VLSI FOR NEURAL NETWORKS AND ARTIFICIAL INTELLIGENCE

VLSI FOR NEURAL NETWORKS AND ARTIFICIAL INTELLIGENCE

Edited by

JOSÉ G. DELGADO-FRIAS
State University of New York at Binghamton
Binghamton, New York

and

WILLIAM R. MOORE
Oxford University
Oxford, United Kingdom

PLENUM PRESS • NEW YORK AND LONDON

Library of Congress Cataloging-in-Publication Data

```
VLSI for neural networks and artificial intelligence / edited by José
  G. Delgado-Frias and William R. Moore.
       p.   cm.
     "Proceedings of an International Workshop on Artificial
  Intelligence and Neural Networks, held September 2-4, 1992, in
  Oxford, United Kingdom"--T.p. verso.
     Includes bibliographical references and index.
     ISBN 0-306-44722-3
     1. Neural networks (Computer science)--Congresses.  2. Artificial
  intelligence--Congresses.  3. Integrated circuits--Large scale
  integration--Congresses.   I. Delgado-Frias, José.  II. Moore,
  William R.   III. International Workshop on Artificial Intelligence
  and Neural Networks (1992 : Oxford, England)
  QA76.87.V58  1994
  006.3--dc20                                                94-31811
                                                                 CIP
```

Proceedings of an International Workshop on Artificial Intelligence and Neural Networks,
held September 2–4, 1992, in Oxford, United Kingdom

ISBN 0-306-44722-3

©1994 Plenum Press, New York
A Division of Plenum Publishing Corporation
233 Spring Street, New York, N.Y. 10013

All rights reserved

No part of this book may be reproduced, stored in a retrieval system, or transmitted in any form or by
any means, electronic, mechanical, photocopying, microfilming, recording, or otherwise, without written
permission from the Publisher

Printed in the United States of America

PROGRAMME COMMITTEE

Nikolaos Bourbakis, *State University of New York at Binghamton, USA*
Howard Card, *University of Manitoba, Canada*
José Delgado-Frias, *State University of New York at Binghamton, USA*
Simon Lavington, *Univeristy of Essex, UK*
Will Moore, *University of Oxford, UK*
Alan Murray, *University of Edinburgh, UK*
Lionel Tarassenko, *University of Oxford, UK*
Stamatis Vassiliadis, *IBM, USA*
Michel Weinfield, *EP Paris, France*

PREFACE

Neural network and artificial intelligence algorithms and computing have increased not only in complexity but also in the number of applications. This in turn has posed a tremendous need for a larger computational power that conventional scalar processors may not be able to deliver efficiently. These processors are oriented towards numeric and data manipulations. Due to the neurocomputing requirements (such as non-programming and learning) and the artificial intelligence requirements (such as symbolic manipulation and knowledge representation) a different set of constraints and demands are imposed on the computer architectures/organizations for these applications.

Research and development of new computer architectures and VLSI circuits for neural networks and artificial intelligence have been increased in order to meet the new performance requirements. This book presents novel approaches and trends on VLSI implementations of machines for these applications. Papers have been drawn from a number of research communities; the subjects span analog and digital VLSI design, computer design, computer architectures, neurocomputing and artificial intelligence techniques.

This book has been organized into four subject areas that cover the two major categories of this book; the areas are: analog circuits for neural networks, digital implementations of neural networks, neural networks on multiprocessor systems and applications, and VLSI machines for artificial intelligence. The topics that are covered in each area are briefly introduced below.

Analog Circuits for Neural Networks

Using analog circuits as a mean to compute neural networks offers a number of advantages such as smaller circuit size, higher computing speed, and lower power dissipation. Neural network learning algorithms may need be implemented directly in analog hardware in order to reduce their computing time. *Card* presents a number of analog circuits for on-chip supervised and unsupervised learning. *Landolt* has developed and implemented in analog CMOS a full Kohonen network with learning capability. *Valle et al* report an implementation of a backpropagation learning algorithm along with some simulation results. *Lafargue et al* address the problem of designing a mixed analog-digital circuit for the Boltzmann machine learning rules. *Brause* introduces the building blocks of a linear neuron that uses anti-Hebb rule and restricted weights. *Raffo et al* present a set of functional primitives that perform synaptic and neural functions which are implemented on an analog circuit.

Digital Implementations of Neural Networks

Digital implementations of neural networks provide some advantages, such as: noise free, programmability, higher precision, reliable storage devices. *Delgado-Frias et al* present a pipelined bit-serial digital organization that is able to implement a backpropagation learning algorithm. *Fornaciari and Salice* propose a tree-structure architecture based on a pseudo-neuron approach. *Viredaz et al* introduce a multi-model neural network computer based on the systolic computing paradigm. *Asonovic et al* present a fully programmable single chip microprocessor that is intended to be used within a Sun Sparcstation. *Ae et al* extend the self-organizing system (or Kohonen network) for semantic applications. *Hui et al* present a probabilistic RAM-based model that incorporates a reward-penalty learning algorithm. *Hurdle et al* propose an asynchronous design approach for neurocomputing; the CMAC neural model is used to demonstrate the approach capabilities.

Neural Networks on Multiprocessor Systems and Applications

König and Glesner present an architecture and implementation of a scalable associative memory system. *Ryan et al* describe how neural network can be mapped onto dataflow computing approach. A custom associative chip that has 64 fully interconnected binary neurons with on-chip learning is presented by *Gascuel et al* The use of RISC processors as implementation of neural network simulation is a study by *Rückert et al;* they report results from eight different processors. *Kolcz and Allinson* have studied the implementation of CMAC, a perceptron-like computational structure, using conventional RAMs. *Wang and Tang* have developed GENET which is a competitive neural network model for AI's constraint satisfaction problems. *Luk et al* introduce a declarative language, called Ruby, that could be used in developing of neural network hardware. *Palm et al* present a study of the integration of connectionist models and symbolic knowledge processing. *Styblinski and Minick* present two Tank/Hopfield-like neural network methods for solving linear equations.

VLSI Machines for Artificial Intelligence

Eight papers address a number of current concerns in the hardware support of artificial intelligence processing. *Lavington et al* present a SIMD approach for pattern matching that is often used in production systems. *Howe and Asonovic* introduce a system that incorporates 148X36-bit content addressable parallel processors. *Cannataro et al* explain a message-passing parallel logic machine that can exploit AND/OR parallelism for logic programs. *Rodohan and Glover* outline an alternative search mechanism that can be implemented on the Associative String Processor. *De Gloria et al* present two performance analysis of a VLIW architecture where global compaction techniques are used. *Civera et al* discuss the application of a VLSI prolog system to real time navigation system; requirements, such as processes, memory bandwidth, and inferences per second, are presented. *Demarchi et al* present the design and a performance evaluation of a parallel architecture for OR-parallel logic programs. *Yokota and Seo* present a RISC Prolog processor which include compound instructions and dynamic switch mechanisms.

ACKNOWLEDGMENTS

This book is an edited selection of the papers presented at the *International Workshop on VLSI for Neural Networks and Artificial Intelligence* which was held at the University of Oxford in September 1992. Our thanks go to all the contributors and especially to the programme committee for all their hard work. Thanks are also due to the ACM-SIGARCH, the IEEE Computer Society, and the IEE for publicizing the event and to the University of Oxford and SUNY-Binghamton for their active support. We are particularly grateful to Laura Duffy, Maureen Doherty and Anna Morris for coping with the administrative problems.

CONTENTS

ANALOG CIRCUITS FOR NEURAL NETWORKS

Analog VLSI Neural Learning Circuits - A Tutorial 1
Howard C. Card

An Analog CMOS Implementation of a Kohonen Network with
Learning Capability 25
Oliver Landolt

Back-Propagation Learning Algorithms for Analog VLSI Implementation 35
Maurizio Valle, Daniele D. Caviglia and Giacomo M. Bisio

An Analog Implementation of the Boltzmann Machine with Programmable
Learning Algorithms 45
V. Lafargue, P. Garda and E. Belhaire

A VLSI Design of the Minimum Entropy Neuron 53
Rüdiger W. Brause

A Multi-Layer Analog VLSI Architecture for Texture Analysis Isomorphic
to Cortical Cells in Mammalian Visual System 61
*Luigi Raffo, Giacomo M. Bisio, Daniele D. Caviglia, Giacomo Indiveri
and Silvio P. Sabatini*

DIGITAL IMPLEMENTATIONS OF NEURAL NETWORKS

A VLSI Pipelined Neuroemulator 71
*José G. Delgado-Frias, Stamatis Vassiliadis, Gerald G. Pechanek, Wei Lin,
Steven M. Barber and Hui Ding*

A Low Latency Digital Neural Network Architecture 81
William Fornaciari and Fabio Salice

MANTRA: A Multi-Model Neural-Network Computer 93
Marc A. Viredaz, Christian Lehmann, François Blayo and Paolo Ienne

SPERT: A Neuro-Microprocessor 103
*Krste Asanović, James Beck, Brian E. D. Kingsbury, Phil Kohn,
Nelson Morgan and John Wawrzynek*

Design of Neural Self-Organization Chips for Semantic Applications 109
Tadashi Ae, Reiji Aibara and Kazumasa Kioi

VLSI Implementation of a Digital Neural Network with Reward-Penalty Learning 119
Terence Hui, Paul Morgan, Hamid Bolouri and Kevin Gurney

Asynchronous VLSI Design for Neural System Implementation 129
John F. Hurdle, Erik L. Brunvand and Lüli Josephson

NEURAL NETWORKS ON MULTIPROCESSOR SYSTEMS AND APPLICATIONS

VLSI-Implementation of Associative Memory Systems for Neural
Information Processing 141
Andreas König and Manfred Glesner

A Dataflow Approach for Neural Networks 151
*Thomas F. Ryan, José G. Delgado-Frias, Stamatis Vassiliadis,
Gerald G. Pechanek and Douglas M. Green*

A Custom Associative Chip Used as a Building Block for a Software
Reconfigurable Multi-Network Simulator 159
*Jean-Dominique Gascuel, Eric Delaunay, Lionel Montoliu, Bahram Moobed
and Michel Weinfeld*

Parallel Implementation of Neural Associative Memories on RISC Processors *U. Rückert, S. Rüping and E. Naroska*	167
Reconfigurable Logic Implementation of Memory-Based Neural Networks: A Case Study of the CMAC Network *Aleksander R. Kolcz and Nigel M. Allinson*	177
A Cascadable VLSI Design for GENET *Chang J. Wang and Edward P. K. Tsang*	187
Parametrised Neural Network Design and Compilation into Hardware *Wayne Luk, Adrian Lawrence, Vincent Lok, Ian Page and Richard Stamper*	197
Knowledge Processing in Neural Architecture *G. Palm, A. Ultsch, K. Goser and U. Rückert*	207
Two Methods for Solving Linear Equations Using Neural Networks *M. A. Styblinski and Jill R. Minick*	217

VLSI MACHINES FOR ARTIFICIAL INTELLIGENCE

Hardware Support for Data Parallelism in Production Systems *S. H. Lavington, C. J. Wang, N. Kasabov and S. Lin*	231
SPACE: Symbolic Processing in Associative Computing Elements *Denis B. Howe and Krste Asanovic*	243
PALM: A Logic Programming System on a Highly Parallel Architecture *Mario Cannataro, Giandomenico Spezzano and Domenico Talia*	253
A Distributed Parallel Associative Processor (DPAP) for the Execution of Logic Programs *Darren P. Rodohan and Raymond J. Glover*	265
Performance Analysis of a Parallel VLSI Architecture for Prolog *Alessandro De Gloria, Paolo Faraboschi and Mauro Olivieri*	275
A Prolog VLSI System for Real Time Applications *Pier Luigi Civera, Guido Masera and Massimo Ruo Roch*	285
An Extended WAM Based Architecture for OR-Parallel Prolog Execution *Danilo Demarchi, Gianluca Piccinini and Maurizio Zamboni*	297
Architecture and VLSI Implementation of a Pegasus-II Prolog Processor *Takashi Yokota and Kazuo Seo*	307
CONTRIBUTORS	317
INDEX	319

ANALOG VLSI NEURAL LEARNING CIRCUITS - A TUTORIAL

Howard C. Card

INTRODUCTION

This paper explains the various models of learning in artificial neural networks which are appropriate for implementation as analog VLSI circuits and systems. We do not cover the wider topic of analog VLSI neural networks in general, but restrict the presentation to circuits which perform in situ learning. Both supervised and unsupervised learning models are included.

There are two principal classes of artificial neural network (ANN) models, as shown in Fig. 1. These may be described as feedforward and feedback models. In the feedforward case, during the classification of an input pattern, information flows unidirectionally from input to output through the various network layers. The best known example of a feedforward ANN is the multilayer perceptron employing the backpropagation of errors, otherwise known as the generalized delta learning rule (Rumelhart *et al* 1986). Similarly, the best known feedback model is the Hopfield network (Hopfield 1984), which generally employs some version of a Hebbian (Hebb, 1949) learning rule. A variation on the feedforward case allows for crosstalk connections within a given layer. This intermediate layer may itself be regarded as a feedback network which is embedded in the otherwise feedforward structure. Networks of this type include competitive learning models (Von der Malsburg 1973) and Kohonen maps (Kohonen 1990). Feedback or recurrent networks exhibit dynamics, and there is a relaxation period after the presentation of a new input pattern. These ANN models and their variations are described in detail in two excellent

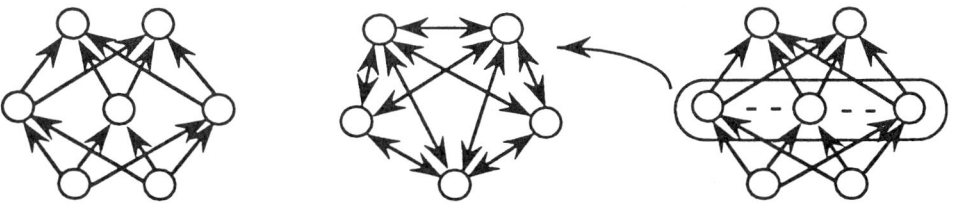

Figure 1 Two classes of artificial neural network (ANN) models are feedforward (left) and feedback (middle) models. The feedforward case with crosstalk connections within a layer (right) is employed in competitive learning models. The crosstalk layer may be regarded as a feedback network within the larger network.

references: the recent text by Hertz et al (1991) and the review of ANN learning by Hinton (1989).

In supervised learning tasks, a set of training patterns is shown repeatedly to the network. The differences between the outputs generated by the present weights and the desired outputs indicated by the training patterns are used to compute weight changes. The hidden units can learn to represent those features of the input which best help the system obtain a low error on the training data. Having learned the training set, the network can generalize to novel patterns, by interpolation on the training examples. In unsupervised learning on the other hand, it is the learning rule itself which completely determines the features which the network extracts from the training data. These features are generally based on correlations inherent in the data, and result in clustering of the input patterns into a limited number of classes.

NEURAL LEARNING MODELS

Analog Neurons And Synapses

In Fig.1, the circles are the artificial neurons, which in analog circuits are usually implemented as nonlinear (saturating) amplifiers. The arrows between the neurons represent weighted connections or synapses. A common form of the saturation characteristic of the neuron is a sigmoid given by either the Glauber logistic function (for $0 < V_i < 1$)

$$V_i(x_i) = [(1 + \exp(-2\beta x_i))]^{-1} \qquad (1)$$

or the tanh function (for $-1 < V_i < 1$)

$$V_i(x_i) = \tanh(\beta x_i) \qquad (2)$$

where the input to the neuron x_i is the weighted sum of its inputs

$$x_i = \Sigma W_{ij} V_j \qquad (3)$$

with W_{ij} the weight of the ij synapse, and with V_i the activation of neuron i, as in Fig.2. Nonzero thresholds Θ_i and external inputs I_i for the neurons (often written as extra terms in the equation) may be included in (3) by considering them as normal connection weights to an extra neuron in the network whose activation is permanently unity. If we regard the activation V_i as the probability of a hypothesis i, then the system of interconnected neurons in Fig. 2 may be regarded as solving an optimization problem with weak constraints. The weights W_{ij} correspond to mutually supporting or conflicting constraints from other hypotheses j (positive or negative weights), the external inputs I_i to external evidence in the present case, and the θ_i to prior probabilities.

In addition to the synaptic multiplication function required by Eqn (3), the synapses must perform in situ learning computations which modify their W_{ij} values. These learning computations generally make gradual changes to the weights. Analog circuits with in situ learning therefore have dynamics on two time scales. They differ from conventional analog signal processors, not only in performing nonlinear transformations of their inputs, but in employing time varying parameters. These adaptive synapses accelerate the learning process as compared with the more common off-chip learning, and can compensate for nonstationary environments. Their disadvantage is that they must commit to a predetermined learning rule and cannot take advantage of progress in learning algorithms subsequent to chip fabrication. In the early stages of analog ANN development this lack of flexibility will be a disadvantage, but in the long term and at least for special purpose appliocations, this approach is promising.

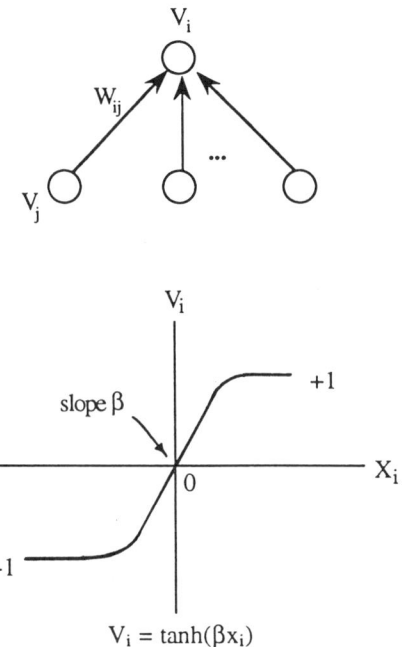

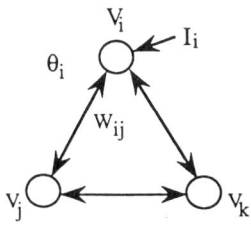

Figure 2 Neurons in ANNs first perform the linear input computation $x_i = \Sigma\ W_{ij}\ V_j + \theta_i + I_i$ and then produce the nonlinear output $V_i = \tanh(x_i)$ where θ_i and I_i are the threshold and the external inputs to the neuron, W_{ij} and V_j are the synaptic weight and activation of neuron j connected to neuron i.

Supervised learning rules

Learning rules are the analytical form which controls the update of the synaptic weights. Two common versions of the learning rule are (i) Hebbian learning (Hebb 1949) which is governed by the correlation of presynaptic and postsynaptic neural activities

$$\Delta W_{ij} = \varepsilon \, V_j \, V_i \tag{4}$$

and (ii) the generalized delta rule (Rumelhart *et al* 1986) which is driven by the anticorrelation of the presynaptic activity and the postsynaptic error

$$\Delta W_{ij} = -\varepsilon \, V_j \, \delta V_i \tag{5}$$

where ε is a learning rate parameter and in the final expression δV_i is the error term. There is often an additional term in these learning rules, proportional to $-W_{ij}$, to represent weight decay (Hertz *et al* 1991, Hinton 1989).

Multilayer feedforward networks. The error term δV_i in Eqn (5) is layer-dependent in a multilayer network (Fig.3). The well-known backpropagation (BP) algorithm (Rumelhart *et al* 1986) for these networks is based on gradient descent in a squared global error measure E

$$E = \Sigma_c \Sigma_i (V_i - V_i^t)^2 \tag{6}$$

with summations over all training cases c and over all output units i. Gradient or steepest descent leads to synaptic weight changes governed by

$$\Delta W_{ij} = -\varepsilon \frac{\partial E}{\partial W_{ij}} \tag{7}$$

By the chain rule

$$\frac{\partial E}{\partial W_{ij}} = \frac{\partial E}{\partial V_i} \frac{\partial V_i}{\partial x_i} \frac{\partial x_i}{\partial W_{ij}} \tag{8}$$

This analysis results in a generalized error term in Eqn (5) for output units governed by a tanh (bx) nonlinearity given by (Hertz *et al* 1991)

$$\delta V_i = \frac{\partial E}{\partial V_i} \frac{\partial V_i}{\partial x_i} = (V_i - V_i^t) \beta (1 - V_i^2) \tag{9}$$

For neurons in the hidden layer below, the contribution of their activations V_j to the total error must be summed over all the output units i

$$\frac{\partial E}{\partial V_j} = \Sigma_i \frac{\partial E}{\partial x_i} \frac{\partial x_i}{\partial V_j} = \Sigma_i \frac{\partial E}{\partial x_i} W_{ij} \tag{10}$$

Analog CMOS networks using an in situ BP learning rule have been reported by (Hinton 1989) among others. The advantage of this learning rule is that it is fairly robust and is based upon a well understood gradient descent procedure. On the other hand it has several shortcomings. Since updates to any weight affect the global error which in turn determine the changes to all other weights, learning becomes very slow in large networks. All weights are mutually dependent. This may be circumvented by hierarchical connections of small networks or by algorithms such as competing experts (Nowlan and Hinton 1991). Another drawback of BP and related rules is their demands on high precision and large dynamic range in the weights, especially of weights in the upper layers of the network.

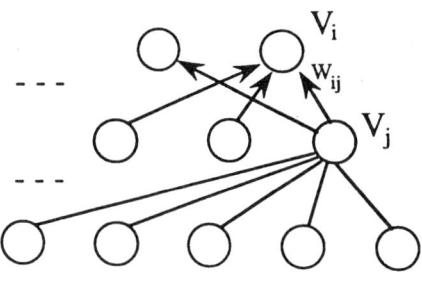

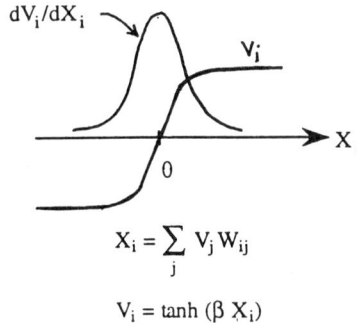

$$X_i = \sum_j V_j W_{ij}$$

$$V_i = \tanh(\beta X_i)$$

Figure 3 Feedforward network or multilayer perceptron as employed with the backpropagation learning rule. The slope of the neuron transfer function at $x_i = 0$ is β.

Fully connected recurrent networks. Recurrent networks have exhibited many computational advantages which result from their dynamics. These networks can solve difficult optimization problems with soft constraints, as they relax to their minimum energy configuration (Hopfield 1984). Even without incorporation of learning these networks have many practical applications in fields such as vision. An algorithm for supervised learning in analog relaxation ANNs called *mean field networks* has been found (Peterson and Anderson 1987) and is based upon the mean field approximation to Boltzmann statistics. The synapses perform what is termed *contrastive Hebbian learning* (CHL)

$$\Delta W_{ij} = \varepsilon \left[(V_i V_j)^+ - (V_i V_j)^- \right] \tag{11}$$

which is an extension of Eqn (4). In this case, Hebbian correlations $V_i V_j$ are determined for two phases. The + phase is termed the teacher phase, in which inputs and outputs are clamped to values determined by the current training pattern. In the - or student phase, only the inputs are clamped, while the outputs are freely determined by the present weights in the network. The weight changes in the + and - phases may also be regarded as learning and unlearning operations. When learning is complete, the outputs agree in the two phases, and the two terms cancel in Eqn (11).

This network (Fig.4) is a deterministic version of a Boltzmann machine (Ackley *et al* 1985), and performs gradient descent in the contrastive function J

$$J = F^+ - F^- \tag{12}$$

where F^+ and F^- are the free energy functions of the network after settling in the respective phases (Movellan 1990). Settling of the neural activations is accompanied by gradually sharpening the tanh nonlinearity by increasing the gain of the neurons. This approximates the decreasing temperature of an annealing process.

The mean field approximation accelerates conventional Boltzmann learning by eliminating the need to obtain statistical correlation of the random binary variables in (Ackley *et al* 1985). Only simple analog multiplications are required in the mean field approximation to Boltzmann statistics. A desirable property of a CHL rule is that weight updates depend only upon quantities local to the individual synapses (the update to W_{ij} depends only on V_i and V_j and, with weight decay, on W_{ij} itself). This type of learning rule scales well to large networks, unlike the BP rule above. Like Boltzmann machines, it is also expected to require less precision in the weights. On the negative side, CHL rules and mean field networks are less well understood than BP rules, and are sensitive to learning rates and to the schedules for annealing of the neural activation dynamics (Card *et al* 1992).

Analog CMOS mean field networks have been reported by several groups with variations in their methods of implementing the synapses and the learning rule of Eqn (11). For

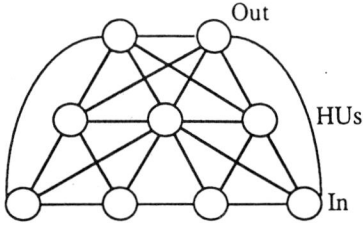

Figure 4 Mean field network employing contrastive Hebbian learning rule $\Delta W_{ij} = \varepsilon [(V_i V_j)^+ - (V_i V_j)^-]$. HUs are hidden units.

example, Alspector *et al* (1991) have used digital weights, whereas we and others (Schneider and Card 1991, 1992, Arima *et al* 1991) have employed capacitive weights in a fully analog implementation. These approaches are compared in a later section.

Unsupervised learning rules

Unsupervised and semi-supervised learning rules circumvent the excessive training times of supervised networks. Competitive learning (CL) models (Von der Malsburg 1973, Hertz *et al* 1991, Hinton 1989) are related to clustering techniques in classical pattern recognition (Fig.5). In 'hard' CL, a single node is the winner on each input pattern - a winner take all (WTA) network. WTA can be implemented by having inhibitory connections among all the competing nodes, i.e. lateral inhibition, and each node must have a self-excitatory connection. The strengths of the inhibitory connections and the nonlinear activation functions of the nodes must be judiciously chosen to avoid oscillation while producing a single excited node. Once the winner is decided its weight is updated according to the standard competitive learning rule

$$\Delta W_{ij} = \varepsilon V_i (V_j - W_{ij}) \tag{13}$$

where V_i is the excitation of the winning node i and V_j is a component of the input vector. This is the Hebbian learning of Eqn (4) with an added term which gives a form of weight decay. Eqn (13) works best (Hertz 1991) for prenormalized inputs and weights

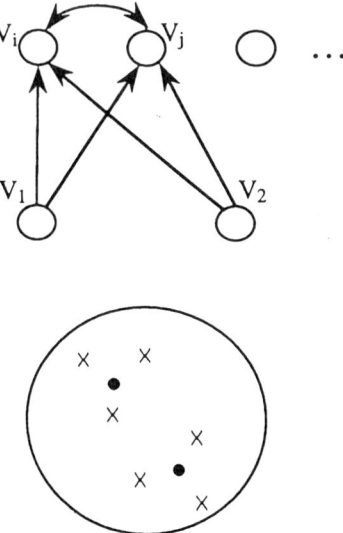

Figure 5 Competitive learning models have mutual interaction which is usually lateral inhibition among the neurons in the hidden layer. This interaction is shown between neurons i and j in the figure but is understood to apply to all neurons in this layer. Lower figure shows normalized input vectors X and weight vectors •.

and
$$\Sigma_j V_j^2 = 1 \qquad (14)$$

$$\Sigma_j W_{ij}^2 = 1 \qquad (15)$$

and in this case learning has, for a given node, the geometric analogy of moving weight vectors on the surface of a unit sphere into the center of gravity of all those input vectors (also on the unit sphere) for which it is the winner (Fig.5).

WTA circuits have natural analog VLSI implementations and have been described by several groups, for example (Lazzaro *et al* 1989). The basic idea of these circuits has been mutual inhibition among competing subthreshold CMOS amplifiers. Also included in (Lazzaro *et al* 1989) were local inhibition circuits based on resistive grids. Mann *et al* (1990) describe analog chips with unary digital weights which implement an approximation to CL. Hochet *et al* (1991) have reported a similar implementation which stores multiple discrete levels of synaptic weights on a capacitor. The weights are refreshed without the need for full A/D or D/A conversion by employing a voltage ramp which is gated by clock signals to charge the capacitor.

Radial basis functions (Moody and Darken 1989) implement a soft form of competitive learning in which each node is activated according to its distance from an input pattern, without mutual inhibition as in WTA discussed above. A common example of a distance function is

$$V_i = C \exp[-(V_j - W_{ij})^2 / 2\sigma^2] \qquad (16)$$

Adaptive resonance models (Carpenter and Grossberg 1987) are another generalization of CL which enforce the formation of stable categories. This is accomplished by having an abundance of node units which are only enlisted as needed. Inputs are accepted by existing nodes according to a vigilance parameter r which determines how close they must be to these nodes. One can obtain coarse or fine categorizations depending on the value of r. ART-2 is the adaptive resonance model employed with analog inputs. ART models have been implemented as analog circuits in (Tsay and Newcomb 1991, Linares-Barranco *et al* 1992).

Feature mapping (Willshaw and von der Malsburg 1976) is a refinement on CL in which the inhibitory WTA connections are replaced by more complex lateral connections. Correlated weight updating by neighbouring nodes can lead to topology-preserving mappings. Kohonen maps are the best known example of these.

Kohonen maps. In Kohonen feature maps (Kohonen 1990) as in Fig. 6, weight vectors for nodes in the neighbourhood of the winner are updated as well as the weight vector of the winner itself. This can be represented by

$$\Delta W_{ij} = \varepsilon N_{xy}(t) (V_j - W_{ij}) \qquad (17)$$

where $N_{xy}(t)$ is a neighbourhood function (NFs) in which the indices x,y determine the displacement of node i from the winning node. The neighbourhood in which $N_{xy}(t)$ is appreciably large typically shrinks with time during the course of learning, as progressively finer topological details are ironed out. The neighbourhood function may be either Gaussian

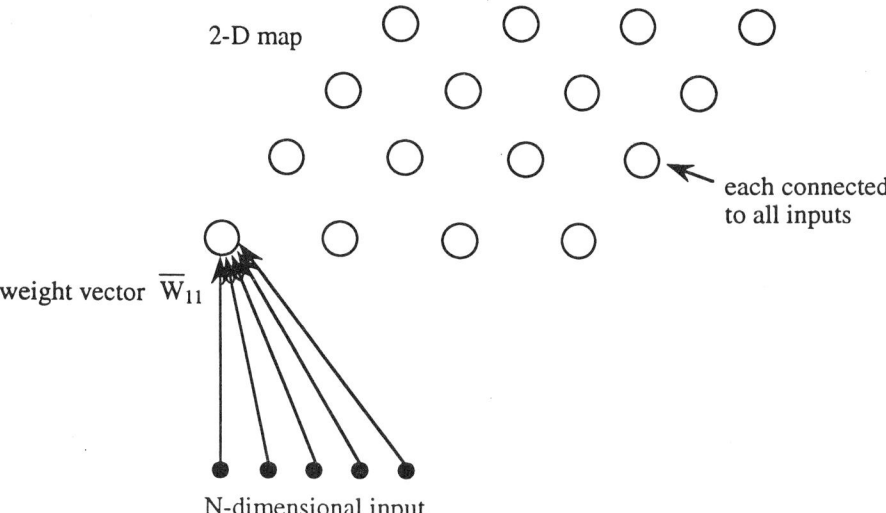

Figure 6 In Kohonen learning models, the neurons in the competitive layer are arranged into a 2-D grid, and the weight updates apply not only to the winning neuron, but to those neurons within a prescribed neighbourhood $N_{xy}(t)$ around the winner.

or step functions of position, and its time dependence may be linearly or exponentially decaying, or inversely proportional to t (Kohonen 1990).

Analog circuits for Kohonen nets have been reported by several groups, for example (Mann and Gilbert 1990, Hochet *et al* 1991). The competitive learning operations are implemented as discussed above. The neighbourhood function of Eqn. (17) has received less attention, but Hochet *et al* (1991) suggest in their paper the use of cellular automata to generate the NFs. One may also employ a diffusive process to obtain the NF, as is used in vision chips with resistive grids mentioned above (Lazzaro *et al* 1989).

Elastic nets (Durbin and Willshaw 1987) can be regarded as an extension of feature maps in which updates to the weights depend upon current values of neighbouring weights as well as the current input. Note that this poses extra geometric constraints in addition to the topological constraints of Kohonen maps. Other unsupervised algorithms are based on statistical procedures such as principal component analysis PCA (Linsker 1988).

Principal component analysis. Unsupervised learning makes use of redundancy in the input data to extract meaningful features, so the information content of the input data stream must be well below the capacity of the input channel. In this case, spatial and temporal correlations in the input data can be discovered and used to reduce the dimensionality of this data. For a single node, a learning rule due to Oja (1989) can be shown to extract the principal component of the input data

$$\Delta W_{ij} = \varepsilon V_i (V_j - V_i W_{ij}) \qquad (18)$$

where the first term is normal Hebbian learning and the second term is a form of weight decay proportional to the square of the output V_i^2. This constrains weight magnitudes, causing the directions of weight vectors to line up with the input data. This rule maximizes the mean squared output $<V_i^2>$ or variance over the input patterns (Hertz *et al* 1991). With M units the first M principal components of the input data can be extracted using the rule (Sanger 1989)

$$\Delta W_{ij} = \varepsilon V \ (V_j - \Sigma_{k=1}^{i} V_k W_{kj}) \tag{19}$$

The nodes or neurons in Eqns (18) and (19) are linear, as opposed to the nonlinear sigmoid units discussed earlier.

Spatial coherence. An alternative unsupervised technique is based on spatial or temporal coherence in the input data (Becker and Hinton 1992). Learning in this case can employ a delta rule as in Eqn (5) but, instead of an external error, neighbouring neurons guide one another's weight updates so as to maximize their mutual information. By extracting features from adjacent patches of the image which have large mutual information, adjacent subnetworks learn sets of weights which correspond to spatially coherent features such as depth, reflectance or surface orientation. Fig. 7 illustrates this process, in which the subnetworks may be regarded as local BP networks with an error measure based on maximizing the mutual information between the outputs a and b from adjacent patches.

$$I(a,b) = H(a) + H(b) - H(a,b) \tag{20}$$

H(a) and H(b) are the entropies of the two patches over the data set, and H(a,b) is the cross entropy. One may employ the variances (Becker and Hinton 1992)

$$I(a,b) = \log \frac{V(a)}{V(a-b)} + \log \frac{V(b)}{V(a-b)} \tag{21}$$

so as to minimize the difference between a and b while at the same time maximizing their individual variation over the data set. Note that the correlation of a and b is not sufficient, as this could be maximized by having all weights learn to become zero.

The output error values are computed in a two pass process. In the first phase, the means

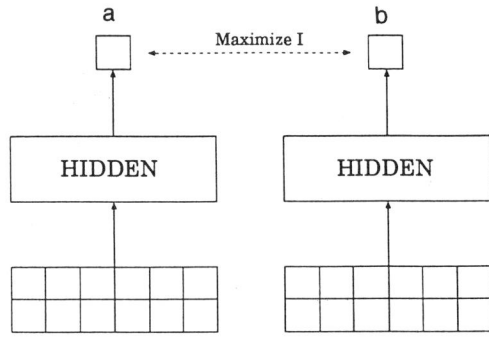

Figure 7 In unsupervised networks based on spatial coherence in the input patterns, a feature is extracted from neighbouring patches a and b in such a way that the mutual information is maximized. After Becker and Hinton (1992).

and variances of a and b are computed over the entire training set. In the second phase the error derivatives (in a similar manner to the BP algorithm) are computed using these variances. We are currently exploring efficient analog VLSI implementations of this and related learning rules.

ANALOG VLSI CIRCUITS FOR ANNs

Analog circuits for neurons

Two basic approaches to VLSI synapses in analog ANNs, to the learning of the synaptic weights, and to their connections to the neurons are shown in Fig. 8. These are discussed

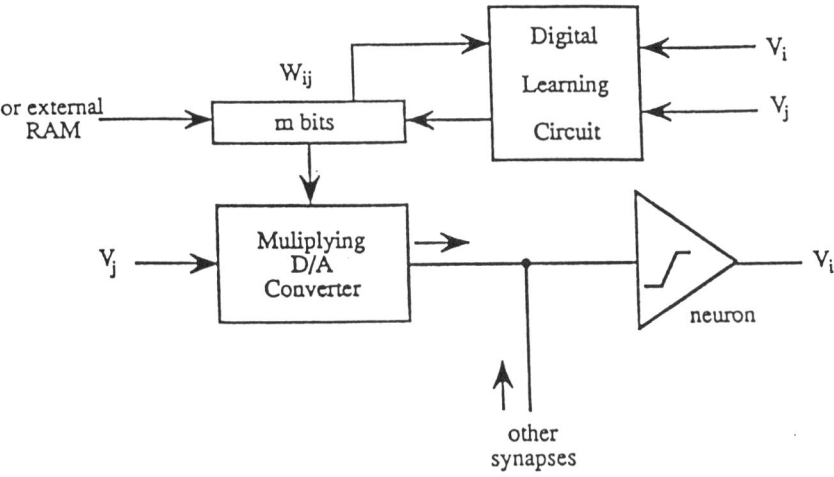

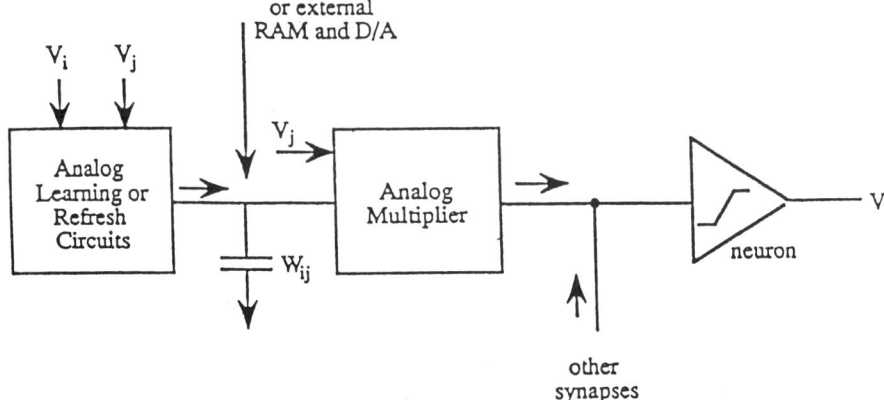

Figure 8 Two approaches to analog artificial neural networks (ANNs). In the upper diagram, the synaptic weights are stored in binary registers, whereas in the lower figure the weights are represented as charge on capacitors. Both in situ learning and the alternative external learning modes are indicated.

further below. There are various types of analog neurons in these networks. The common feature of these methods is that the neuron should exhibit a sigmoid nonlinearity, that it should sum the contributions from its synapses using charge accumulation (Kirchoff's current law), and that it should have the capability to drive the many synapses to which it is connected.

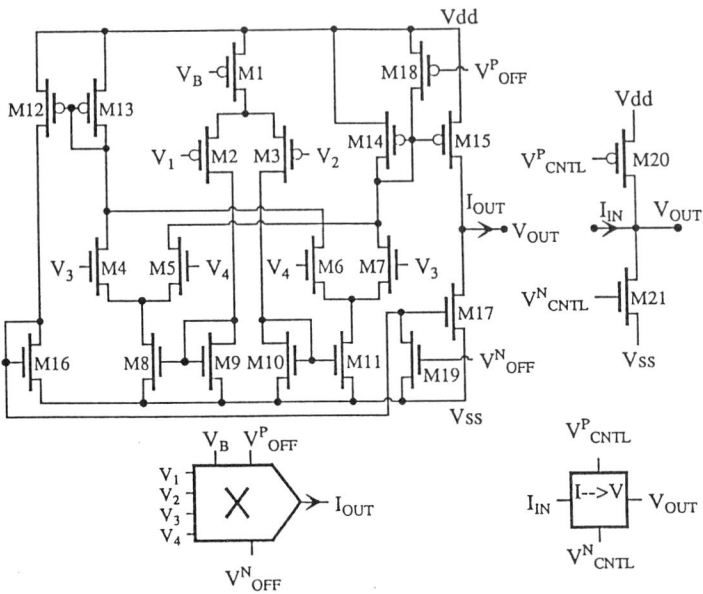

Figure 9 Schematic diagrams of typical CMOS wide range Gilbert multiplier and current-voltage converter circuits (Schneider and Card 1991).

Analog inverters. The simplest version of a neuron is the analog CMOS inverter circuit consisting of two transistors. The gain of the neuron is controlled by the conduction parameters of the p and n channel devices, in other words by their W/L ratios. There is no mechanism to control the gain of this neuron electronically.

Neurons based on analog multipliers. A much more flexible analog neuron can be obtained using a simple transconductance amplifier/multiplier or a Gilbert multiplier. Mead (1988) has described versions of these circuits in which the transistors operate in the subthreshold region, to obtain minimal power dissipation for sensory preprocessing, especially in vision and audition tasks. Other implementations have been described in which the transistors operate above threshold, and improved speed is obtained at the cost of increased power. Power dissipation remains however well within acceptable limits. Since there are typically very few neurons as compared to the number of synapses in ANNs the area minimization of neurons is not a major issue, and in the interest of flexibility it is advantageous to employ more elaborate circuits for neurons, such as the Gilbert wide-range analog multipliers of Fig. 9. The output current of these multipliers exhibits a good approximation to a tanh nonlinearity, the resultant neurons have adjustable gain, and are excellent current sources as shown in Fig 10. They also provide sufficient drive capability, even for fully-connected

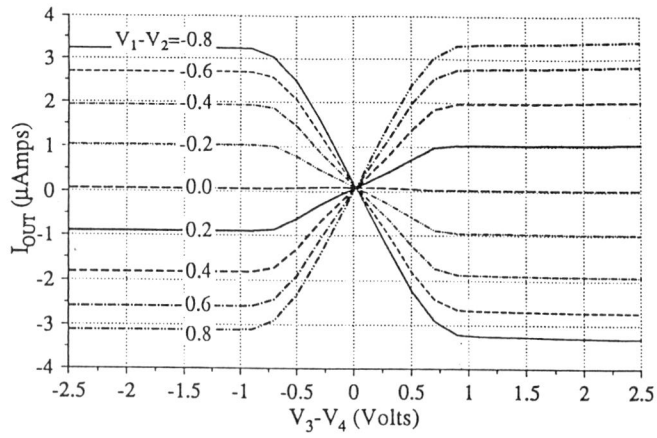
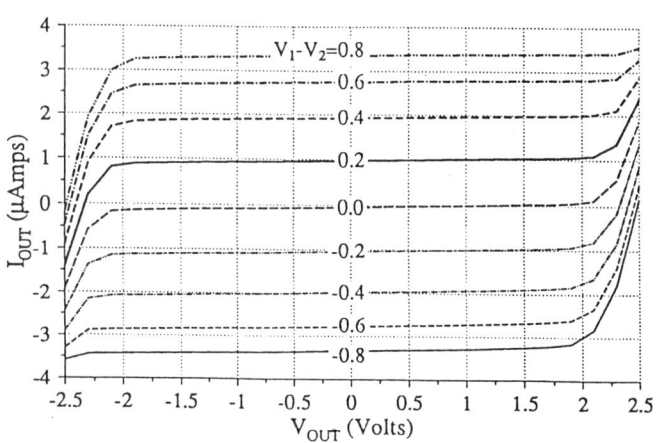

Figure 10 Measured response of Gilbert multipliers. Upper figure shows multiplier characteristics and lower figure behaviour as a current source (Schneider and Card 1991).

networks. Note the wide voltage range of these characteristics which is one benefit of operating the transistors above threshold.

Neurons for pulse stream circuits. In some implementations of ANNs (Murray *et al* 1991), the neural signals are represented by streams of pulses, the frequency of which encodes the value of the signal. This is similar to biological signal processing. Digital versions of this approach (Tomlinson *et al* 1990) employ synchronous pulses. Summation of multiple streams may be obtained simply using OR gates and multiplication may be compactly achieved in certain cases using AND gates. The compression of the summation at the top end of the pulse frequency range gives a squashing function which resembles the desired neural sigmoidal nonlinearity. In analog systems the pulse streams are asynchronous, so that pulse overlap interferes with OR gate operation. Because of this, and the fact that the useful fanin to an OR gate is limited, analog pulse stream circuits have accumulated charge from these pulses onto a capacitor. Alternatively, neurons are integrator circuits based on operational amplifiers (or multipliers for gain control) in which case the synapses first convert pulse-coded information into analog currents whose direction and magnitude is controlled by the weights. These currents are integrated by the neuron to produce a voltage which then drives a voltage-controlled oscillator. VCOs of various types have been reported, both linear and nonlinear (with sigmoid transfer functions) with typically 20 transistors. CMOS neurons with functional characteristics much more faithful to mammalian biological (neocortical pyramidal) neurons were also recently reported by Mahowald and Douglas (1991).

Switched capacitor circuits. SC circuits for the implementation of ANNs are closely related to the pulse stream circuits discussed above. An example is shown of the simplest form of SC circuits for ANNs in Fig 11 (Maundy and El-Masry 1990). The synaptic weight W_{ij} is an analog voltage and the neural activation V_j is a clocked pulse stream as in (Murray *et al* 1991). Increasing frequency of the pulse stream reduces the effective resistance at the input to the integrator by simulating the resistor by a switched capacitor, as in switched capacitor filters. The input current is integrated and converted to a voltage V given by

$$V = \frac{C_2}{C_1} W_{ij} V_j \qquad (22)$$

This voltage V drives a VCO to generate a pulse stream representing the neural activation V_i of neuron i. Among the virtues of this approach are the dependence of V on a capacitance ratio rather than on absolute capacitance values, and the relative economy of transistors per synapse. Caution must however be used in employing synchronous clocks in large analog systems. Note also that the weights may be bipolar by choosing appropriate thresholds but that the neural activations in the simple scheme of Fig 11 are unipolar.

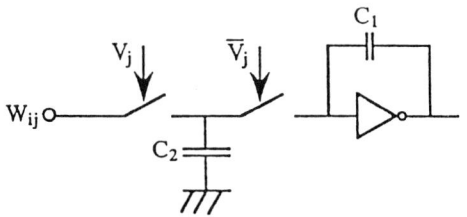

Figure 11 Basic switched capacitor synapse (Maundy and El-Masry 1990).

Analog circuits for synapses

The basis of both synaptic weighting of neural activations as in Eqn (3) and learning computations as in Eqns (4) and (5) is an analog multiplication. This section therefore initially discusses circuit implementations of these multipliers.

Simple multipliers. There have been a number of circuits suggested and tested as synaptic multipliers, which trade off complexity (number of transistors) for accuracy. The simplest approximations to a multiplier may be realized using characteristics of individual MOS transistors. This is the approach taken for example in (Walker *et al* 1989, Card *et al* 1991). The current in an MOS transistor with drain voltage a V_1 where a is a small constant can be written

$$I = 2aK V_1 V_2 \qquad (23)$$

where $V_2 = V_g - V_t$ with V_t the threshold voltage and K the conduction parameter of the transistor $K = \mu C_o (W/L)$. This approach may be directly employed in circuits with unipolar neural activations (analog voltages in the range $0 < V_i < 1$) and unipolar weights ($0 < W_{ij} < 1$). For bipolar activations and weights several transistors per synapse are required for the multiplication operation in Eqn (3). An EEPROM technology (see also synapse section) which provides magnitude and polarity control of V_t permits the design of a bipolar multiplier with reasonably linear characteristics using only two transistors (Shoemaker *et al* 1988).

Gilbert multipliers. In order to obtain good approximations to a linear multiplication over an appreciable voltage range one usually employs CMOS versions of Gilbert multipliers such as that of Fig. 9. Their characteristics (Schneider and Card 1991) are reasonably linear near the origin and saturate for large voltage excursions. Note that the characteristics also remain flat over a wide output voltage range as required for a current source to provide the learning currents of Eqns (4) or (5) independent of present values of the weights.

Weights stored in binary registers. In this case (upper diagram in Fig. 8) the weight is stored in a binary register at the synaptic site, and the contents of this register are multiplied by a neural activation, for one term in Eqn (3), by a multiplying D/A converter which produces an analog current (Raffel *et al* 1987). The multiplying D/A converter employs transistors of progressively increasing size for the more significant bits so that their contributions to the total current double with each bit, as in Fig 12. This means that the area of the synapse grows exponentially with the resolution of the weights. An alternative is to employ a D/A converter to produce an output analog voltage on a capacitor which may then be treated, for example during learning, as an analog weight.

An advantage with this approach is that a digital host computer has access at any time during learning to the synaptic weight matrix. Alspector and his colleagues at Bellcore have designed and fabricated several analog chips with in situ learning of digital weights (Alspector *et al* 1991). Digital processors local to the synapses perform the weight updates. The limited weight resolution with register based weights places a lower bound on the magnitude of the weight changes that may be learned. This limitation may be offset to an extent by the ability of some chips of this type to perform stochastic weight updates. 5 bit weights (4 bits + sign bit) were employed by the Bellcore group on a chip with 32 neurons

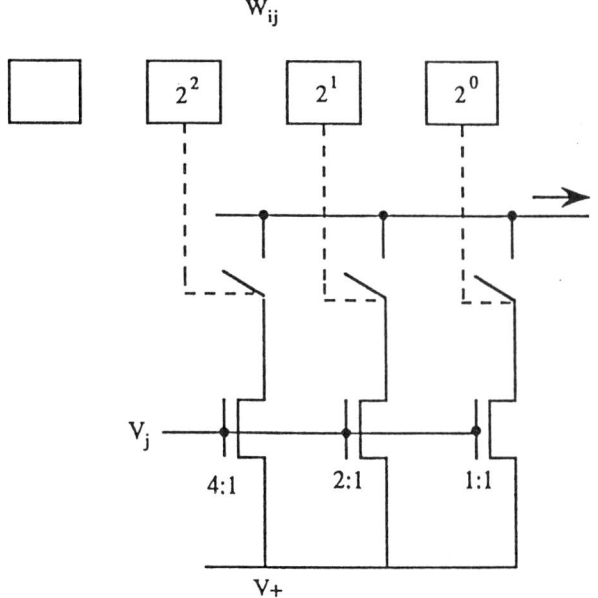

Figure 12 Multiplying D/A converters (Raffel *et al* 1987).

and 496 bidirectional connections. They have implemented both stochastic and deterministic Boltzmann machines, with digital weight processors to perform learning computations.

Stochastic neurons are driven in (Alspector *et al* 1991) by input signals together with zero mean Gaussian noise. In recent chips, the analog noise is generated by low-pass filtering high frequency binary outputs from on-chip LFSR random number generators. In the deterministic case, contrastive Hebbian learning is approximated by Manhattan weight updating. Their work also included simulations of backprop, Boltzmann and mean field machines learning parity, replication, and NetTalk, with comparable results for similar numbers of training cycles. It was observed that in the presence of a global weight decay, the limited dynamic range of 5 bit weights did not generally compromise learning. An exception was noted for large replication problems in mean field networks. This supports the observations in our laboratory that CHL, when using hidden units, requires appreciable dynamic range in the weights, the degree depending on the problem or dataset.

Weights stored as charge on capacitors. This method is shown in the lower diagram of Fig. 8, and in more detail in Fig. 13. The blocks P1,P2 and P4 correspond to the Gilbert multipliers, and P3,P5 to the current-voltage converters in Fig. 9 (Schneider and Card 1991). Small synaptic area ($3 \times 10^4 mm^2$ which includes the in situ learning circuits) is possible with this approach. Another virtue is that weight updating of capacitors employs relatively simple analog circuitry, leading to compact fully-analog ANNs. There are some concerns with capacitive weights, however. The voltage on capacitors decays with time in the presence of leakage currents, and the weights must be refreshed periodically. Our solution to this problem is to periodically refresh the weights via the learning circuitry, using a repetition of

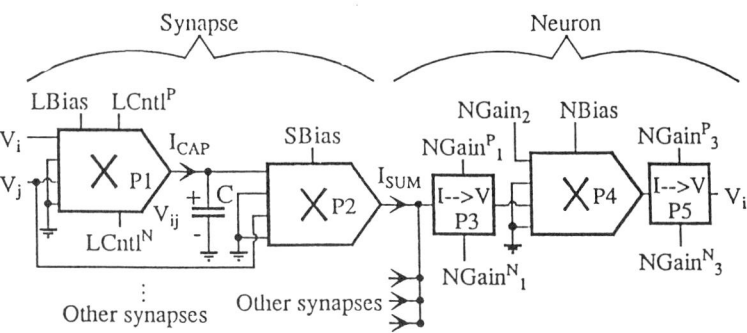

Figure 13 Schematic diagram of neuron and synapse with in situ Hebbian learning. The block P1 is slightly more complicated for contrastive Hebbian learning in mean field networks (Schneider and Card 1991, 1992).

the training data. This has the added benefit of compensating for drifts in the analog circuitry due for example to temperature changes. Novel data to be classified is then interspersed with the training data (with learning switched off if desired).

One can also control the decay rate by modest cooling of the entire chip and at the same time obtain a desirable rate of weight decay which optimizes the learning algorithm. A potential objection to capacitive weights is that circuit operation may be too fast for an external host to maintain the required I/O data rates. Although this will eventually limit the performance, one can alternatively use these circuits in systems with analog (sensory) inputs and analog (motor) outputs, such as in fully-analog autonomous robots.

Fig. 13 actually corresponds to a capacitive synapse which implements in situ Hebbian learning. Analog multiplication in these circuits may also be used to perform delta learning rules as in Eqn (5). These synapses are shown in a fully-connected CMOS network in Fig. 14. The power dissipation is 0.34 Wcm^{-2}. This approach has been extended to mean field synapses, with the contrastive Hebbian learning (CHL) rule of Eqn (11), using a circuit similar to Fig. 13. The output $(V_iV_j)+ - (V_iV_j)-$ of this circuit is used to drive Manhattan updating circuits (Schneider and Card 1992) resulting in synaptic weight changes of

$$\Delta W_{ij} = \varepsilon \ \text{sgn} \ [(V_iV_j)^+ - (V_iV_j)^-] \tag{24}$$

Test results for 1.2μm CMOS synapses with CHL indicate that learning pulses as short as 50 ns may be employed to update the weights, and that updates as small as 10^{-4} to 10^{-3} are reliably made using these circuits. This implies a dynamic range in excess of 10 bits for these capacitive synaptic weights. Of course the precision as measured by intrachip variation is much lower, but in situ learning makes this precision less important. Intrachip variation in device characteristics was measured to be 3-5% and interchip variation to be 30-40% in our 1.2 μm CMOS test chips.

Simulations of mean field learning in networks based upon parameters measured from these CMOS circuits are presented in Fig. 15 for a pattern matching problem. The effects of imperfect components and of a systematic mismatch between the two phases of the mean field learning algorithm are shown. The system clearly tolerates the former but not the latter. Offsets in the multiplication operation cause no problems. Offsets encountered in *subtracting* the two terms in Eqn (24) are important however and may drive the weights to extreme values. These are the offsets responsible for the mismatch effects in Fig. 15. These effects may be curtailed by thresholding the output of the CHL circuits before performing the

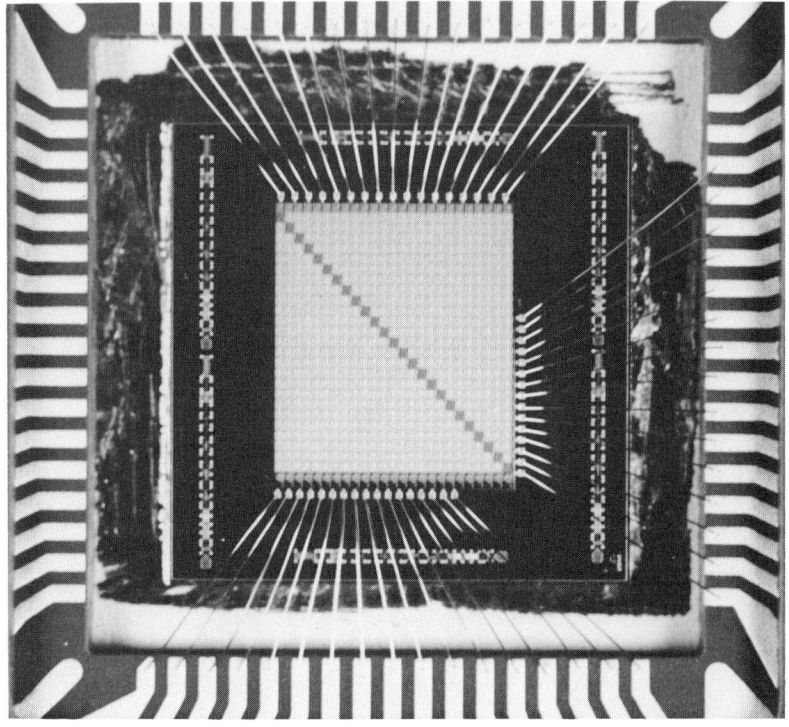

Figure 14 Analog ANN implementation as 3µm CMOS chip with in situ Hebbian learning (25 neurons, 600 synapses). 1.2µm versions of mean field networks have also been fabricated and tested (Schneider and Card 1991, 1992).

Manhattan updates, at the expense of small increases in classification error. It was also found that, when hidden units were present, a small learning step size was required to obtain stable learning. These and other simulations of mean field networks based on our measured neuron and synapse properties (multiplier characteristics, interchip variations, etc) suggest that these systems learn in spite of nonideal components. It turns out that there are certain tasks which mean field networks have difficulty learning and some with unstable learning behaviour. Networks of nonideal components seem however to be able to solve the same problems as those with ideal components.

Arima *et al* of Mitsubishi (1991) have also reported analog learning of capacitive weights in CMOS networks. In their designs, charge pump circuits implement the learning algorithm. Weights are changed in rather large increments of 10% of their full-range values. Chips containing up to 400 neurons and 40K synapses in 1.0 and 0.8 μm CMOS technologies for 1 bit neural activations and 5 bit weights perform 10^{12} CPS and 3×10^{10} CUPS (learning mode). The capacitive weights are updated at 100 ms intervals. The large neural integration levels are a result of employing only 0 or 1 values of activations, and of using coarse learning updates. Learning is similar to that of Boltzmann machines, but without statistical correlations. It also lacks the full multiplicative correlations of mean field approximations.

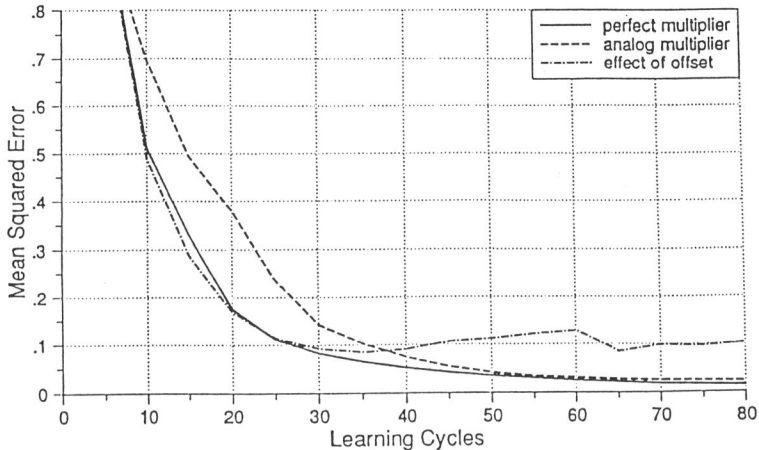

Figure 15 Simulation of supervised learning in mean field network based on measured neuron and synapse parameters (Card *et al*, in press). The network gracefully tolerates component inaccuracy in multiplication but not offsets in addition operations.

True correlations of even single bit activations require +1 and -1 activations. Even Manhattan updates in mean field networks occur after the sign of the difference of multiplicative correlations in the teacher and student phases are determined. The chips of Arima et al may be cascaded and have been shown to learn multiple binary patterns. Weights are refreshed by periodic repetition of the training set. An important contribution of this work is the demonstration that high density analog chips with in situ learning of capacitive weights can perform classification or pattern completion tasks. These chips can been configured to have hidden units. However learning tasks reported using these chips did not involve hidden units, but were restricted to a single layer of weights. Our simulations show that, with tasks requiring hidden units, in other words with difficult learning tasks, fine control of learning updates is necessary.

EEPROM-based weights. An advantage of capacitive weights is in the compact layouts that result. A disadvantage may be in the weight decay due to leakage from the capacitor. Weight decay can be advantageous, but one would like to be able to control its magnitude. In other words one wants storage elements that have little or no inherent decay, and to add active decay circuitry if desirable. For this purpose, one may wish to replace the synaptic capacitors above with EEPROMs or other non-volatile storage devices. EEPROMs (electrically erasable/programmable read-only memories) have traditionally been binary storage devices, and it is only recently that they have been employed as analog storage elements in ANNs (Holler et al 1989).

EEPROMs employ a floating gate FG to store charge in a nonvolatile way. The FG charge is usually modified using tunneling currents, but it may also be induced to exhibit bidirectional analog weight changes using ultraviolet light. Two current problems with EEPROMs are (i) their requirement for relatively large programming voltages (progress is being made to reduce these) and (ii) the dependence of programming voltages, for a specified weight increment, upon the FG charge, i.e. upon the present weight value. There may also be significant variations in these parameters among devices, and unwanted effects due to modulation of their characterisitics with repeated write/erase cycles.

Other implementations of synapses. Among the early reported implementations of ANNs as analog systems were the MNOS/CCD circuits of Sage *et al* (1986), who combined the analog storage of MNOS devices, which are one type of EEPROM, with the analog signal processing capabilities of CCDs (charge coupled devices). This approach was based on work in the 1970s on adaptive signal processing systems, which are closely related to ANNs, and may be regarded as their linear precursors. A further advantage of CCDs is that these devices have been investigated intensively as imagers, with input as 2-D arrays of photodetectors, as for example in solid state television cameras. We do not however pursue these circuits here since they have not to this point included in situ learning.

Compact multipliers have also been developed for synapses in analog pulse stream networks (Murray *et al* 1991). A pulse stream representation of the neural activation switches a current source on and off, either supplying or withdrawing a packet of charge to/from the output node. This results in very compact layouts. So far these synapses have also not incorporated learning mechanisms. An alternative approach with pulse stream systems is to use analog/digital hybrid circuits based on pulse width modulation to perform the multiplication operation. Note that weight storage in these designs is capacitive as in the multipliers described above. This means that if the weights are to be obtained from digital registers for loading/unloading (since no learning is present on chip) one must employ D/A converters. Weights are in external RAM and can share D/A facilities for serial downloading. Pulse coded competitive learning (Watola *et al* 1992) and other adaptive algorithms are just now beginning to be explored by several groups.

Summary

Analog CMOS circuits with in situ learning mechanisms appropriate to a variety of ANN models have been demonstrated. Most of this work is at the level of very small networks, as the embedded learning mechanisms make these time variant as well as nonlinear circuits, and their performance, stability, and other properties are complex. A considerable amount of work has been performed on circuits with lateral inhibition for use in competitive learning in unsupervised systems. In supervised tasks, algorithms such as mean field learning have advantages over the more common delta rules such as backpropagation, as a consequence of learning rules which are dependent only upon information local to the synapses.

Acknowledgements

The author wishes to thank his graduate students Chris Schneider, Roland Schneider, Dean McNeill, Brion Dolenko, and Jeff Dickson of the University of Manitoba, and also Geoffrey Hinton of the University of Toronto for discussions on the topics of this paper. The financial support of NSERC and Micronet is also gratefully acknowledged.

References

Ackley, D.H., Hinton, G.E. and Sejnowski, T.J. "A learning algorithm for Boltzmann machines", *Cognitive Science*, Vol. 9, pp.147-169, 1985.
Alspector, J., Allen, R.B., Jayakumar, A., Zeppenfeld, T., and Meir, R., "Relaxation networks for large supervised learning problems", in *Advances in Neural Information Processing Systems 3*, D. S. Touretzky, J. Moody, and R. Lippmann, editors, San Mateo, CA: Morgan Kaufmann, pp. 1015-1021, Apr. 1991.
Arima, Y.,. Mashiko, K., Okada, K., Yamada, T., Maeda, A., Notani, H., Kondoh, H., and Kayano, S., "A 336-Neuron, 28K Synapse, Self-Learning Neural Network Chip with Branch-Neuron-Unit Architecture", *IEEE J. Solid St. Ccts.*, Vol. 26, pp. 1637-1644, Nov. 1991.
Becker. S. and Hinton, G. E., "Self-Organizing Neural Network that Discovers Surfaces in Random-Dot Stereograms", *Nature*, Vol. 355, pp. 161-163, 1992.
Card, H.C., Schneider, C.R., and Schneider, R.S., "Learning Capacitive Weights in Analog CMOS Neural Networks", *J. VLSI Signal Proc.*, in press.
Card, H. C., Schneider, C., and Moore, W.R., "Hebbian plasticity in MOS synapses", *IEE Proc. F*, Vol. 138, pp. 13-16, Feb. 1991.
Carpenter, G.A. and Grossberg, S., "ART-2: Self Organization of Stable Category Recognition Codes for Analog Input Patterns", *Appl. Optics*, Vol., 26, pp. 4919-4930, 1987.
Durbin, R. and Willshaw, D., "An Analog Approach to the Travelling Salesman Problem using an Elastic Net Method", *Nature*, Vol. 326, pp. 689-691, 1987.
Hebb, D.O. , *Organization of Behaviour*. New York: John Wiley, 1949.
Hertz, J., Krogh, A., and Palmer, R. G., *Introduction to the Theory of Neural Computation*. Menlo Park, CA: Addison-Wesley, 1991.
Hinton, G. E., "Connectionist learning procedures", *Artificial Intelligence*, Vol. 40, pp. 185-234, 1989.
Hochet, B., Peiris, V., Abdo, S., and Declercq, M.J., "Implementation of a Learning Kohonen Neuron based on a new Multilevel Storage Technique", *IEEE J. Solid St. Ccts.*, Vol. 26, pp. 262-267, Mar 1991.
Holler, M., Tam, S., Castro, H., and Benson, R., "An electrically trainable artificial neural network with 10240 floating gate synapses", *Proc. IJCNN-89*, Part II, pp, 191-196, 1989.
Hopfield, J., "Neurons with Graded Response have Collective Computational Properties like those of Two-State Neurons", *Proc. Natl. Acad. Sci.*, Vol. 81, pp. 3088-3092, 1984.
Kohonen, T., "The Self Organizing Map", *Proc. IEEE,* Vol. 78, pp. 1464-1480, Sept 1990.
Lazzaro, J., Ryckebusch, S., Mahowald, M.A., and Mead, C.A., "Winner-Take-All Networks of O(N) Complexity", in *Advances in Neural Information Processing Systems I*, D.S. Touretzky, editor, San Mateo, CA: Morgan Kaufmann , pp. 703-711, 1989.

Linares-Barranco, R., Sanchez-Sinencio, E., Rodriguez-Vazquez, A., and Huertas, J.L., "A Modular T-Mode Design Approach for Analog Neural Network Hardware Implementations", *IEEE J. Solid St. Ccts.*, Vol. 27, pp. 701-713, 1992.

Linsker, R., "Self Organization in a Perceptual Network", *IEEE Computer*, Vol. 21, pp. 105-117, 1988.

Mahowald, M. and Douglas, R., "A Silicon Neuron", *Nature*, Vol. 354, pp. 515-518, Dec 1991.

Mann, J. R. and Gilbert, S., "An Analog Self-Organizing Neural Network Chip", in *Advances in Neural Information Processing Systems 1*, D.S. Touretzky, editor, San Mateo, CA: Morgan Kaufmann, pp. 739-747, 1990.

Maundy, B. and El-Masry, E., "Switched Capacitor Neural Networks using Pulse Based Arithmetic", *Elect. Lett.*, Vol. 26, pp. 1118-1119, 1990.

Mead, C.A., *Analog VLSI and Neural Systems*. Reading: Addison-Wesley, 1988.

Moody, J. and Darken, C.J., "Fast Learning in Networks of Locally Tuned Processing Units", *Neural Computation*, Vol. 1, pp. 281-294, 1989.

Movellan, J. R., "Contrastive Hebbian Learning in the Continuous Hopfield Model", *Connectionist Models: Proceedings of the 1990 Summer School*, D.S. Touretzky et al, editors, pp. 10-17, 1990.

Murray, A. F., Del Corso, D., and Tarassenko, L., "Pulse Stream VLSI Neural Networks Mixing Analog and Digital Techniques", *IEEE Trans. on Neural Networks*, Vol. 2, pp. 193-204, Mar. 1991.

Nowlan, S.J., and Hinton, G.E., "Evaluation of Adaptive Mixtures of Competing Experts", in *Advances in Neural Information Processing Systems 3*, D. S. Touretzky, J. Moody, and R. Lippmann, editors, San Mateo, CA: Morgan Kaufmann, pp. 774-780, 1991.

Oja, E., "Neural Networks, Principal Components, and Subspaces", *Int. J. Neural Systems*, Vol. 1, pp. 61-68, 1989.

Peterson, C. and Anderson, J.R., "A mean field theory learning algorithm for neural networks", *Complex Systems*, Vol. 1, pp. 995-1019, 1987.

Raffel, J., Mann, J., Berger, R., Soares, A., and Gilbert, S., "A Generic Architecture for Wafer Scale Neuromorphic Systems", *Proc. IEEE Int. Conf. Neural Networks*, Vol.III, pp. 501- 513, 1987.

Rumelhart, D.E., Hinton, G.E. and Williams, R.W., "Learning Representations by Back Propagating Errors", *Nature*, Vol. 323, pp.533-536, 1986.

Sage, J.P., Thompson, K., and Withers, R.S., "An Artificial Neural Network Integrated Circuit based on MNOS/CCD Principles", In *AIP Conf. Proc. 151, Neural Networks for Computing*, Snowbird, ed. J.S. Denker, American Institute of Physics, New York, pp. 381-385, 1986.

Sanger, T.D., "Optimal Unsupervised Learning in a Single Layer Linear Feedforward Neural Network", *Neural Networks*, Vol. 2, pp. 459-473, 1989.

Schneider, C. and Card, H.C., "Analog CMOS Contrastive Hebbian Networks", SPIE Proc. 1709, *Applications of Artificial Neural Networks III*, Orlando, Florida, Apr 21-24, pp. 726-735, 1992.

Schneider, C. R. and Card, H. C., "Analog CMOS Hebbian Synapses", *Elec. Lett.*, Vol 27, pp. 785-786, 1991.

Shoemaker, P.A., Lagnado, I., and Shimabukuro, R., "Artificial neural network implementation with floating gate MOS devices", in *Hardware Implementations of Neuron Nets and Synapses*, NSF/ONR Workshop, San Diego, CA, P. Mueller, editor, pp.114-119, Jan 14-15, 1988.

Tomlinson, M.S., Walker, D.J., and Sivilotti, M.A. (1990), "A Digital Neural Network Architecture for VLSI", *Int. Joint Conf. on Neural Networks*, San Diego, Vol. II, pp. 545-550, 1990.

Tsay, S.W. and Newcomb, R.W., "VLSI Implementation of ART1 Memories", *IEEE Trans. Neural Networks*, Vol. 2, pp. 214-221, 1991.

Von der Malsburg, C., "Self Organization of Orientation Selective Cells in Striate Cortex", *Kybernetik*, Vol. 14, pp. 85-100, 1973.

Walker, M., Hasler, P., and Akers, L., "A CMOS Neural Network for Pattern Association", *IEEE Micro*, pp. 68-74, Oct. 1989.

Watola, D., Gembala, D., and Meador, J., "Competitive Learning in Asynchronous Pulse Density Integrated Circuits", *IEEE Int. Symp. on Ccts. and Systems*, San Diego, CA, May 10-13, 1992.

Willshaw, D.J. and von der Malsburg, C., "How Patterned Neural Connections can be Set Up by Self Organization", *Proc. Royal Soc. London*, Vol. B 194, pp. 431-445, 1976.

AN ANALOG CMOS IMPLEMENTATION OF A KOHONEN NETWORK WITH LEARNING CAPABILITY

Oliver Landolt

INTRODUCTION

Kohonen (1988) introduced a new type of neural network that exhibits very interesting pattern classification and clustering properties. So far, they have been successfully used to solve various problems in fields like image processing or robot control. Unfortunately, although the task each cell has to accomplish is rather simple, computing time of software simulations becomes prohibitive as the size of a network increases. For real-time applications, parallel processing is required. Analog circuits are believed to allow potentially highest density of integration, and highest speed. However, since density is essential, only poor accuracy may be achieved. It is therefore necessary to select architectures that guarantee proper behaviour of the network despite of all error sources that may affect it.

A circuit has been designed that implements a one dimensional Kohonen network, the inputs of which being two-dimensional vectors. Those two dimensions were chosen to meet the requirements of a particular application (Sorouchyari 1992), and are not set by any fundamental limitation. The main goals considered include full on-chip learning, and multichip functioning capability. The latter property should allow splitting of very large networks into several chips.

The architecture of the network is being described in the following. Several issues are discussed regarding particular constraints, set by the multichip functioning capability. Then the principle of the original synapse circuit is presented, as well as a quick review of the other parts of the cell. Last, several limits of this circuit are being discussed.

KOHONEN'S ALGORITHM

A Kohonen network is built upon one layer of cells arranged in a string or a lattice fashion. Each cell stores one N-dimensional vector M_i, representing N synaptic weights. A vector of inputs X is distributed to all cells. Kohonen's so-called simplified learning algorithm consists of iterating the following operations:

- provide one input vector X selected at random in the input space;
- determine which cell in the network stores the M_i vector most similar to X; this best matching cell will be further called the *winner*;
- select a group of cells centered around the winner; this group is called the *activity bubble*;
- update the synaptic vector of each cell in this group according to following equation:

$$M_i(k+1) = M_i(k) + \alpha (X(k) - M_i(k)) \tag{1}$$

where k denotes a discrete time index, and α is a parameter ranging from 0 to 1.

The parameter α and the size of the activity bubble, i.e. the number of cells which are being updated during each cycle, may vary in time. Large values of those parameters confers the network high flexibility, and are advisable at the beginning of learning, whereas low values make it more slow and stable, allowing good expansion during the final phase of learning.

The similarity between the input vector X and a synaptic vector M_i defines the *activity* of the cell. It may be measured by various types of criteria, the most popular of which is the Euclidean distance. A cell is said to be sensitive to a given input vector if it shows high activity on presentation of this particular vector.

By repeating the learning cycle many times, the synaptic vectors in the network tend to cover the input space in such a way that neighbouring cells in the network become sensitive to neighbouring vectors in the input space. Moreover, the distribution of the synaptic vectors reproduces the density function of the input vectors during training.

NETWORK ARCHITECTURE

Cell structure

The Kohonen network circuit is made of a chain of identical cells. The block scheme of one cell is drawn in Figure 1. An input vector X, the two components of which are denoted X_1 and X_2, is applied to the cell. In each synapse, one component of the synaptic vector M_i is being stored. The activity of the cell, which is measured by the Euclidean distance between X and M_i, is computed by a dedicated circuit (Landolt *et al* 1992).

A so-called Winner Take All (WTA) selects the most active cell in the network. It is a distributed circuit, i.e. each cell contains one elementary part of it. The collective action of all elementary parts results in the selection of the winning cell.

An activity bubble is determined around the winner using a non-linear diffusion network (Heim *et al* 1991). Each cell is being associated one node of this network. The most active cell injects a signal onto its own node, pulling up the nodes associated to its immediate neighbours. Every cell the node of which reflects sufficient activity is said to belong to the activity bubble.

In order to achieve learning, the synaptic vectors of the cells that belong to the bubble are being updated according to Equation 1. This computation is being done by the synapses, the principle of which will be described in a further section.

Note that the process of generating a bubble for a given input vector doesn't need to be a sequential, discrete-time process, but may occur continually. However, updating the synaptic vector M_i has to be done in a second time phase according to Kohonen's original algorithm, because M_i has to remain constant while determining the bubble.

Reference problem

As mentioned in the introduction, one particular requirement to this design is the possibility splitting a large network into several chips. A network is being built upon cells which should ideally be all identical. Moreover, they should all be provided with exactly the same input signals. In reality, even if all the cells are fabricated from the same layout, characteristics may vary from cell to cell due to the spread of technological parameters. Matching between cells belonging to the same chip may usually be considered as satisfactory, and thus most operations may be considered to perform exactly the same way on the whole chip. This is no longer true between cells on two different chips, for which parameters such as threshold voltage or gain factor cannot be guaranteed to match better than within 20% or 30%. The consequence of this is that any operator dependent on those parameters, such as most voltage to current convertors, will yield an output scaled to a different reference in each chip. This may or may not be acceptable depending on the application. If it is not, then the circuit should be designed either by avoiding any operation dependent on the process parameters, or by performing those operations according to some reference element or signal distributed exactly to all the chips.

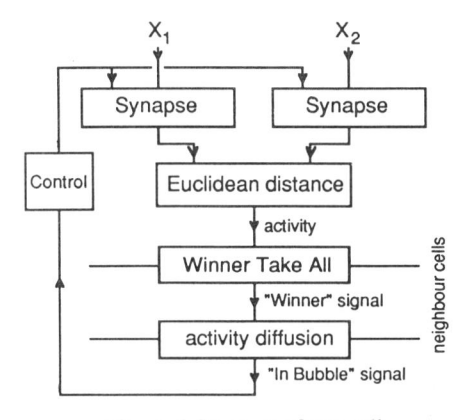

Figure 1 Structure of one cell

In the case of a Kohonen network, it is not a fundamental problem having signals scaled to various references *as long as they are internal to a cell*, and do not interact with signals provided by other cells. For instance, if the input vector **X** is not distributed exactly to all the cells, but is affected by offset or scaling, the internal synaptic vector of each cell won't take the same value as if **X** would be ideal, but *each cell will still be most sensitive to the same input vector* as in the perfect case, if the imperfections are constant in time.

However, all global signals or information exchange between neighbour cells have to be represented according to some common reference. The application this circuit was primarily designed for (Sorouchyari 1992) requires computing the length of the network, i.e. the sum of the distances between the points each pair of adjacent cells are most sensitive on. Computing these distances is easiest done by using the synaptic vectors themselves. Thus, a linear correspondence is required between the space of input vectors and the space of synaptic weights.

The global signals distributed to the network include the input vector **X**, and one signal for the Winner Take All. Local information exchange is provided by the diffusion network that propagates activity from one cell to its neighbours. To allow distributing the global signals to the whole network, they must be represented by voltages, since it is difficult providing exact copies of a current to several separate chips. On the other hand, current-mode circuits were found much more convenient to implement a cell. Thus, the input vector has to be converted to currents, and the activity of each cell has to be converted back to a voltage in the Winner Take All. In this implementation, the first conversion is done using external precision resistors as a reference. The second uses normalization to a current delivered by a special reference circuit (Heim 1991) compensating for the dependence of the convertor on the process parameters. All other operations in the cells are independent on those parameters.

Chip structure

A block schematic of one chip building the network is shown in Figure 2. Only three cells are drawn for convenience, but there is no fundamental limit to the number of cells one single chip may contain.

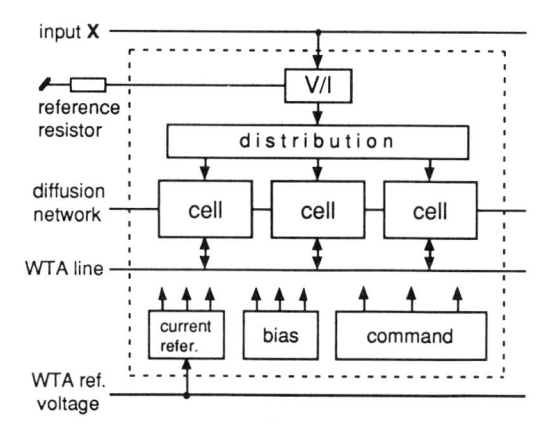

Figure 2 Block schematic of one chip

In addition to the cells themselves, the common blocks and reference elements are represented. The V/I conversion for the input vector is shown explicitly; the corresponding circuit doesn't require much surface, since the conversion has to be performed only once in each chip. The current signal may then be replicated easily by a multiple output current mirror, with acceptable accuracy. On the other side, an I/V conversion of the activity is done in each cell as part of the Winner Take All block. Reducing the need for surface expensive normalization circuits is achieved by a hierarchical approach. Details will be given in the next section.

MAJOR BUILDING BLOCKS

Several building blocks used in this architecture are completely new designs. These include the synapse, the Euclidean distance circuit, and the architecture of the Winner Take All. Other solutions, such as the diffusion network (Heim *et al* 1991) or the implementation of the process-independant Winner Take All (Heim 1991) were adapted from previous ideas. The explanations given below focus on the original and unpublished work. They intentionally stay on the level of principles, and don't include details about the actual circuit implementation. Interested readers should refer to the given references to get more information about such aspects.

Synapse

Each synapse is expected to perform two major functions. The first is storing one component of the synaptic vector, or synaptic weight. The second is being able to update this weight according to Kohonen's learning algorithm (Equation 1). The synapse circuit, whose principle is illustrated in Figure 3, is built upon an integrator and a set of switches.

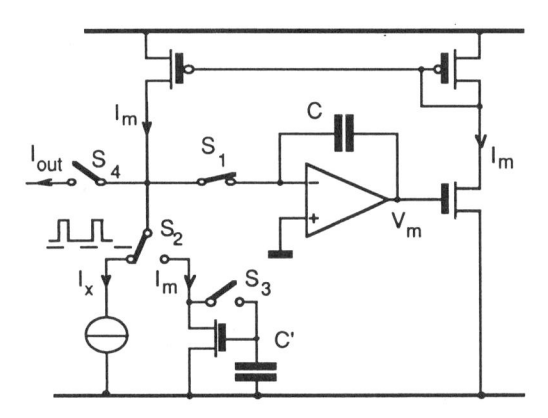

Figure 3 Principle of the synapse circuit

Memory. The integrator is being used as a low leakage capacitive memory (Vittoz *et al* 1990). This configuration allows holding the stored value for several minutes with acceptable loss of charge. This is sufficient for using the network under continuous training. When one-time learning is required, the weights should be stored in long-term memory devices such as floating gates (Säckinger and Guggenbühl 1988) after training.

The output voltage V_m of the integrator is converted to a current I_m which is fed back to the input. Writing the memory is done with the switch S_1 closed, by integrating the difference between the input current and I_m. The feedback ensures that I_m converges to the input current without offset. Although the physical type of element that is stored is a charge on the capacitor, the actual content of the memory is considered to be the current I_m.

The response of this structure to a small input current step is that of a first-order low pass filter, assuming that the change in V_m remains small enough to consider the V_m to I_m conversion as linear. The time constant τ of the response depends on the operating point V_m, and is given by Equation 2. This filtering property is being exploited to update the memory.

$$\tau(V_m) = C / \frac{dI_m}{dV_m} \qquad (2)$$

The content of the memory is being hold by opening the switch S_1. Since the ideal switch represented in Figure 3 is implemented by a MOS transistor, some charge will be dumped from its channel to the storage capacitor as it will be cut off. Although compensation techniques might be used (Vittoz and Wegmann 1990), this charge injection phenomena cannot be totally avoided, and will set a limit to the accuracy that may be expected from this circuit. In practise, the error at each cycle can be kept very low, but its effect over a large number of cycles may not be negligible, and will be discussed later on.

Learning. A temporary memory and two switches S_2 and S_3 are needed to implement Kohonen's algorithm. The problem consists of determining the next I_m to load into the memory, given the present value of I_m, the input current I_x, and the learning coefficient α. Using these symbols to represent one component of M_i and X, Equation 1 may be rewritten as

$$I_m(k+1) = \alpha\, I_x(k) + (1-\alpha)\, I_m(k) \tag{3}$$

Updating the memory may thus be achieved by holding $I_m(k)$ on the auxiliary memory, and by switching periodically between this copy of $I_m(k)$ and the input current $I_x(k)$ by means of switch S_2. If T represents the period of the switch control signal, S_2 is positioned towards I_x during a fraction of time αT, and towards I_m the remaining of the period $(1-\alpha)T$. The resulting signal is a *pulsed current with a DC component equal to $I_m(k+1)$*. If the period T is much smaller than the time constant τ of the memory (Equation 2), the AC components are filtered out, and the content of the memory converges to the correct value.

The auxiliary memory needs to hold I_m only for a very short time. A simple current copier (Vittoz and Wegmann 1990), as drawn in Figure 3, is sufficient to meet the storage requirements.

This way of updating the weights provides a few important advantages. First, it allows very easy and flexible control of the parameter α by adjusting the duty cycle of a periodic, binary control signal. Moreover, α can be guarantied to be exactly the same in all synapses driven by the same control signal, in particular for all components of a given synaptic vector. Last, the new value stored onto the memory is fundamentally guarantied to stay between I_m and I_x. In other words, the parameter α ranges strictly from 0 to 1. This is important to ensure convergence of the Kohonen algorithm.

Output. The weights stored in the synapses of a cell are being used for two different purposes. The first is to compute the activity of that cell, i.e. the Euclidean distance between X and M_i. Hence, the first operation to be done with each component I_m is subtracting it from the corresponding I_x. The second purpose is to determine the new values for the updating of the memory.

It would be easy to add an extra output to the current mirror of the memory circuit to get an extra copy of I_m. However, this copy would be accurate only to the matching of the two transistors involved. Another solution is taking advantage of the fact that computing the activity and updating the weights are done in two different time phases. Since two copies of I_m are never required simultaneously, *the original and accurate I_m may be switched* between one task and the other.

It may be seen in Figure 3 that the synapse circuit allows achieving both functions easily and precisely. By sending the current I_x to the output of the memory via S_2, a current I_{out} equal to $I_m - I_x$ may be extracted through an output switch S_4, and may be sent directly to the distance circuit. On the other hand, holding an exact copy of I_m may be achieved by sending the original I_m to the current copier by setting S_2 accordingly.

Operation. As described above, this synapse circuit is being used successively in various modes of operation, which are selected by setting the switches properly. Its major functions are listed in Table 1. In this table, the status of switch S_2 is given by "left" and "right", which refer to the drawing in Figure 3.

Table 1 Synapse modes of operation

Operation	S_1	S_2	S_3	S_4
send I_m-I_x to distance circuit	off	left	don't care	on
store I_m into temporary memory	off	right	on	off
update main memory	on	alternate	off	off
store I_x into main memory	on	left	don't care	off

The first three operations listed in Table 1 are necessary for the learning cycles. The last one has been added for initialization purposes. In addition to these basic operations, several other intermediate states should be added to ensure a proper switching sequence during transitions from one basic state to another. Furthermore, each application for which this circuit would be used, might require its own set of extra operations depending on some specific circuitry added for this application, and on the actual output expected from the network.

The flow of operations the neurons have to perform, has to be controlled by a digital sequencer. It might either be implemented on chip to get a fixed sequence, or controlled more flexibily by an external processor through a decoder. This implementation of a Kohonen network is therefore an example of *analog processing unit*, the operation of which is being controlled by a digital command unit.

Winner Take All

Lazzaro *et al* (1988) proposed a very efficient Winner Take All circuit to find out the largest of a set of currents. It is made of a chain of elementary parts taking one input current each. All of these elements are connected to a unique global wire, the voltage of which is being adjusted to represent the maximum input value. The relation between this voltage and the current it represents depends on process parameters. Thus, this solution cannot be used in its basic form for a network distributed on multiple chips.

As mentioned earlier, this problem may be overcome by normalizing the input currents properly in order their voltage representation to match some common reference. Unfortunately, such normalization circuits are quite surface expensive, and including one of them in each cell of the network would increase its size sensibly. This is why a two-stage hierarchical Winner-Take-All circuit was chosen. The first stage selects the most active cell of the chip, making use of the simple version of the circuit. The second stage selects the most active chip in the network, thus only one normalization circuit is required for each chip. The price to pay for this economy of surface is some loss of accuracy of the Winner Take All, and some increase of complexity of the circuit.

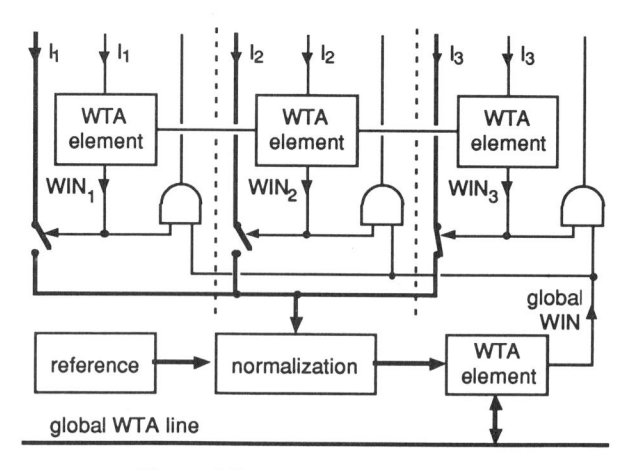

Figure 4 Two stage Winner Take All

A block scheme of this two stage solution is depicted in Figure 4. Two distinct copies of each input current is required. The local Winner Take All is based on Lazzaro's circuit. Access to the global level is given to the second copy of the largest current by means of a switch. The second stage is being implemented by an enhanced version of Lazzaro's circuit. Determining which cell won is done by combining the local and global binary outputs with an AND gate. This signal will be used for generating the activity bubble by means of the diffusion network.

LIMITS

Any practical implementation of a system suffers some restrictions compared to its theoretical, ideal definition. Quantization is the typical limitation of software or digital implementations. Offsets or other accuracy limits is the main trouble of analog circuits. In any case, it's important evaluating the impact of those limits, and determining under which conditions the real system may be considered as equivalent to its theoretical definition. The specific limitations of the Kohonen circuit presented above will be discussed in the following.

Charge injection

Each time the access transistor to a capacitor is blocked (switch S_1 in Figure 3), a fixed amount of charge is dumped from its channel to the capacitor. This charge injection phenomena induces a small systematic error on the value of each synaptic weight, every time it is being updated. In the space of synaptic vectors, this might be considered as a fixed direction and magnitude error vector ΔM_ε which is added to M_i every time it is modified.

The global tendency of the synaptic vectors due to such a systematic effect is sliding progressively towards one corner of the space. The Kohonen algorithm may intrinsically compensate for this tendency, since the mean correction made to weights should grow towards the empty space if the density of synapses is higher in some region. This can be true only if *the mean correction made to the synaptic vectors remains much larger than the systematic error*. The mean magnitude of updates increases with α and decreases with the density of synaptic vectors in space. This density is indirectly related to the number of cells in the network.

Charge leakage

Leakage of charge from the capacitive memories is another systematic effect, the consequences of which are similar to those of charge injection. However, the conditions on the circuit operation differ since this phenomena depends on time, whereas charge injection is bound to the number of learning iterations. Again, a necessary condition for the network to possibly behave correctly is that *the mean speed of each synaptic vector in space is much larger than the systematic drift* which is applied to it. The speed of the synaptic vectors increases with the cadence of the iteration loops, with α and with the size of the bubble.

The two conditions on charge injection and leakage are clearly necessary. It is not clear whether they are *sufficient*, i.e. if the network remains in some stable state with such a systematic error as time and the number of training iterations grow to infinity. By this time, no theoretical work on the Kohonen network provides a definite answer to such questions.

Other limits

More bounds are set to the performance of this implementation by the limited accuracy of analog circuits. In particular, the finite discrimination capability of the Winner Take All might induce errors if the local density of synaptic vectors gets too high around a given input vector. Such situations relate indirectly to the number of cells in the network, and to the topology of the space of input vectors. Hence, it is very difficult identifying clear criteria to distinguish what is acceptable and what is not. However, such problems are far less important than the systematic effects affecting the synaptic weights since they won't be accumulated over many iterations. Thus, the classification properties of the network might suffer, but the convergence should not be affected.

CONCLUSION

The architecture presented here to implement a Kohonen network by an analog circuit leads to a realistic solution. The size, accuracy and other expected performance should make it possible integrating large networks for practical applications. If no more than about 200 cells are required for a given application, a one chip implementation may be considered, allowing some simplifications to the design.

A test chip has been designed including six cells and the related common circuitry. It has been integrated using a double-metal, double-poly 1.2µm process. Each cell occupies only about $0.1mm^2$ on the chip. The network is expected to learn at rates over 10000 iterations a second, independently of the number of cells. Measurement results are not available yet.

One particular problem met in this work requires some attention. Software simulations showed the non-negligible influence of systematic perturbations in the training process of the Kohonen network. Minimising such errors was one key goal for this design, and required much effort until a suitable structure was found. This need for accuracy in implementing the learning algorithm is quite unexpected, since neural networks are generally said to be very robust against imprecision or individual failures of their cells. Biological networks show a high structural robustness even if the cells they are made of perform a very imprecise calculation. This divergence may lead to the conclusion that *some key feature allowing potential robustness is not exploited* by Kohonen's simplified algorithm.

A major source of the robustness of neural networks is the fact that their output is the result of a collective computation involving a large number of units. Hence, a change in the behaviour of an individual cell doesn't affect much the overall result. One reason for the observed sensitivity in the Kohonen network might be that none of the steps involved in the learning process is the result of a collective computation. Hence, the algorithm fits well to software implementations (which is probably the purpose it was developed for) that are almost immune to uncontrolled perturbations, but has no fundamental properties anymore that may compensate efficiently for imperfections encountered in analog chips.

Therefore, another approach for implementing a Kohonen network by an analog circuit might be stepping back to the continuous-time differential equation from which the simple algorithm was deduced (Kohonen 1988). Using the same symbols as in Equation 1, this equation is given by

$$\frac{dM_i}{dt} = \alpha \eta_i(t) (X(t) - M_i(t)) \qquad (4)$$

where $\eta_i(t)$ denotes the activity of cell i at time t. This activity is the collective result of the activity of that cell and the activities of its neighbours through a set of excitatory and inhibitory connections. It is believed that a network built upon this more "biological" equation might show better properties than the simplified algorithm for an analog circuit implementation.

Acknowledgment

The author would like to thank E. Vittoz at CSEM, and P. Heim at the Swiss Federal Institute of Technology (EPFL), for their continuous support throughout the work. This research was supported by the ESPRIT-BRA.3049 *NERVES* project ("Innovative Architectures for Neuro-Computing Machines and VLSI Neural Networks"), and the Fonds national de la recherche scientifique (FNRS), project ref. #21-25622.88.

References

Heim, P., Hochet, B.and Vittoz, E., "Generation of Learning Neighbourhood in Kohonen Feature Map by Means of Simple Nonlinear Network", Electronics Letters, January 1991, Vol. 27, No 3, pp. 275-277

Heim, P., private communication, 1991

Kohonen, T., *Self-Organisation and Associative Memory*, Springer Verlag 1988, ch. 5, pp. 119-157

Lazzaro, J., Ryckebush, S., Mahowald, M. A. and Mead, C., "Winner-Take-All networks of order N complexity", Proc. 1988 IEEE conference on neural information processing - natural and synthetic, Denver, 1988, pp. 703-711

Landolt, O., Vittoz, E. and Heim, P., "CMOS self-biased Euclidean distance computing circuit with high dynamic range", Electronics Letters, February 1992, Vol. 28, No 4, pp. 352-354

Säckinger, E. and Guggenbühl, W., "An analog trimming circuit based on a floating-gate device", IEEE Journal of Solid State Circuits, Vol. SC-23, No. 6, December 1988, pp. 1437-1440

Sorouchyari, E., "Fractal dimension computation based on a self-organizing neural network", to be published

Vittoz, E.*et al*, "Analog storage of adjustable synaptic weights", Proc. ITG/IEEE workshop on microelectronics for neural networks, Dortmund, Germany, 1990

BACK-PROPAGATION LEARNING ALGORITHMS FOR ANALOG VLSI IMPLEMENTATION

Maurizio Valle, Daniele D. Caviglia and Giacomo M. Bisio

INTRODUCTION

Many different VLSI implementations of Neural Networks (NNs) have been proposed: digital, analog and mixed-mode (IEEE Micro ,1989, and Murray, 1991). Advocates of digital VLSI NNs (DVNNs) (Ramacher, 1991) claim the higher computational accuracy, noise immunity and speed of digital implementations. On the other hand, analog VLSI NNs (AVNNs) are more closely related to the biological models and are better suited to exploit innovative circuit solutions. The discussion about the best implementation approach for VLSI NNs is still open. Anyway, the choice (digital versus analog) strongly depends on the application.

Main features of AVNNs, as outlined by Mead (1989, 1990) and Vittoz (1990), are:

- the information is represented by the relative values of electric analog signals;
- the computation is carried out in an analog way;
- elementary physical phenomena are used as computational primitives.

Advantages of AVNNs are:

- real time processing;
- low power consumption (circuits working in non-conventional regions of operation of devices, e.g. subthreshold);
- high density; i.e., simple, compact and small area elementary circuits (neurons and synapses).

Nevertheless, potential disadvantages are represented by devices imperfections and non idealities, that strongly affect the performances of common analog circuits (e.g., converters, filters, etc.). Moreover in AVNNs, to obtain an efficient (low area) implementation, synaptic multipliers are implemented using analog circuits operating in the strong (Long and Guggenbühl, 1992) or weak inversion region (Mead, 1990), though they feature strong non linearities. To answer these questions, it is necessary to investigate the relationships between the neural algorithm and the non ideal computational behavior of neurons and synapses, that reflects their circuit implementation.

In this paper, we will investigate these questions with references to the case of Multi Layer Perceptron networks trained by the Back Propagation (BP) learning algorithm. The BP algorithm (Rumelhart and McClelland, 1986, and Vogl et al, 1988) is a supervised learning algorithm. The learning refers to the problem of finding a network function that approximates some desired target function. Usually the network function is an input-output mapping on a set of examples, the training set. The learning is an iteration procedure: at each iteration step the synaptic weights are adjusted to minimize an error measure. Early attempts have been made to implement the BP algorithm with analog VLSI circuits (Paulos and Hollis, 1988, Furman et al, 1988, Shima et al, 1991). It has been observed (Widrow and Lehr, 1990, Jabri and Flower 1992) that the BP algorithm is not suited for the analog hardware implementation because it needs high computational accuracy particularly in the learning phase. To solve this problem, two approaches can be followed. One (Reyneri and Filippi, 1991) tries to attain good performances searching for a near-optimum value for the learning rate. Otherwise, one can analyses the features of VLSI analog circuits to derive a constraint-driven formulation of the BP algorithm. Thus the precision requirements of the analog implementation are mitigated by exploiting the heuristics in the formulation of the algorithm to increase the adaptativity and locality of computation. In this way, it is possible to use simple circuits at the expense of a slower learning procedure. In this paper we illustrate this last approach presenting a new formulation of the Back-Propagation algorithm. The variants respect to the standard algorithm are introduced to satisfy the constraints imposed by the analog VLSI implementation. The paper demonstrates, through functional simulations, that the new formulation performs efficiently even with nonlinear synapses.

This paper is organized as follows. Section II discusses the design and verification methodology for AVNNs. Section III presents the analog VLSI constraint-driven version of the BP algorithm. Section IV illustrates the circuit implementation of synaptic multipliers and its features. Section V describes the simulation results.

DESIGN AND VERIFICATION METHODOLOGY

An AVNN system, like any other complex computational system, can be decomposed hierarchically into a set of simpler components: system, module, macrocell, basic cell, device. Each level of the hierarchy is characterized by a particular type of description that is identified by the primitives used to describe and verify the system: behavioral (algorithm + application), architectural, circuit macromodel, circuit. Usually, the design process follows a top-down flow that explores the hierarchy: each level of description is detailed in the implementation level immediately below.

In AVNNs computations are performed using analog circuits that feature non ideal and non linear behavior. If an efficient implementation (small area and low power consumption circuits) is pursued, the non linearities strongly affect the performances of the system. Consequently, the top-down design flow must be slightly modified. *The non linearities of neuron and synapse circuits are viewed no longer as imperfections but as specific features of a computational device.* The macromodel description (Casinovi and Sangiovanni-Vincentelli, 1991) of the circuits is then propagated upwards to the previous description levels until the behavioral level is reached. *At this level the constraints of the circuit implementation force the designer to reformulate the neural algorithm.* In this paper we focus our attention on the behavioral simulations, that must consider the analog macromodels of the basic circuit blocks. The macromodels are abstract mathematical models that hide the details of circuit implementation, and captures not only the input/output function of the circuit is to perform, but also the non idealities intrinsic to analog operation.

AN ANALOG VLSI CONSTRAINT-DRIVEN VERSION OF BACK-PROPAGATION

In a learning problem the network and the node function in general are fixed while the interconnection weight vector $\overline{w}$ (where w_{ji} is the synaptic weight from neuron i to neuron j) is the only free parameter of the system. Learning refers to the problem of finding a network function $O(\cdot)$ that approximates some desired target function $T(\cdot)$, defined over the same set $\overline{x}$ of input vectors as the network function, and it is accomplished by iteratively adjusting the weight vector to minimize an error measure. The most common error measure is the sum-squared error E over all the patterns p:

$$E(\overline{w}) = \sum_{p \in\, trainingset} E(p, \overline{w}) = \frac{1}{2} \sum_{p \in\, trainingset} \sum_j (T_j(p) - O_j(p, \overline{w}))^2$$

The learning problem is thus reduced to find a weight vector $\overline{w}$ for which $E(\overline{w})$ is minimized. At each iteration step k, the values of the weights are modified moving the weight vector in the direction opposite of the gradient (*gradient descent*):

$$\Delta W_{ji}^{(k)} = -\eta \frac{\partial E}{\partial W_{ji}} = -\eta\, \delta_{pj}\, o_{pi}$$

where η is the learning rate or step size, δ_{pj} is the error contribution from neuron j after presentation of pattern p and o_{pi} is the output of neuron i.

On this basis, the standard update formulation of the BP algorithm is:

$$\Delta W_{ji}^{(k+1)} = -\eta\, \delta_{pj} o_{pi} + \alpha\, \Delta W_{ji}^{(k)}$$

where α is the momentum factor.

Many different versions of the standard algorithm have been proposed in the literature to improve the convergence properties of the algorithm (Vogl et al, 1988, Jacobs 1988, Battiti, 1989, Tollenaere, 1990, Sontag 1991). Vogl et al (1988) introduced the adaptation over time of the learning rate η that varies accordingly with the shape of the local error surface. A more adaptive and local algorithm was introduced first by Jacobs (1988) and subsequently improved by Tollenaere (1990): i) the learning rate η is local at each synapse site (η_{ji}) and varies over time following the computation of the local error derivative ($\frac{\partial E}{\partial W_{ji}}$); ii) weights are updated since the error derivative doesn't change sign; when a change in the sign is detected, the previous weight updates (which caused the change in the derivative sign) are undone.

The momentum factor α low-pass filters the weights updates (Widrow, 1990) but it needs extra-circuitry for the storage of the weight update at the previous iteration: we set $\alpha = 0$ (in general this results in longer learning times, Tollenaere, 1990).

As we will show in the next section, the circuits that update the synaptic weights, memorize them, and perform the ensuing multiplication with the input pattern, can be characterized as follows:

- the weight factor of the feedforward multiplication is controlled by a voltage, through a non-linear dependence;
- such voltage is updated through the BP method and is stored dynamically on a capacitor.

In formal terms, let W be the weight value, V the voltage control value, $W = \Psi(V)$ the functional relation between V and W. If we consider the inverse function $\zeta = \Psi^{-1}$, i.e. $V = \zeta(W)$, the voltage variation can be related to the weight steps of the BP rule (see eq. 1 with $\alpha = 0$), as follows:

$$\Delta V_{ji} = \frac{\partial \zeta}{\partial W_{ji}} \cdot \Delta W_{ji} = -\overline{\eta}_{ji} \cdot \delta_{pj} o_{pi}$$

where

$$\overline{\eta}_{ji} = \eta \cdot \frac{\partial \zeta}{\partial W_{ji}}$$

represents a local learning rate dependent on the bias point of the synaptic multiplier.

This form of the learning rule evidences that, if we operate on the voltage, we should use a learning rate $\overline{\eta}_{ji}$ specific for each synaptic site. However, it is not necessary that the control of the learning rate be able to produce variations of the same extent the function Ψ does. In fact, if one considers the adaptive nature of this type of algorithm, one could develop criteria for determining adaptatively the local learning rates, thus non requiring the explicit knowledge of the bias point of each synaptic circuit. In Jacobs (1988) and Tollenaere (1990) the rate of convergence of BP was increased by providing each weight of the network of its own adaptative learning η_{ji}. We can extend the heuristics of Tollenaere (1990) for our circuit implementation in the following way:

```
do {
```
$$\Delta V_{ji}^{(k)} = - \sum_{p \in \text{trainingset}} \left(\overline{\eta}_{ji}^{(k)} \cdot \delta_j^{(p)} \cdot x_i^{(p)} \right)$$
$$V_{ji}^{(k)} = V_{ji}^{(k-1)} + \Delta V_{ji}^{(k)}$$
$$\text{if } \text{sgn}\left(\frac{\partial E}{\partial W_{ji}}\right)^{(k)} = \text{sgn}\left(\frac{\partial E}{\partial W_{ji}}\right)^{(k-1)} \text{ then}$$
$$\overline{\eta}_{ji}^{(k+1)} = A \cdot \overline{\eta}_{ji}^{(k)} \qquad \{\text{where } A > 1\}$$
else
$$\overline{\eta}_{ji}^{(k+1)} = B \cdot \overline{\eta}_{ji}^{(k)} \qquad \{\text{where } B < 1\}$$
endif
$k \Leftarrow k + 1$
} until Error(k) $\leq \varepsilon$ OR $k \geq k_{\max}$

The decision of increasing or decreasing the learning rate is taken by comparing the sign of the gradient components at two successive iterations. This situation has the consequence that the learning rate can be adapted to overcome the non linearities of the circuit in the exact point they occur, thus ensuring better performances.

Let us consider the circuit implication of this local approach. Firstly, the only thing to be maintained from one iteration to the following is the sign of $\frac{\partial E}{\partial W_{ji}}$, (which is a digital information), and not the actual analog value of ΔW_{ji}. Furthermore, the monitoring of the learning process is distributed over all synaptic sites.

BEHAVIORAL MODELING OF THE SYNAPSE MULTIPLIER

The circuit behavior

We have proposed (Valle et al, 1992) an analog VLSI architecture, whose circuit implementation strongly recalls continuous-time filter design (Tsividis et al, 1987, Johns et al, 1989, Bibyk and Ismail, 1989). The architecture is based on two analog computational primitives: i) the neuron module implemented through a simple amplifier with a circuit for the calculation of the local error; ii) the synapse module implemented through analog multipliers with a local weight-storage circuit.

The synapse circuit (Figure 1) is made by two complementary transconductance amplifiers (OTAs T1 and T2), each composed of a differential stage and a current mirror stage. The bias and differential stages are local at each synapse module while there is only one current mirror stage for each synaptic row. Synaptic inputs are the chip input voltages V_{in} and the signal ground voltage V_{ref}, usually set at 2.5V. The output node of the current mirror stage is connected to the neuron input which behaves as a virtual ground close to 2.5V.

The value of $G_{m1,2}$ (transconductance of the stages T1 and T2) depends on the current in the bias transistors M1 and M2 (Gray and Meyer, 1984, Vittoz, 1985). This current is determined by the current I_{bji} that flows in the transistor M_{b1} or M_{b2}, of the bias stage. I_{bji} is set by the control voltage V_{bji} stored dynamically on the gate capacitance of M_{b1} and M_{b2}. Since transistors M_{b1} and M_{b2} polarize the two OTAs alternatively according to the sign of $(V_{bji} - V_{ref})$, the synaptic output current I_{ji} can be expressed as follows:

$$I_{ji} = G_{m1}(V_{in} - V_{ref}) + G_{m2}(V_{ref} - V_{in}) = (G_{m1} - G_{m2})(V_{in} - V_{ref}) = G_m(V_{in} - V_{ref})$$

To increase the dynamic range of the weight value G_m, the synapse multiplier extends its operation also in the subthreshold region. In the subthreshold region of operation the currents I_{bji}^1 and I_{bji}^2 in transistors M_{b1} and M_{b2} depends exponentially on the control voltage V_{bji} (Vittoz, 1985, Mead, 1989):

$$I_{bji}^1 = I_{D0}\, e^{\frac{V_{GS}}{nV_\theta}}(1 - e^{-\frac{V_{DS1}}{V_\theta}}) \approx I_{D0}\, e^{\frac{V_{GS}}{nV_\theta}} = I_{D0}\, e^{\frac{V_{bji} - V_{ref}}{nV_\theta}}$$

$$I_{bji}^2 = I_{D0}\, e^{-\frac{V_{GS}}{nV_\theta}}(1 - e^{-\frac{V_{DS2}}{V_\theta}}) \approx I_{D0}\, e^{-\frac{V_{GS}}{nV_\theta}} = I_{D0}\, e^{-\frac{V_{bji} - V_{ref}}{nV_\theta}}$$

where V_{GS} and V_{DS} are respectively the gate to source and the drain to source voltages, n is a technological parameter, in the range [1.3 : 2], I_{D0} plays a role similar to saturation current in bipolar transistor and V_θ is the thermal voltage. When transistors enter the strong inversion region, the dependence of the OTA transconductance G_m on V_{bji} slows down to a linear dependence (Gray and Meyer, 1984). Experimental data about the dependence of the output current from the control voltage are presented in Figure 2.

The macromodel

The operations performed by the neuron j^{th} and the N synapses connected to it, are summarized in Figure 3.

A behavioral model that approximates the circuit multiplier can be expressed as follows, where V_{TH} represents the boundary between weak and strong inversion regimes:

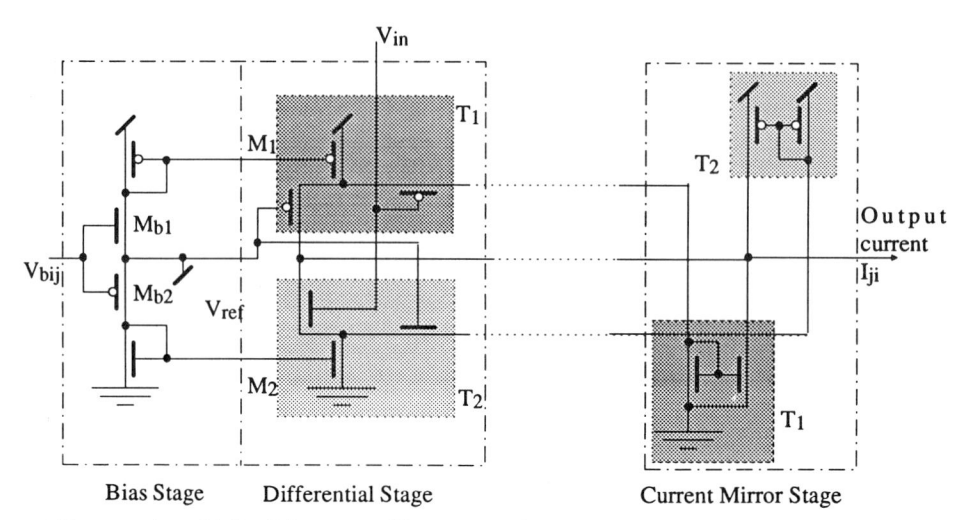

Figure 1 The synaptic multiplier. It is composed by two complementary transconductance amplifiers (T1 and T2): the bias and differential stages are local at each synapse module; for each synaptic row there one current mirror stage. Synaptic inputs are the chip input voltages V_{in} and the signal ground voltage V_{ref} usually set to 2.5V. The output node of the current mirror stage is connected to the neuron input which behaves as a virtual ground close to 2.5V.

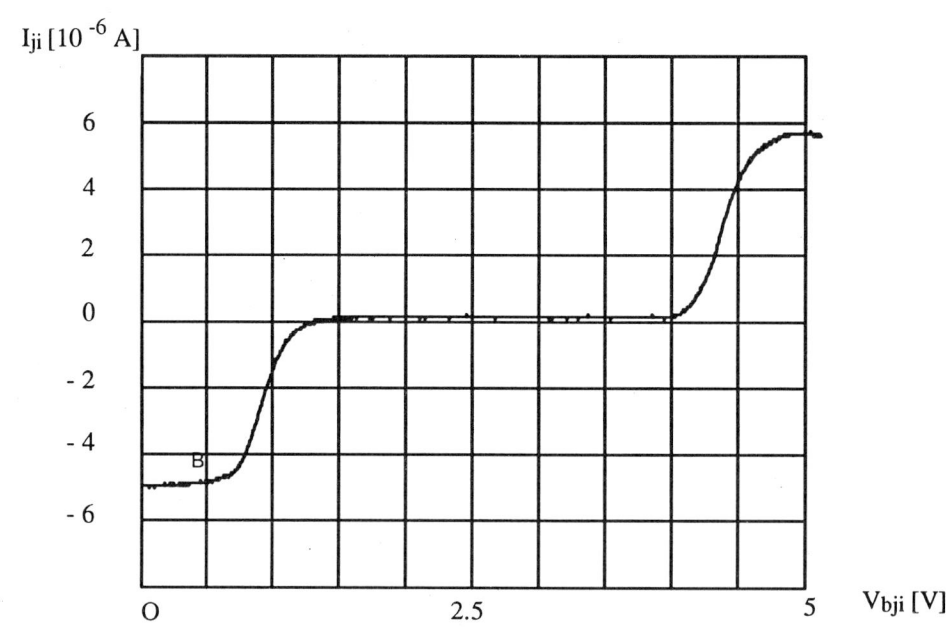

Figure 2 Experimental transfer characteristic of the synaptic multiplier. V_{in} has been kept fixed at 2.7 V, the weight voltage V_{bij} has been varied in the range [0V : 5V]. The output current is in the range [-6 µA : + 6 µA].

if $|V_{bji} - V_{ref}| < V_{TH}$
$$I_{bji} = I_{D0} \cdot (e^{\frac{V_{bji} - V_{ref}}{nV_\theta}} - e^{-\frac{V_{bji} - V_{ref}}{nV_\theta}})$$
else
$$I_{bji} = sgn(V_{bji} - V_{ref}) \cdot I_{sat}$$
endif

The effects of the non linearities on the performances of the algorithm can be studied at behavioral level through functional simulations, choosing for $W = \Psi(V)$ the following *normalized* expression:

if $|V| < V_T$
$$W = K \cdot (e^{20 \cdot V} - e^{-20 \cdot V})$$
else
$$W = sgn(V) \cdot W_M$$
endif

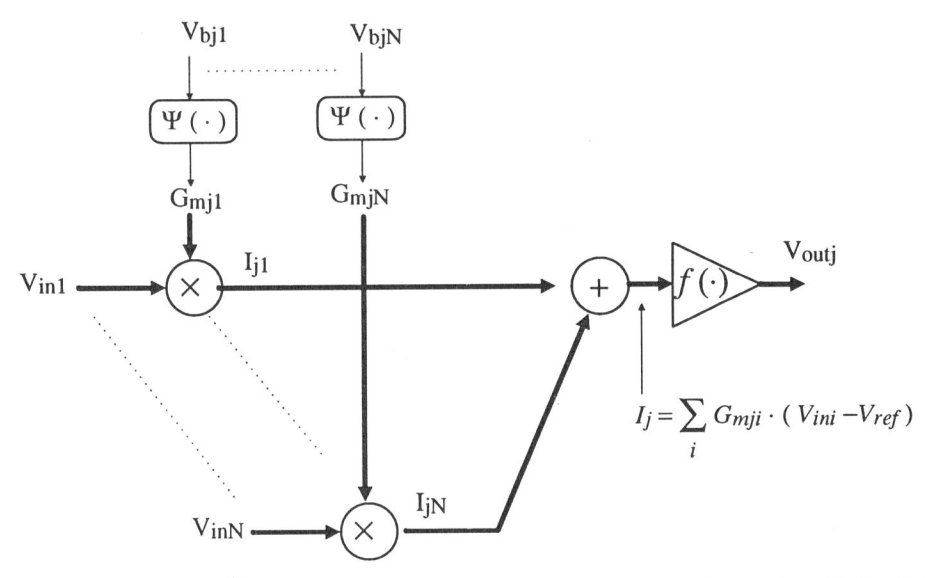

Figure 3 Macromodel of the jth neuron and of the N synapses connected to it. For each synaptic multiplier, the input voltage V_{ini} is multiplied by weight value G_{mji} (tranconductance controlled by the voltage V_{bji}) to produce the synaptic output current I_{ji}. These currents are summed in input to the neuron, characterized by the quasi linear transfer function $f(\cdot)$. The neuron output voltage is V_{outj}: $V_{outj} = f(\sum_i G_{mji} \cdot (V_{ini} - V_{ref}))$.

RESULTS

This section illustrates the performances of the learning algorithm proposed above for three test problems: XOR, parity and character recognition (ten numerals). Various configurations (number of hidden neurons) of the Multi Layer Perceptron architecture are considered.

The neuron transfer function is a sigmoid with values in the range [0 : 1] and the corresponding target values are 0.1 and 0.9. The convergence criterion adopted is:

Table 1 Simulation results for the XOR problem

dB [bit]	N	C
∞	17	100
16	18	97
15	15	97
14	16	100
13	15	100
12	14	98
11	18	97
10	14	69
9	19	66
8	31	66

Table 2 Simulation results for the parity problem with 5 input neurons.

Hidden neurons	10		15		18		20	
dB [bit]	C	N	C	N	C	N	C	N
∞	51	163	83	120	89	96	79	90
16	45	175	82	115	88	92	84	96
15	52	153	86	124	80	81	78	104
14	45	170	84	109	91	96	71	106
13	48	183	83	115	80	96	68	85
12	45	178	79	113	80	88	72	86
11	44	168	84	108	73	97	68	77
10	40	182	74	98	55	90	52	76
9	35	170	60	131	53	97	51	80
8	18	181	55	143	51	121	53	104

Table 3 Simulations results for the character recognition problem. Comparisons are made between the BP algorihms with the global and local and adaptive learning rate.

	C		Number of epochs to convergence when successful (N)					
			average		minimum		maximum	
	global η	local η_{ji}	global η	local η_{ji}	global η	local η_{ji}	global η	local η_{ji}
Numerals	72.5	100	16114	150	2027	50	65911	488

max $(|T_j - O_j|) < 0.4$ (maximum error allowed for all outputs j) for each patterns belonging to the training set for the numerals and max $(|T_j - O_j|) < 0.25$ for the XOR and parity problem. An iteration limit (number of input presentations) of 10000 has been used for the XOR and parity, while a 200000 limit has been used for the character recognition. The parameters for the normalized synaptic relation $W = \Psi(V)$ are the following: $V_T = 1.05$, $W_M = 3$ and $K = 2.275 \cdot 10^{-9}$. Initial values for the synaptic control variables V's were chosen randomly in the range $\pm [0.95 : 1.05]$. These variables were allowed to vary in the range $\pm [0.01 : 5]$. Different logarithmic resolutions for the control variable were considered $(d_B = \log_2 \frac{(V_{max} - V_{min})}{\Delta V_{min}}$, Reyneri and Filippi, 1991).

For each problem 100 experiments were run with different random initial configuration. Two performance parameters are considered: i) the average number of epochs for successful runs, N; ii), the average number of converged runs, C.

The experiment results for the XOR problem, with 4 hidden neurons, are summarized in Table 1. Table 2 illustrates the experiment results for the parity problem with 5 input neurons and various network configurations (10, 15, 18 and 20 hidden neurons).

In the character recognition problem, we considered the ten numerals coded by a 8×8 pixel matrix. Table 3 summarizes the simulation results for this problem comparing the performances of BP algorithms with global and local and adaptive learning rates.

CONCLUSIONS

This paper has proposed and evaluated a new Back Propagation algorithm designed to be implemented through analog CMOS circuits operating in subthreshold regime. The algorithm considers some major features of the analog VLSI implementation: the non ideal and non linear behavior of synaptic multipliers. The approach followed privileges the adaptativity and locality of the neural computation resulting in a formulation very similar to the SSAB algorithm (Tollenaere, 1990).

The approach followed in this paper conforms to constraints of locality, since the BP algorithm so modified performs only local computations. Computational models with local features can be viewed as metaphors for biological networks (Mead, 1989, Mead 1990), and can be easily implemented with scalable parallel architectures (Seitz, 1984).

REFERENCES

Battiti, R., "Accelerated Back Propagation Learning: Two Optimization Methods", *Complex Systems*, Vol. 3, 1989.

Bibyk, S. and Ismail, M., "Issues in Analog VLSI and MOS Techniques for Neural Computing", in *Analog VLSI Implementation of Neural Systems*, C. Mead and M. Ismail (ed), Kluwer Academic Publishers, 1989.

Casinovi, G. and Sangiovanni-Vincentelli, A., "A Macromodeling Algorithm for Analog Circuits", *IEEE Trans. on Computer-Aided Design*, Vol. 10, No 2, February 1991.

Furman, B., White, J. and Abidi, A.A., CMOS Analog IC Implementing the Back Propagation Algorithm, in *Proc. of the INNS-88*, 1988.

Gray, R.P. and Meyer, R. G., *Analysis and Design of Analog Integrated Circuits*, John Wiley & Sons, 1984.

Jabri, M. and Flower, B., "Weight Perturbation: an Optimal Architecture and Learning Technique for Analog VLSI Feedforward and Recurrent Multilayer Networks", *IEEE Trans. on Neural Networks*, Vol. 3, No. 1, January 1992.

Jacobs, R.A., "Increased Rates of Convergence through Learning Rate Adaptation", *Neural Networks*, Vol. 1, No. 4, 1988.

Johns, D. A., Snelgrove, W. M. and Sedra, A.S.," Continuous-Time Analog Adaptive Recursive Filters",in *Proc. of IEEE ISCAS*, 1989.

Lont, J.B. and Guggenbühl, W., "Analog CMOS Implementation of a Multilayer Perceptron with Nonlinear Synapses", *IEEE Trans. on Neural Networks*, Vol. 3, No. 3, May 1992.

Mead, C., *Analog VLSI and neural Systems*, Addison Wesley, Reading 1989.

Mead, C., "Neuromorphic Electronic Systems", *Proc. of the IEEE*, Vol. 78, No. 10, October 1990.

IEEE MICRO, special issue on Silicon Neural Networks, Dec. 1989.

Murray, A. F., "Silicon Implementations of Neural Networks", *IEE Proceedings-F*, Vol. 138, No. 1, Feb. 1991.

Paulos, J.J. and Hollis, P.W., "A VLSI Architecture for Feedforward Networks with Integral Back-Propagation", in *Proc. of the INNS-88*, 1988.

Ramacher, U., "Guide Lines to VLSI Design of Neural Nets", in *VLSI Design of Neural Networks*, U. Ramacher and U. Ruckert (ed), Kluwer Academic Publishers, 1991.

Reyneri, L.M. and Filippi, E., "An Analysis on the Performance of Silicon Implementations of Backpropagation Algorithms for Artificial Neural Networks", *IEEE Trans. on Computers*, Vol. 40, No. 12, December 1991.

Parallel Distribuited Processing,Rumelhart, D.E. and Mc Clelland, J.L. (ed), MIT Press, Cambridge, Mass., 1986.

Seitz, C.L., "Concurrent VLSI Architectures", *IEEE Trans. on Computers*, Vol. C-33, No. 12, Dec. 1984, pp. 1247-1265.

Shima, T., Kimura, T., Katamani, Y., Itakura, T., Fujita, Y. and Iida, T., "Neuro Chips with On-chip BackPropagation and/or Hebbian Learning", in *Proc. of 1992 IEEE International Solid State Circuits Conference.*

Sonteg, E.D. and Sussmann, H.J., "Back Propagation Separates where Perceptrons Do", *Neural Networks*, Vol. 4, No. 2, 1991.

Tollenaere,T., "SuperSAB: Fast Adaptive Back Propagation with Good Scaling Properties", *Neural Networks*, Vol. 3, 1990.

Tsividis, Y., Banu, M. and Khoury, J., "Continuous-Time MOSFET-C Filters in VLSI", *IEEE Trans. on Circuits and Systems*, Vol. CAS-33, N0. 2, February 1986.

Tsividis, Y., "Analog MOS Integrated Circuits - Certain New Ideas, Trends and Obstacles",*IEEE Journal of Solid State Circuits*, Vol. SC-22, N0. 3, June 1987.

Valle, M., Caviglia, D.D. and Bisio, G. M., "An Experimental Analog VLSI Neural Chip with On-Chip Back-Propagation Learning", *Proc. of ESSCIRC '92*, pp. 203 - 206, 1992.

Vittoz, E.A., "Micropower Techniques", in *Design of MOS VLSI Circuits for Telecommunications*, Y. Tsividis and P. Antognetti (ed), Prentice-Hall Publisher, 1985.

Vittoz, E.A., "Analog VLSI Implementation of Neural Networks", *Proc. ISCAS 90*, New Orleans, 1990.

Vittoz, E.A., "Future of Analog in the VLSI Enviroment", in *Proc. ISCAS 90*, New Orleans, 1990.

Vogl, T.P., Mangis, J.K., Rigler, A.K., Zink, W.T. and Alkon, D.L., "Accelerating the Convergence of the Back-Propagation Method", *Biological Cybernetics*, 59, 1988.

Widrow,B. and Lehr, M.A., "30 Years of Adaptive Neural Networks: Perceptron, Madaline, and Back-Propagation", *Proc. of the IEEE*, Vol. 78, No. 9, September 1990.

AN ANALOG IMPLEMENTATION OF THE BOLTZMANN MACHINE WITH PROGRAMMABLE LEARNING ALGORITHMS

V. Lafargue, P. Garda and E. Belhaire

INTRODUCTION

Most current neural nets experiments use the MultiLayer Perceptrons and the Backpropagation learning algorithms. Whereas the experimentation speed is insufficient, few analog integrated circuits have been realized for these algorithms, because neither their implementation nor their parallelization are obvious. On the other hand, Boltzmann Machines (Hinton et Sejnowski 1984) show a number of very attractive features, including high recognition rates, but their simulations are desperately slow. Therefore mixed analog/digital implementations have been described (Alspector *et al* 1987a, Alspector *et al* 1987b, Kreuzer *et al* 1988), whose learning algorithm is hardwired.

However, several learning algorithms were recently proposed by Azencott (1989) and (1990). Whereas they have a more general scope than the older algorithms, their effectiveness for the application of analog circuits has to be assessed. For this purpose, we present in this paper a new analog/digital implementation of the Boltzmann Machine whose learning algorithm is programmable.

We present the Boltzmann Machine and its different learning algorithms in the first section. We describe a faithful analog implementation of the Boltzmann machine relaxation in the second section. We then introduce a new architecture for the programmable synaptic cell in the third section and a prototype implementation in the fourth section.

BOLTZMANN MACHINES

The Boltzmann Machine model was introduced by Hinton and Sejnowski in 1984 (Hinton and Sejnowski 1984). It is an asynchronous model, where a single neuron updates its state at each iteration. However, in a parallel hardware implementation, it is natural for all the neurons to update their state simultaneously at each iteration. Therefore, a new model was introduced by Azencott in 1989 (Azencott 1989) to deal with this case, which is called the Synchronous Boltzmann Machine. We describe now the operation of these models.

The Asynchronous Boltzmann Machine, introduced by Hinton and Sejnowski in (Hinton *et al* 1984), is operated as follows. Let $(u_i)_i$ be a set of *neurons*, x_i^n the *neuron state* of

u_i at instant n (which may have values 1 or 0) and w_{ij} the *synaptic weight* between neurons u_i and u_j. Let V_i^n be the *network contribution* to u_i after instant n, computed according to the equation (1):

$$V_i^n = \sum_{j \neq i} w_{ij} x_j^n \tag{1}$$

Then the activation state of the neuron u_i which is updated at discrete time step (n+1) is tossed at random with the probability given by the equation (2):

$$P(x_i^{n+1} = 0) = \frac{1}{1 + \exp(V_i^n/T)} \tag{2}$$

Now let us consider the learning process. For this we have to choose input, output and hidden neurons among the network of neurons. The weight update process is repeatedly performed for all the pattern associations, and for each of them, it consists in a clamped and a free phase. During each phase, the neurons update their state, and a cooccurence counter is associated to each weight. It is computed according to equation (3) for asynchronous networks, and according to equation (4) for synchronous networks (Azencott 90):

$$p_{ij} = \frac{1}{N} \sum_{n=M}^{N+M} x_i^n . x_j^n \tag{3}$$

$$p_{ij} = \frac{1}{2N} \sum_{n=M}^{N+M} x_i^n . x_j^{n-1} + x_i^{n-1} . x_j^n \tag{4}$$

During the clamped phase, a pattern is imposed both on the input and output neuron states while the hidden neurons are left free to change their states, and the cooccurence p_{ij}^+ is computed. Whereas during the free phase, the input pattern is presented to the input neuron states while the output and hidden neurons are left free to choose their states, and the cooccurence p_{ij}^- is computed. After these clamped and free phases, the weights are updated according to one of the rules introduced in (Hinton et al 1984), (Azencott 90) and (Lacaille 92). They are given by equations (5), (6) or (7), where the gain η and the temperature T are two positive parameters.

Linear rule: $$\Delta w_{ij} = \frac{\eta}{T} \cdot \left(P_{ij}^+ - P_{ij}^- \right) \tag{5}$$

Threshold rule: $$\Delta W_{ij} = \eta \cdot \text{sgn}\left(P_{ij}^+ - P_{ij}^- \right) \tag{6}$$

Momentum rule: $$\Delta W_{ij}^n = \alpha . \Delta W_{ij}^{n-1} + (1-\alpha).\eta.\left(P_{ij}^{n+} - P_{ij}^{n-} \right) \tag{7}$$

Several mixed analog/digital implementations were described by (Alspector et al 1987) (Alspector et al 1989) (Arima et al 1991). They all included a hardwired learning algorithm with the thresholded weight update rules of asynchronous network according to the equations (3) and (6).

It is not proved that these rules are the best suited to an analog implementation. Whereas software simulations were performed to assess this point, they do not capture enough characteristics of the analog circuits to lead to a definite conclusion. Therefore it is interesting to provide the possibility of experimenting different learning algorithms in an analog circuit and we propose a new architecture for this purpose. We describe in the two following sections firstly its analog part and then its programmable part.

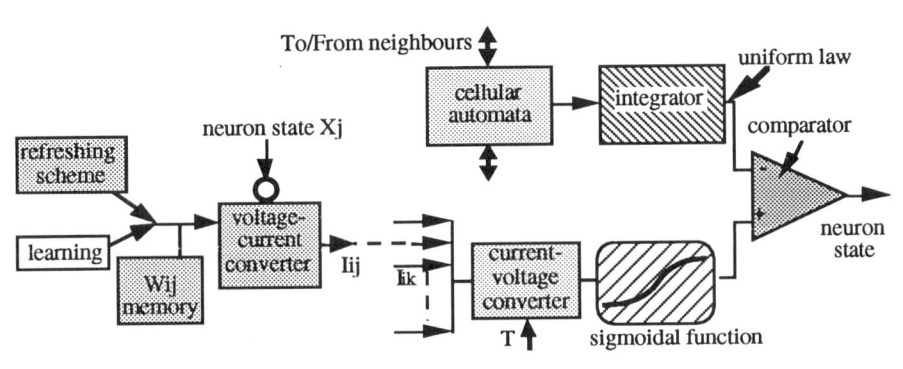

Figure 1. synaptic and neuron cells

ANALOG CELLS

The functional analog cells implement the computations required by the equations (1), (2) and store the variables they require.

Firstly, the synaptic cell is in charge of the storage of the weight, its learning or refreshing, and the computation of one $w_{ij} x_j^n$ product. The synaptic cell is divided into two subcells. The first one is in charge of the relaxation of the network: it is an analog cell and it is described in this section. The second one is a digital processor used for the weight learning and it will be described in the next section.

The network contribution is computed according to equation (1) by a single rail current summator. The analog part of the synaptic cell includes a linear voltage-to-current converter driven by the weight capacitor (Belhaire *et al.* 1991). The converter output is connected to a net N_i in order to provide on N_i a current I_{ij} representing the contribution $w_{ij} x_j^n$. This current has an order of magnitude of one micro-ampere (see Figure 1).

The neuron cell is in charge of the update of the neuron state according to equation (2). It takes as input the sum I_i of the currents I_{ij} which represents the network contribution V_i. This current is then converted into a voltage representing $\frac{V_i}{T}$, so that the converter gain implements the temperature T of equation (2). A sigmoidal function is then applied to the output of this converter. It is built out of compatible lateral bipolar transistors available as parasitic devices on CMOS technologies but which are efficient enough for our purpose. The output of this function consists of a differential current and it is then compared to a random differential current. The latter is derived from the conversion of a voltage following a uniform law. The generation of this random voltage is described below.

Finally the neuron cell has two inputs, the random voltage and the current I_i, and one output, the state x_i. It consists (cf Figure 1) in a current to voltage converter, a sigmoidal function, an integrator and a comparator. The schematic of this cell is represented in Figure 2. The minimum number of transistors for this cell is fifteen.

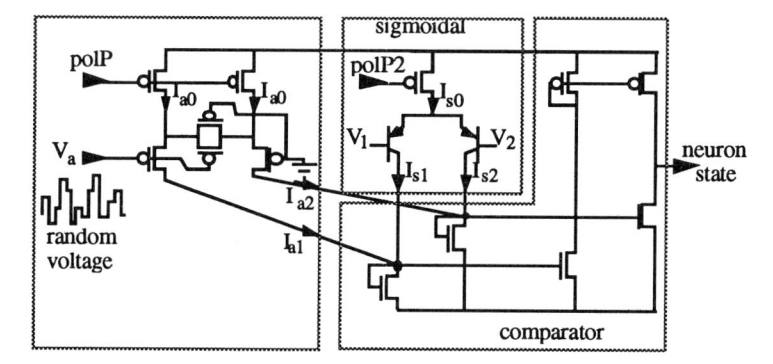

Figure 2. schematic of neuron cell.

The random tossing required by the Boltzmann machine results in some troubles: the use of resistor thermal noise (Alspector et al 87) leads practically to correlated generators, as pointed out in (Kreuzer et al 1988) and confirmed experimentally by (Alspector et al 1991). We investigated an original solution based on a uniform law random number generator: the binary activation state is resulting from the comparison of the output v_i of this uniform law generator to the sigmoidal function $F(V_i/T)$.

The uniform law v_i is built out of the integration of a sequence of unbiased independent binary values b_i^n. This integration is performed by a switched-capacitor filter shown in the

Figure 3 (as in a serial DAC). The binary random numbers are generated by a 1-D chaotic cellular automaton introduced in (Wolfram 1986). This cellular automaton uses the rule 30 given by:

$$x_i^n = x_{i-1}^{n-1} \text{ XOR } (x_i^{n-1} \text{ OR } x_{i+1}^{n-1}) \tag{8}$$

As the states sequences of distant cells are uncorrelated, we use site spacing in order to generate in parallel the binary inputs of the neuron cells (Hortensius 89). Actually each neuron cell includes 8 automata, and the automata of all the neuron cells are linked as a single 1-D cellular automaton. Therefore its period is about $2^{0.61(8N+1)}$ for a system including N neuron cells (Wolfram 1986), which is very large even for small systems and increases with the number of neurons. This is a major advantage over the shift register implementation described in (Alspector et al 1991). Moreover the modular approach of this realization and its connection scheme are well suited to an integrated electronic realization.

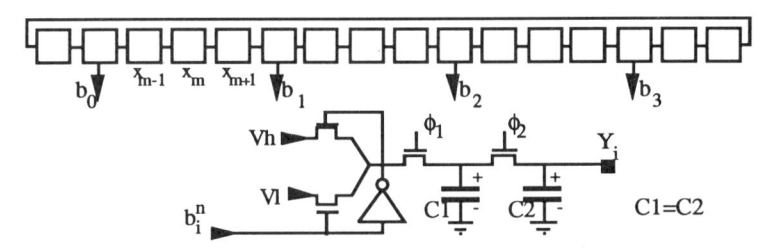

Figure 3. Cellular automata based random generator.

ARCHITECTURE OF THE PROGRAMMABLE SYNAPTIC CELL

We propose a new architecture for the synaptic cell, depicted in Figure 4. It is built out of a bit-serial digital processor, called hereafter the processor, a digital-to-analog converter and the analog cells included in the voltage to current converter described in the previous section.

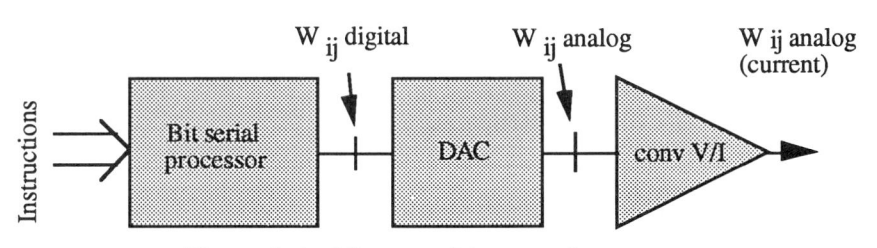

Figure 4. Architecture of the synaptic neurons

The bit-serial processor is programmable and it includes a digital memory where the integer representation of R_{ij}, ΔW_{ij}, and W_{ij} are stored for the synaptic cell located at (i,j). It performs the computations required by equations (3) or (4), and (5), (6), or (7) respectively. For this purpose, the states X_i an X_j of the neurons U_i and U_j are fed into the processor and stored into its memory. All the processors are operated in SIMD mode under the control of a single common sequencer.

IMPLEMENTATION

The implementation of the processor is depicted in Figure 5. It is similar to a usual bit-serial processor such as the GAPP from NCR. The processor includes a 1-bit wide digital SRAM, 3 1-bit registers and an ALU. This ALU is 1-bit wide also, and it performs all the computation serially. It is very efficient for medium precision additions, subtractions, and scaling by powers of 2. The weight w_{ij} is output from the digital SRAM in order to be used by the analog voltage to current converter described in the previous section. For this purpose a serial digital to analog converter has been used, and this is homogeneous with the architecture of the bit serial processor. The digital to analog converter is depicted on Figure 6 and simulations results are presented in Figure 7. They show a good linearity over a 6 to 7 bit range.

A prototype circuit was designed and realised in the MIETEC/ALCATEL 2.4µm DLM DLP CMOS process. It included an array of 4x4 synaptic cell in a 3.9x4.4 µm^2 chip. This chip was simulated with a 20 MHz clock and it was successfully tested with a 11 MHz clock. This speed is compatible with that of the analog update of the neurons states which was performed at 300 iterations per millisecond.

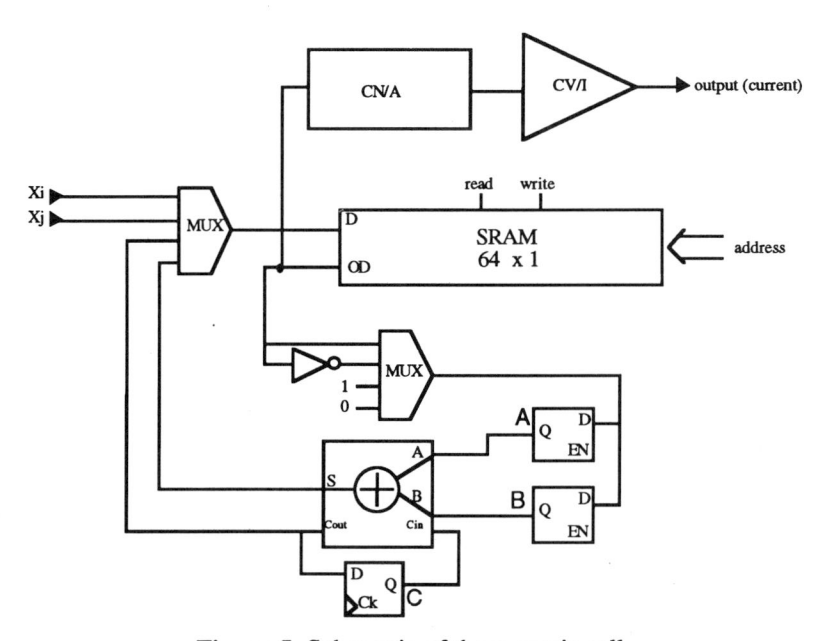

Figure 5. Schematic of the synaptic cell

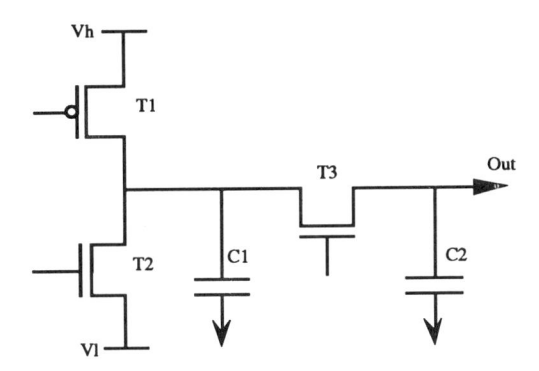

Figure 6. Schematic of the DAC

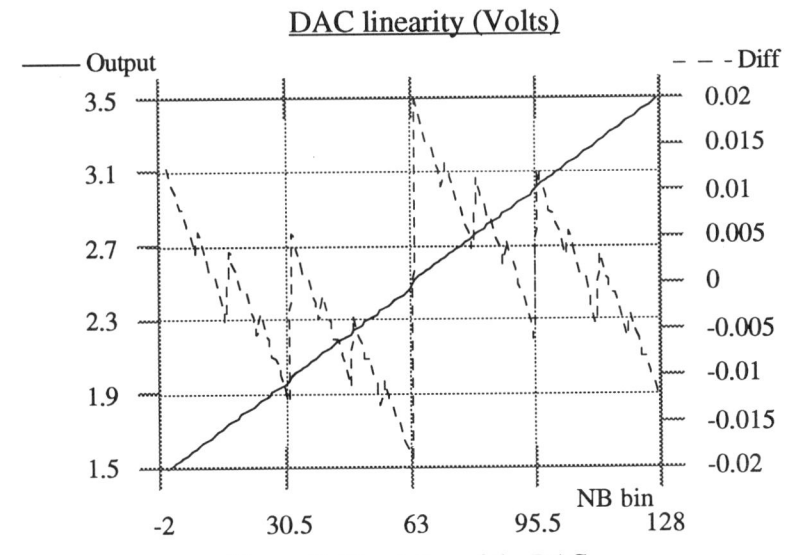

Figure 7. Simulation of the DAC

CONCLUSION

In this paper, we have described the architecture of a new synaptic cell for the mixed digital/analog implementation of the Boltzmann Machines, and we introduced several original points. This synaptic cell is programmable, and this allows to tailor the learning algorithm for the network and the application under study. Simple algorithms may be used for small networks learning easy tasks, whereas complex algorithms may be preferred for large networks whose learning is difficult. As the algorithms are performed in a bit-serial way, their duration is proportional to the complexity of the algorithm. Finally the width of the variables which are used for the learning, such as the cooccurrences or the weight, may easily be adapted to the task under learning.

Acknowledgments

This work has been supported by D.R.E.T. under Contract 87/292 and by C.N.R.S., M.R.T., M.C.I. and M.E.N. under P.R.C.-A.M.N., Cognisciences and G.C.I.S. project RA. We thank Robert Azencott and Jérome Lacaille, from D.I.A.M.E.N.S. in Paris, Hubert Pujol and Zhu Yi Min from I.E.F., for fruitful discussions.

References

Alspector J. & al., "Stochastic learning network and their electronic implementation", *Procs N.I.P.S.*, A.I.P., 1987

Alspector J. et al., "A neuromorphic V.L.S.I. learning system", *Stanford Conference on VLSI*, M.I.T. Press, 1987

Alspector J. et al., "Relaxation Networks for Large Supervised Learning Problems", *Procs N.I.P.S.*, 90

Alspector J. et al., "A VLSI-efficient Technique for Generating Multiple Uncorrelated Noise Sources and its Application to Stochastic Neural Network", *JSSC*, vol. SC-38, pp. 109-123, 1991

Arima et al, "A 336-Neuron, 28K-Synapse, Self-Learning Neural Network Chip with Branch-Neuron-Unit Architecture",*JSSC*, vol. SC-26, pp. 1637-1644, 91

Azencott R., "Synchronous Boltzmann Machines and their learning algorithms", *Neurocomputing*, Les Arcs, 1989, Springer-Verlag NATO ASI Series, Vol. F-68

Azencott R., "Boltzmann Machines: high-order interactions and Synchronous learning", *Procs Stochastic models, statistical methods and algorithms in image analysis,* Ed. by P.Barone and A.Frigessi, Lecture Notes in Statistics, Springer-Verlag, 1990

Belhaire E. & al., "A linear "sum-of-product" circuit for Boltzmann machines", *ISCAS 91*, Singapore, June 11-14 1991, pp. 1299 - 1302

Belhaire E. & al., "A faithful analog implementation of the Boltzmann Machine", *ICANN 92*, Brighton, September 4-7 1992

Garda P.& al., "An analog circuit with digital I/O for Synchronous Boltzmann Machines", *VLSI for Artificial Intelligence and Neural Networks*, Oxford, September 2-4 1990, Ed. by J.Delgado-Frias et W. Moore, Plenum Publishing Corp., 1991, pp. 245-253

Hinton G. *et al.*, "Boltzmann Machines", *C.M.U. Technical Report CMU-CS-84-119*, Carnegie Mellon University, 1984

Hortensius P. *et al.*, "Cellular Automata-Based Pseudorandom Number Generators for Built-In Self Test", IEEE Trans. on CAD, Vol. 8, pp. 842-859, 89

Kreuzer I. *et al.*, "A modified model of Boltzmann Machines for W.S.I. realization", *Signal Processing IV,* 1988

Lacaille J., "Machines de Boltzmann, Théorie et Applications", Ph.D. Thesis, June 25th 1992, University of Paris Sud

Wolfram S., "Random Sequence Generation by Cellular Automata", Advances in Applied Mathematics, 7, pp. 123-169, 1986

A VLSI DESIGN OF THE MINIMUM ENTROPY NEURON

Rüdiger W. Brause

INTRODUCTION

For many purposes the necessary processing of sensor input signals is realized by using a system which implements the maximization of the transinformation from the input to the output of the system. For deterministic systems, this corresponds to the maximization of the output entropy (maximum entropy principle). In pattern recognition theory, it is well known that for Gaussian distributed sources this corresponds to the minimization of the mean square error of the output. For linear systems, this is done by a linear transformation to base of the eigenvectors of the autocorrelation matrix (Fukunaka 1972). Furthermore, we can compress (encode) the the input information by using only the base vectors (eigenvectors) with the biggest eigenvalues. Neglecting the ones with the smallest eigenvalues (m<n, see Fig.1) results in the smallest reconstruction error on the encoded input (Fukunaga 1972). Generally, this approach can be used for sensor signal coding such as picture encoding, see e.g. (Jayant and Noll 1984).

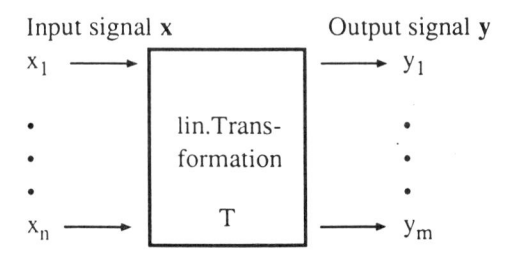

Figure 1 The feature signal transformation

The neural network models of this approach use linear neurons, where each neural weight vector corresponds to one eigenvector. Examples of those architectures are the Oja subspace network (Oja 1989, Williams 1985), the Sanger decomposition network (Sanger 1989) and the lateral inhibition network of Földiák (1989) or Rubner and Tavan (1989). The first mentioned networks decompose sequentially the input vector **x**, see figure 2. They use as a basic building block the linear correlation neuron which learns the input weights by a Hebb-rule, restricting the weights $w_1,..,w_n$. The input for each stage is obtained by subtracting sequentially all the projections of the weight vectors on the input pattern

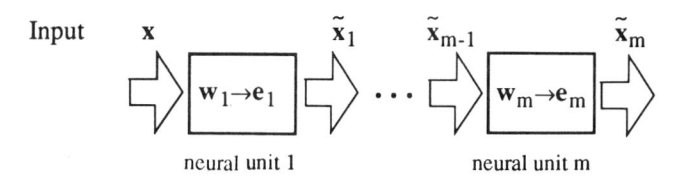

Figure 2 The sequential learning of the eigenvectors

$$\tilde{x}_i = \tilde{x}_{i-1} - y_i w_i \qquad \tilde{x}_0 = x$$

As Oja showed (Oja 1982), this learning rule let the weight vector of the first neuron converge to the eigenvector of the expected autocorrelation matrix C of the input patterns $\tilde{x}$ with the biggest eigenvalue λ_{max}:

$$w_1 \rightarrow e_k \quad \text{with} \quad \lambda_k = \max_i \lambda_i \qquad \text{and} \quad C_i e_i = \lambda_i e_i \qquad \text{with} \quad C_i = \langle \tilde{x}_{i-1} \tilde{x}_{i-1}^T \rangle$$

The subsequent neuron learns the eigenvector with the biggest eigenvalue of the autocorrelation matrix of the remaining input: The eigenvectors remain the same as before, but in this system λ_k becomes zero and the second eigenvalue becomes in fact the biggest one.

The whole network is used in two modes: in the learning mode, where the input is propagated sequentially through the units, and in the transforming, filtering mode, where the input is presented to the neurons in parallel, see figure 3. The basic building block of these networks is the linear neuron, learning with the Oja learning rule (Oja 1982, Sanger 1989); the same as in other asymmetric networks with additional lateral inhibition (Rubner and Tavan 1989, Földiák 1989).

Nevertheless, some researchers in pattern recognition theory (Tou and Gonzales 1974) claim that for pattern recognition and classification clustering transformations are needed which reduce the intra-class entropy. This leads to stable, reliable features and is implemented for Gaussian sources by a linear transformation using the eigenvectors with the *smallest* eigenvalues.

Additionally, the well-known, important problem of approximating randomized, disturbed measures of hypersurfaces can be solved by minimizing the total least square error (TLSE) of an approximation surface. It can be shown that the approximated surface coefficients can be obtained by the minor components (eigenvector with the smallest eigenvalue) of the observed data (Xu,Oja and Suen 1992).

In another paper (Brause 1992) it is shown that the basic building block for such a transformation can be implemented by a linear neuron using an Anti-Hebb rule and restricted weights. More formally, the transformation of n input variables $(x_1,..,x_n) = x$ to m output variables $(y_1,...,y_m) = y$ is made by a linear transformation. For one output variable y_i the transformation can be implemented by a formal neuron. For each neuron, the input is weighted by the weights $w=(w_1,...,w_n)$ and summed up to the activation z of the neuron

$$y_i = z(t) = \sum_j w_j x_j = w^T x \qquad \textit{linear activation} \qquad (1)$$

which is expressed as the scalar product of x and the transpose of w. Additionally, the transformation coefficients w_j are locally learned on chip by the stochastic version of the Hebbian law and $\hat{w}(t+1) = w(t) - \gamma(t)x(t)y(t)$ *Anti-Hebb-Rule*   (2)

$w(t+1) = \hat{w}(t+1) / (\Sigma_j \hat{w}_j(t+1)^2)^{1/2}$ *Normalization*   (3)

This paper shows the VLSI design for such a building block, using standard modules of multiplication and addition for Eqs(1) and (2). Additionally, the normalization of Eq. (3) is covered by special circuitry, described in the last section.

THE SYSTEM DESIGN

One of the most interesting objectives for a VLSI design is a fast, real-time oriented architecture design. The neural network paradigm of small, parallel processors allows an efficient, parallel implementation of the proposed transformation. For such a system, the layout in figure 3 shows how the input vector of n sensor lines is fed in parallel to all m neurons via an input bus.

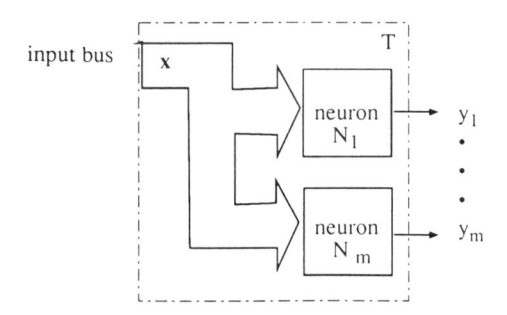

Figure 3 Signal transformation by formal neurons

Each neuron performs a projection of the input to a weight vector by equation (1). The output bus consists of the m output lines of the m neurons.

Additionally, the weights w_{ij} are updated according to Eqs. (2) and (3) in such a way that they will converge to the eigenvector of the autocorrelation matrix with the smallest eigenvalue. This is reflected by the principal system design for one neuron in figure 4.

Nevertheless, to achieve a complete eigenvector decomposition necessarily an interaction between the neurons has to exist. There are several possibilities to construct such a network, see Brause (1992). Since the interactions consist of simple multiplications and additions (see e.g. first section), they are not shown in figure 3 and we concentrate on the basic implementation features for one neuron.

Modeling the weights by voltages w_{ij} at capacitors c_{ij}, each input signal x_j is first multiplied by the weight w_{ij} in the MUL module and the resulting current is summed up in SUM, generating the output signal after Eq. (1). The learning according to Eq. (2) takes place in module HEBB, whereas the normalization of Eq. (3) is done in NORM.

Note that we do not use EEPROMs, floating gate transistors or other permanent means for the storage of the weight values. Thus, the learning takes place in parallel to the normal signal processing of Eq. (1) and reflects therefore the real statistics of a short time period of the input signal, not the prefixed ones of one application. Therefore, the design is self-adaptive in respect to the application problem statistics. In the case where you do not know the application properties (one chip for all applications) this yields an

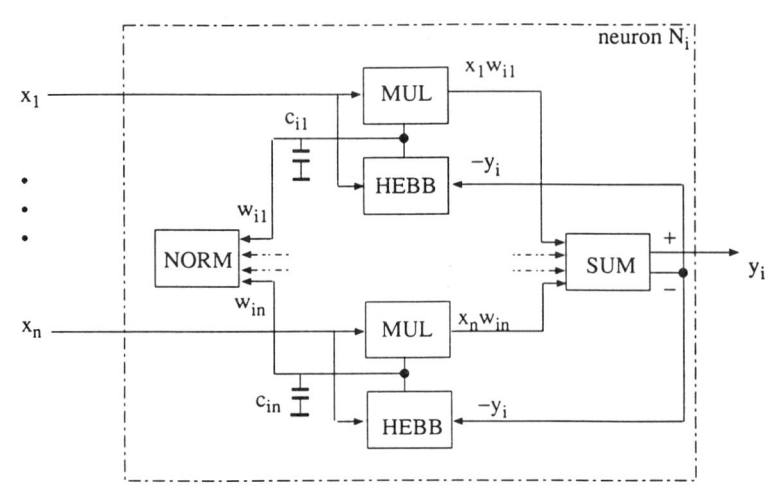

Figure 4 The principal system structure of one neuron

optimal transformation. Nevertheless, when the statistics are well known the design can be simplified by canceling the learning mechanism (HEBB and NORM modules) and implementing the well known, constant weights directly. Let us discuss now the implementation of the various modules.

THE BUILDING BLOCKS

There already do exist some "standard" building blocks for the purpose of neural networks which are cited in several papers (Card and Moore 1989, Mead 1989, Vittoz 1989). Nevertheless, we do not use some of them for certain reasons; others are modified and there are some new building blocks in this paper. In this section, all building blocks are discussed in detail.

The standard building blocks

One of the best known building blocks in figure 4 is the multiplication module MUL. When the input voltage x_j is smaller than the weight voltage w_{ij}, the circuit for positive and negative multiplication might consists of just one FET (linear region) (Weste and Eshraghian 1985, Card and Moore 1989), see figure 5a on the left hand side.

Very often, the input signal is bigger than the weight signal and the transistors leaves the state of weak inversion, resulting in non-linear current. Therefore, we propose the linear region of the well-known Gilbert four-quadrant multiplier circuit (Gilbert 1968, Gilbert 1974) or a wide-range version of it (e.g. Mead 1989) shown in figure 5b on the right hand side for the multiplication $w_{ij}x_j$ in the MUL building block. This is also true for the other multiplication x_jy_i of Eq. (2) in the HEBB building block. Here, we charge the capacitor c_{ij} by a current proportional to the product $w_{ij}x_j$ (see e.g. Card and Moore 1989).

Another important action is the general SUM building block, which exists only once per neuron and implements the sum of Eq.(1). Here we do not use the virtual ground circuit, proposed for instance by Vittoz (1989), because it also automatically scales the input which leads to false results in our case. Instead, we relay on a simple current adder, followed by an amplifier in figure 6.

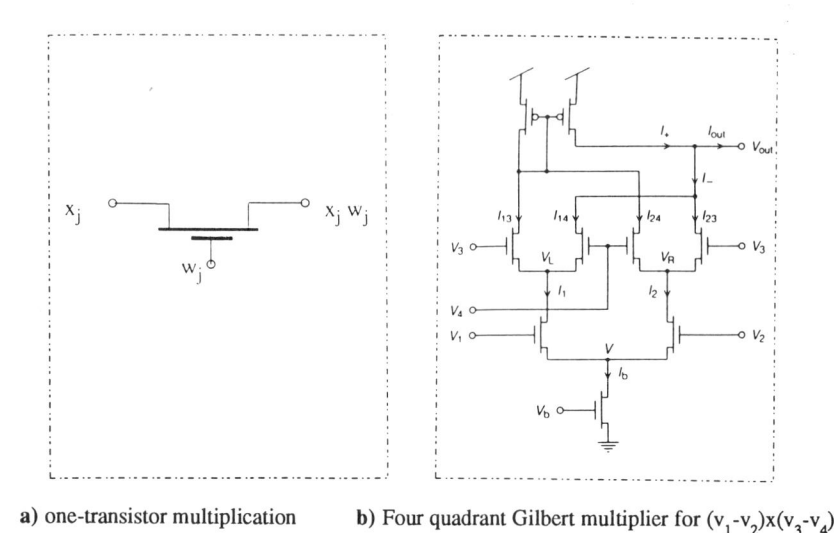

a) one-transistor multiplication   **b)** Four quadrant Gilbert multiplier for $(v_1-v_2) \times (v_3-v_4)$

Figure 5 Multiplication circuits

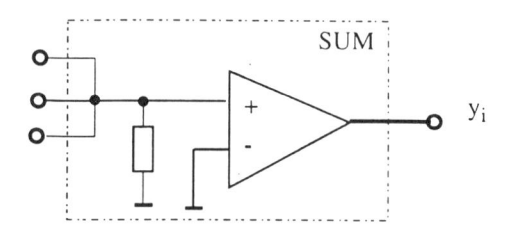

Figure 6 The sum circuit

There remains the most interesting building block, the normalization block which implements Eq.(3). Up to my knowledge, there do not yet exist propositions for an implementation of it. This is covered by the next section.

The normalization block

The basic idea for the mechanism of the learning algorithm used in this paper is the restriction of the resources, the weights. The restriction or side-condition consists of a constant length of the weight vector which is fixed at all iteration steps according to eq. (1.3). To achieve this, the normalization block uses three non-linear circuits.

The most tedious problem in this VLSI-application is the design of this vector normalization circuitry. It can be shown that the standard approaches of weight normalization (Mead 1989, Vittoz 1989, Card and Moore 1989)

$$\mathbf{w}(t+1) = \hat{\mathbf{w}}(t+1) / (\Sigma_j \hat{w}_j(t+1)) \qquad (4)$$

will *not* give the convergence to the eigenvectors for a proper feature transformation because the weights are only normalized in the L_1 norm sense. To avoid this problem, our design differs significantly from the standard approaches by computing the L_2 norm, the real quadratic Euclidean norm of Eq.(3).

The weights signal levels are transformed to squared signals (see figure 7) and then,

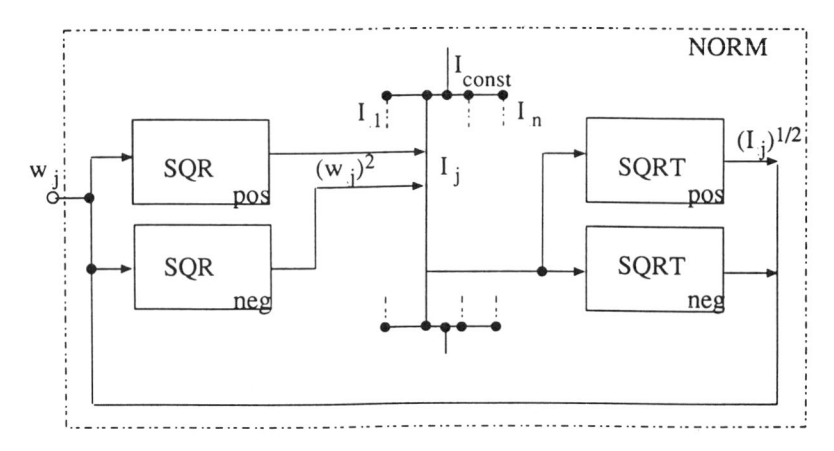

Figure 7 The normalization block schema

by a current mirror, induce a proportional current in a Kirchhoff network which is supplied by a constant current. The relative currents in all branches will be balanced according to the relative values of the squared input signals, such implementing the normalization relations. The normalized currents are then read out by a current mirror, squeezed by a root circuit and fed back to the input where they control the factor γ of Eq. (2) by regulating the output resistance of the root amplifier building block. Thus, the weight signals are normalized. More formally, this can be shown as follows.

Assume that for a certain weight vector w all currents of the constant current network have a certain value I_j such that

$$I_{const} = \Sigma_j I_j \quad \text{and} \quad I_j \sim w_j^2 \tag{5}$$

Now, let us further assume that in time step t+1 each component of w has changed independently by a factor of a_j

$$\hat{w}_j(t+1) = a_j w_j(t) \tag{6}$$

The signal $\hat{w}_j(t+1)$ is converted to its square $\hat{w}_j(t+1)^2$ and induces proportionally a conductance $L_j=1/R_j$ in the branch I_j. With the increase of each conductance

$$L_j(t+1) = a_j^2 L_j(t) \tag{7}$$

the conductance L of all branches becomes

$$L(t+1) = \Sigma_j L_j(t+1) = \Sigma_j a_j^2 L_j(t) = \alpha L(t) \tag{8}$$
$$\text{with } \alpha = (\Sigma_j a_j^2 L_j(t)) / L(t)$$

Since the product of the voltage V and the conductance L, the current source $I_{const} = V(t) L(t)$, is constant, we know with (8) and

$$I_{const} = V(t+1) L(t+1) = V(t+1) \alpha L(t) = V(t) L(t) \tag{9}$$

that the voltage V(t) becomes

$$V(t+1) = V(t)/\alpha \tag{10}$$

By (7) and (10), the current in the j-th branch becomes

$$I_j(t+1) = V(t+1)L_j(t+1) = V(t)a_j^2 L_j(t)/\alpha = I_j(t)a_j^2/\alpha \tag{11}$$

and therefore, the j-component of the weight vector

$$w_j(t+1)^2 = w_j(t)^2 a_j^2/\alpha \tag{12}$$

becomes scaled by the factor of Eq.(6). Additionally, the whole vector is scaled in length by a factor of α which results in a length of

$$w(t+1)^2 = \sum_j w_j(t+1)^2 = \sum_j w_j(t)^2 a_j^2/\alpha = \sum_j w_j(t)^2 a_j^2 L(t) / (\sum_j a_j^2 L_j(t))$$

and with $L_j(t) = c w_j(t)^2$, $L(t) = \sum_j L_j(t) = \sum_j c w_j(t)^2 = c\, w(t)^2$ with a proportion factor c

we get $\quad w(t+1)^2 = c\, w(t)^2 (\sum_j w_j(t)^2 a_j^2) / (\sum_j a_j^2 c\, w_j(t)^2) = w(t)^2$

Thus, the length of the vector **w** will remain fixed, only determined by the fixed value of the constant current.

The next figures will cover the rest of the non-linear circuits. The non-linear, over-proportional square effect is achieved by a kind of feed-back circuit, originally proposed by Degrauwe et al. (1982), shown in figure 8 on the left hand side.

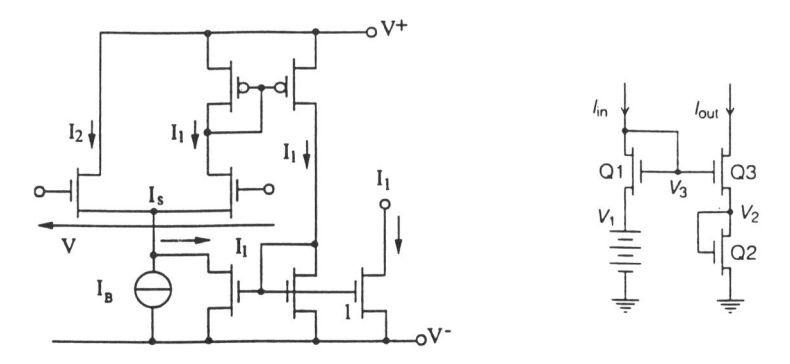

a) The square circuit (after Vittoz 1989)   b) the square root circuit (after Mead 1989)
Figure 8 Expanding and compressing circuits

The non-linear square root circuit is covered by a simple current-mirror design, described for instance in Mead (1989) and shown on the right hand side of figure 8.

Nevertheless, for a real VLSI implementation of this approach some analog detail problems must be solved. One of them is the question, how the coupling of the non-linear normalization feed-back block to the Hebb-learning unit should be designed.

On the one hand, when a normalization of the weight signal is obtained, the ratio of the output resistance of the Hebb-circuit to the output resistance of the root-circuit must be high to highly influence the weight signal. Obviously, a small error due to the finite ratio will be left.

On the other hand, this error voltage is just the one which can influence the normalization circuit and will change the direction of the weight vector. So, the error should not be made too small. Current research is done to overcome this problem.

CONCLUSION

The paper shows how the linear transformation of multidimensional signal features to the optimal feature set, the base of eigenvectors implemented by formal neurons, can be accomplished. The main contribution of this paper to the analog VLSI design of formal neurons consists in a system concept for such a transformation and an analog vector normalization mechanism proposed for the first time. The desired properties of this mechanism is proven on the macro unit level.

A neuron with such a normalization mechanism can be used in a network either for coding purposes and signal decorrelation (maximum entropy network) or for cluster transformation, classification preprocessing or surface fitting (minimum entropy network).

References

Brause, R., "The Minimum Entropy Neuron- a basic Building Block for Clustering Transformations", in *Artificial Neural Networks 2*, I.Aleksander and J.Taylor (Eds.), Elsevier Sc.Publ., pp.1095-1098, 1992

Card, H. C. and Moore, W. R., "VLSI Devices and Circuits for Neural Networks", *Int. J. Neural Systems*, Vol 1/2, pp. 149-165, 1989

Degrauwe, M. et al., "Adaptive biasing CMOS amplifiers", *IEEE Journal of Solid-State Circuits*, SC-17, p. 522, 1982

Földiák, P., "Adaptive network for optimal linear feature extraction", in *Proc. IEEE Int. Joint Conf. on Neural Networks*, pp. I 405-405, San Diego 1989, CA

Fukunaga, K., *Introduction to Statistical Pattern Recognition*, Academic Press, New York 1972.

Gilbert, B., "A precise four-quadrant multiplier with subnanosecond response", *IEEE Journal of Solid-State Circuits*, SC-3:365, 1968

Gilbert, B., "A high performance monolithic multiplier using active feedback", *IEEE Journal of Solid State Circuits*, SC-9 364-373, 1974

Jayant, N.S. and Noll, P., *Digital Coding of waveforms*, Prentice Hall 1984

Mead, C., *Analog VLSI and Neural Systems*, Addison-Wesley 1989

Oja, E., "A Simplified Neuron Model as a Principal Component Analyzer", *J. Math. Biol.* vol 13, pp.267-273, 1982

Rubner, J. and Tavan, P., "A Self-Organizing Network for Principal-Component Analysis", *Europhys.Lett.*, 10(7), pp. 693-698, 1989

Sanger, T., "Optimal unsupervised Learning in a Single-Layer Linear Feedforward Neural Network", *Neural Networks*. Vol 2, pp.459-473, 1989

Tou, J.T. and Gonzales, R.C., *Pattern Recognition Principles*, Addison-Wesley Publ. Comp., 1974

Vittoz, E., "Analog VLSI Implementation of Neural Networks", *Proc. Journées d' électronique*, EPFL, Lausanne, pp.223-250, 1989

Weste, N. and Eshraghian, K., *Principles of CMOS design*, Addison Wesley, 1985

Williams, R.J., "Feature discovery through error-correction learning", *Technical Report 8501*, San Diego, Inst. of Cognitive Science, University of California UCSD, 1985

Xu, L., Oja, E. and Suen, Ch., "Modified Hebbian Learning for Curve and Surface Fitting", *Neural Networks*, Vol. 5, pp.441-457, 1992

A MULTI-LAYER ANALOG VLSI ARCHITECTURE FOR TEXTURE ANALYSIS ISOMORPHIC TO CORTICAL CELLS IN MAMMALIAN VISUAL SYSTEM

Luigi Raffo, Giacomo M. Bisio, Daniele D. Caviglia, Giacomo Indiveri
and Silvio P. Sabatini

INTRODUCTION

VLSI Artificial Neural Networks (ANNs) can be implemented with digital or analog technology. Digital implementations allow high precision and flexibility but to build effective neuromorphic systems it is necessary to exploit all available features and possible modes of operation of the MOS transistor in analog VLSI implementations: the neural computation is mapped directly on the electrical variables that describe the state of the network. In this way the physical restriction on the density of wires, the low power consumption of CMOS circuits (e.g. in the subthreshold domain), the limited precision and the cost of communication imposed by the spatial layout of electronic circuits are similar to the constraints imposed on biological networks (Mead 1989, Schwartz et al 1989, Mead 1990). This choice in favor of analog neural networks is also favored by the development of computational paradigms in computer vision research based on analog models (e.g. minimization of a functional that characterize the degree of acceptability of a solution according to the existing constraints) (Poggio et al 1985, Bertero et al 1988).

One of the most important task performed by natural visual systems is the perception of textural difference and consistency; indeed the ability to distinguish among different objects in the visual world and to separate figures from background depends upon an effective texture discrimination. A great contribution to the comprehension of the mechanisms involved in human texture discrimination was given by Bela Julesz and his colleagues (Caelli and Julesz 1978, Caelli et al 1978) who carried on in the seventies an intensive research on human preattentive vision.

The results of numerous psychophysical experiments led Julesz to the formulation of the "texton theory" (Julesz 1986), according to which texture differences can be perceived instantaneously in terms of the differences between micropatterns (i.e. textons), without a detailed scrutiny of the image by focal attention. Thus texture discrimination is a preattentive (i.e. parallel) process which involves local feature detectors.

In this paper we present an analog VLSI architecture for texture discrimination. The schema of this architecture is derived on the basis of a (lower level) formal neuron model of cortical cells, that can be mapped directly into silicon with high efficiency. Moreover, the organization of our perceptual elements and their interactions resembles those observed in the mammalian visual cortex. In the next Section we will present a model for cortical feature detectors, then, we will propose a neural architecture for texture segmentation. Concluding remarks and perspectives will be discussed in the closing Sections.

MODELING CORTICAL CELLS

From a neurophysiological point of view, Julesz's theory is supported by the fact that cortical cells act as texton detectors since they respond optimally to oriented elongated blobs and their terminators (Hubel and Wiesel 1962) and to oriented gratings of appropriate spatial frequency (Campbell and Robson 1968, Webster and De Valois 1985). The properties of a cell can be characterized by its response to specific visual stimuli. The area of the visual field in which stimuli best elicit the neural response is called the *receptive field* of the neuron. Similarly, the 2D spatial response profile of the receptive field (called briefly, receptive field profile) can be defined, as the distribution of neuron sensitivity to visual stimulation. More specifically, the linear 2D response profile refers to a spatial weighting function by which a local region of the image is multiplied and integrated to generate that neuron's response (see Fig.1). According to this definition, if we consider the cell in (u,v) on the cortical plane that has a receptive field A centered around (i,j) in image plane, the excitation e_{uv} of such cell is:

$$e_{uv} = \sum_{A} w_{uv}(h,k) x(i-h, j-k) \tag{1}$$

where w is the receptive field profile.

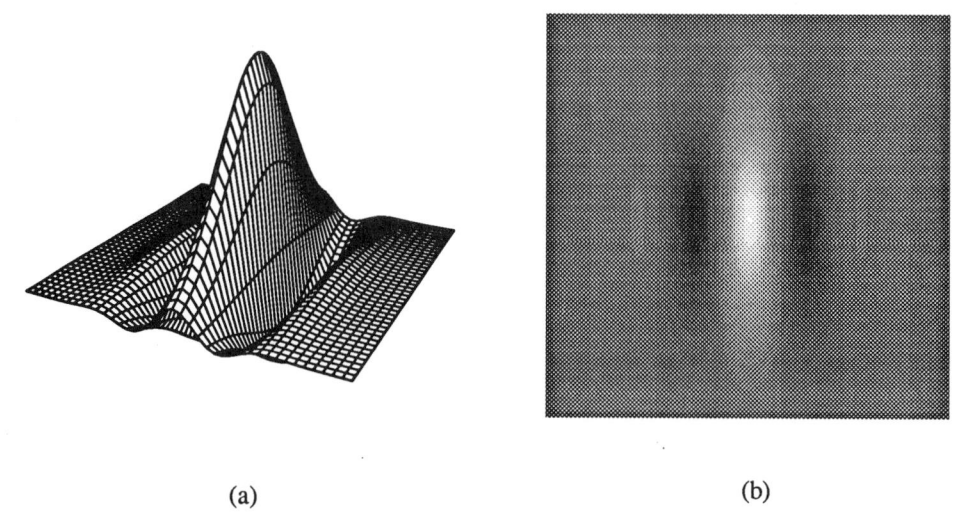

(a) (b)

Figure 1 (a) Example of spatial 2D weighting function describing the receptive field profile; (b) Density plot of the same receptive field profile.

The input-output relation (1) implements a convolution operation on the input image x(i,j). In this case convolution operations are realized by the neuron-synapsis schema, where synapses don't have to change their strenghts because this neural network needs no adaptation as it happens in the adult's visual system.

The observation of elongated Gabor-like receptive fields' profiles in a wide class of cortical cells (Kulikowski and Bishop 1981, Jones and Palmer 1987b) led to the diffusion of a great number of algorithms that perform preattentive texture analysis using two-dimensional Gabor functions as convolution kernels (Turner 1986) or as basic functions for image transformations (Bovik *et al* 1990). Such computational properties justify the attention paid to cortical cells of area 17 and specifically those of the layer IV, that receive the main sensory inputs from the lateral geniculate nucleus (LGN).

The transition from LGN cells with circular receptive fields to cortical ones with elongated receptive fields is a consequence of two main factors: functional projections of LGN afferent connections to cortical cells (Hubel and Wiesel 1962) and intracortical processing (Sillito 1977, Morrone *et al* 1982, Worgotter *et al* 1991). Separating the two contributes, we model the receptive fields, due to the solely LGN projections, with elongated DOOG functions (Differences Of Offset Gaussians) (Malik and Perona 1990) and let intracortical processing mould the definitive shape of the receptive fields. The receptive fields so obtained (Bisio *et al* 1992), in respect to DOOG functions, include additional sidebands similar to the ones found experimentally (Mullikin *et al* 1984, Webster and De Valois 1985) and can be properly fitted by Gabor functions whose processing features have been well characterized in literature (Daugman 1985).

THE ANALOG VLSI ARCHITECTURE

Our aim is to conceive and specify an analog VLSI neural system inspired by the structure of the visual system.

The mammalian visual system fulfills complex tasks such as image segmentation and recognition by a sequel of operation performed in a hierarchical way by different subunits from retina to visual cortex. These subunits perform different levels of computation (Hubel and Wiesel 1962, Albus 1975, Worgotter *et al* 1991). The basic schema of the analog architecture is composed of two layers of computational units: the first one is the input-and-preprocessing layer, the second one is the processing layer. The input layer performs the first step in the chain of visual processing; the input image is pre-processed with operations similar to those accomplished by the retina. At this level the local mean intensity is subtracted to render the process independent on variations of local luminosity. Thus the input layer is organized with a layer of mean extractors. A network of photoreceptors determines the spatial quantization of the image; the pre-processing layer is used for compensation for different illumination conditions. This operation can be realized with elements similar to those used in the subsequent process, because also this operation can be realized with a convolution operation.

The processing layer (whose conception is the specific contribution of this paper) is structured as shown in Fig.2a. The computational units are arranged in two-dimensional regular arrays. Neighboring units are connected with different areas (slightly shifted and mostly overlapped) of the input layer, according to the spatial organization of their receptive fields so that any portion of the image is analyzed by a complete set of orientations (*hypercolumn*); see Fig.2b.

(a)

Outputs

Sensory Input to an hypercolumn

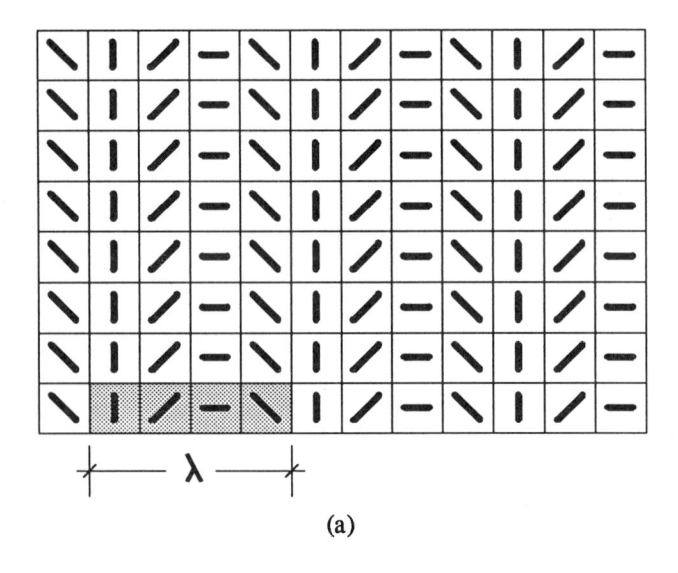

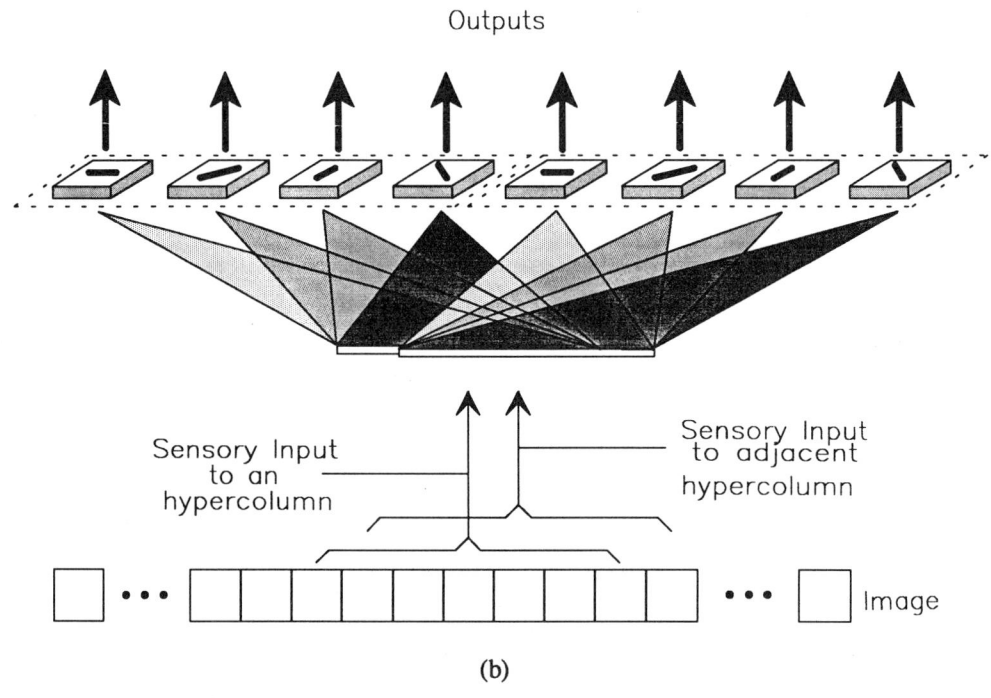

Sensory Input to adjacent hypercolumn

Image

(b)

Figure 2 (a) Schema of the processing layer. Each box, corresponding to an elementary *computational unit* (see Fig.4), is composed of four cells with receptive fields oriented along the line marked on each box (four orientations have been considered in this example). The computational units can be organized in a 2D regular array, according to various schemata. In this case the cells are organized in stripe-like *hypercolumns* of width λ (Worgotter *et al* 1991); (b) Each portion of the image is analyzed by a complete set of orientations (*hypercolumn*). The receptive fields of nearby *hypercolumns* overlap sligthly.

Neighboring units, belonging to different *hypercolumns* are linked with excitatory connections (the output of a unit is added to the input of other units), while the links between units belonging to the same *hypercolumn* are inhibitory (output is subtracted); see Fig.3. Inhibitory and excitatory connections enhance the difference between different oriented patterns. This operation can be realized directly using synapses of fixed strenghts of both positive and negative values.

Each box (i.e. a computational unit) is composed of four cells with DOOG receptive fields oriented along the marked orientation (see Fig.4). Two cells have odd symmetry, two even symmetry; the four receptive fields of each box model the functional projection of LGN afferent neurons to cortical cells (see above). The outputs of the four cells are compared and the maximum value is chosen as the output of the unit to which they belong. Resolution in the spatial frequency domain imposes some constraints on the dimensions of the receptive fields. Implementation efficiency would demand low dimension receptive fields to reduce the number of connections, while an adequate quantization of the receptive field functions imposes a minimum number of connections, that is however not critical for DOOG fields. Typical receptive fields have 25, 81, 225 connections.

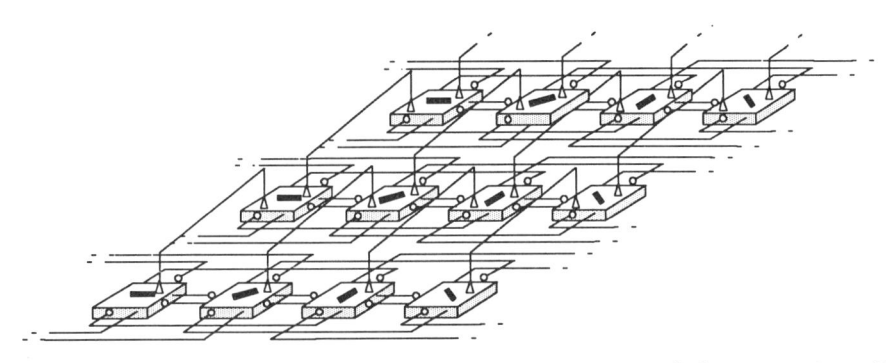

Figure 3 A schematic representation of the connections among *computational units* in three *hypercolumns*. Circles stay for inhibitory synapses whereas triangles stay for excitatory synapses.

When considering a distributed architecture one should qualify it both respect to the computing elements (nodes) and the communication planes (interconnection network). In our architecture, based on receptive fields (i.e. every node is interconnected with all the nodes belonging to the neighborhood of limited extension), communications are localized, thus allowing to achieve good cost and performance metrics in its VLSI implementation (Seitz 1984, Mead 1990). Global information transfer is achieved through the multy-layer structure that makes use of hierarchy of cells (i.e. cells of simple and complex receptive fields). The whole set of objects, action and/or capabilities of the processing layer can be efficiently implemented with analog VLSI circuits (Mead 1989, Vittoz 1990).

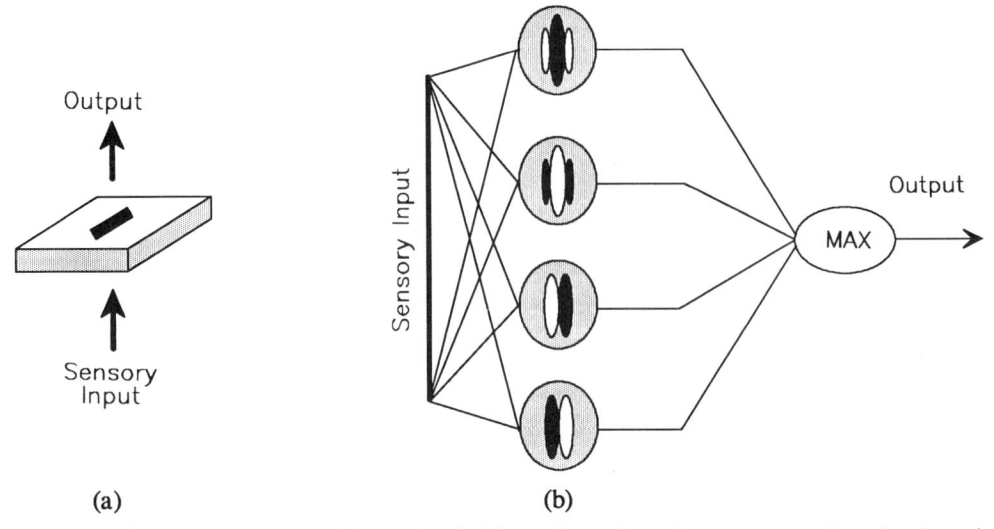

Figure 4 Each *computational unit* (a) is constitued of four cells receiving the same sensory input (as shown in (b)). The cells have the same orientation preference but different contrast sensitivity (ON and OFF) and different spatial phase (EVEN and ODD).

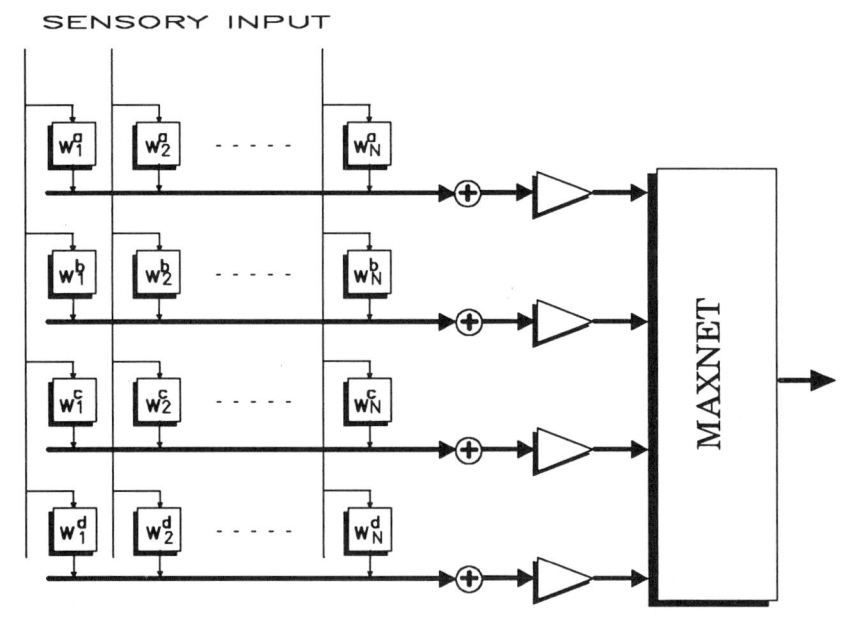

Figure 5 Detailed neural architecture of the cells of the computational unit described in Fig.4. The N wires propagate sensory input across the four rows of synaptic wheights to realize the convolution operation for the four different receptive fields characterizing the computational unit. The four sets of weights used are: *a* for OFF-EVEN cell, *b* for ON-EVEN cell, *c* for ON-ODD and *d* for OFF-ODD. The output of the adders are applied to the MAXNET network that extracts the maximum value in input.

As described in Fig.4b the nodes of our architecture can be identified with the computational cells that perform a convolution on the sensory input, see eq.(1). The coefficients w in eq.(1) determine the orientation of convolution filters. The convolution computation is susceptible of a neural interpretation in which a neuron is activated by the sum of the weighted signals brought in by its input synapses (see Fig.5). Various analog VLSI implementations of neural networks have been considered and realized. (Valle *et al* 1992, Vittoz 1990). The non-linear functionality of analog transistors is applied for implementing complex circuit components (e.g. multipliers). This drastically reduces the size of the network and also decreases power consumption. By example, the maximum operation can be realized with a compact and efficient implementation of the MAXNET network (Lazaro *et al* 1989). Following this analog approach, large networks can be realized if the architectural schema is suitable for VLSI, as it is the one proposed in this paper.

RESULTS

We checked, by computer simulations, the computational capabilities of our model in one basic visual task: texture discrimination. The outputs of the processing layer are used to segment textures: any variation in the texture of the image reflects on a variation of the output of neurons. We considered pictures in which different textures, taken from Brodatz collection (Brodatz 1966), are grouped together. By example, textures with different spatial and spatial-frequency characteristics are represented in the quartered picture shown in Fig.6a. The activities of the four cell populations, corresponding to the four orientations considered (0°, 45°, 90°, 135° from left to right) are shown in Fig.6b. Each cell in the computational unit has a receptive field profile with $N=81$ (see Fig.5). The results are comparable, if not better, than those obtained using an equivalent set of Gabor functions (Turner 1986, Bovik *et al* 1990, Raffo 1991).

DISCUSSION

For our neural architecture it has been possible to choose appropriate connection strengths on the basis of a twofold consideration: the knowledge about the organization and functionalities of cortical cells, and the investigation of features of Gabor and gaussian 2D filters.

Architectures, similar to the one here presented for texture information, based on the concept of receptive field could be devised to elaborate other processes in low-level vision such as edge detection, the extraction of stereo disparity, and the analysis of motion. In perspective, information processing capabilities at intermediate and high-level could be obtained from our architecture adding further processing layers of similar cells. The process can be recursively applied to consider operation like segmentation or recognition of simple objects.

The studies on the structure of mammalian visual systems could further suggest architectural models based on simple computational units, locally strongly interconnected (Koch 1989), apt to perform complex visual tasks in machine vision also for real word

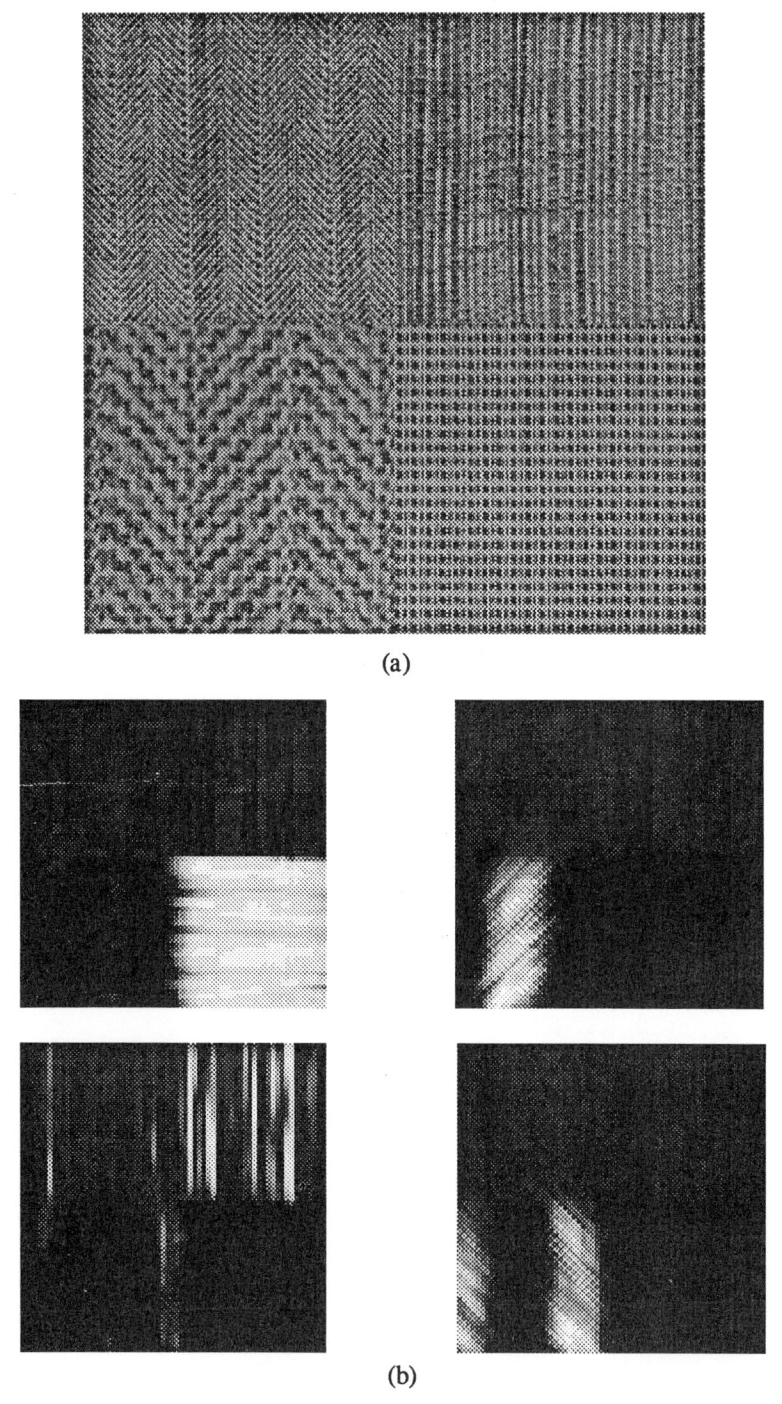

Figure 6 (a) Image of Brodatz collection used for test; (b) The responses of the computational units selective to different orientations (0, 45, 90, 135 degrees from left to right and top to bottom) obtained in the simulations of our architectural model are shown. The textures that can be distinguished better (for this choice of receptive field dimensions) are the ones in the lower quarters, that have low spatial frequency.

scenes, and possibly to overcome the limitations of many AI algorithms strongly oriented to specific applications. Specifically, if we deal with analog computation, where the physical process *is* the computation, a close attention must be paid to the physical structure of neocortex since it is tightly related to its computational properties. The specific organization of neocortex (Douglas and Martin 1992), its economic wiring (Mitchison 1992) and the relations to the properties of cortical neurons justify, indeed, the efforts to understand in what sense cortical microcircuits transform their biology into computation.

REFERENCES

Albus, K., "A Quantitative Study of the Projection Area of the Central and Paracentral Visual Field in Area 17 of the Cat", *Exp. Brain Res.*, vol. 24, pp. 159-202, 1975.

Bertero, M., Poggio, T. and Torre, V., "Ill-Posed Problems in Early Vision", *Proceeding of the IEEE*, vol. 76, pp. 869-889, 1988.

Bisio, G.M., Caviglia, D.D., Indiveri, G., Raffo, L. and Sabatini, S.P., "A neural model of Cortical Cells Characterized by Gabor-like Receptive Fields-Application to Texture Segmentation-", *ICANN '92 Brighton-UK*, 1992.

Bovik, A.C., Clark, M. and Geisler, W.S., "Multichannel Texture Analysis Using Localized Spatial Filters", *IEEE Trans. on PAMI*, vol. 12, pp. 55-73, 1990.

Brodatz, P., *Textures: A Photographic Album for Artists and Designers*, New York, NY: Dover, 1966.

Campbel, F.W. and Robson, J.G., "Application of Fourier Analysis to the Visibility of Gratings", *J. Physiol.*, vol. 197, pp. 551-566, 1968.

Caelli, T. and Julesz, B., "On Perceptual Analyzers Underlyng Visual Texture Discrimination: Part I", *Biol. Cybern.*, vol. 28, pp. 167-175, 1978.

Caelli, T., Julesz, B. and Gilbert, E., "On Perceptual Analyzers Underlyng Visual Texture Discrimination: Part II", *Biol. Cybern.*, vol. 29, pp. 201-214, 1978.

Daugman, J.G., "Uncertainty Relation for Resolution in Space, Spatial Frequency, and Orientation Optimized by Two-Dimensional Visual Cortical Filters", *J. Opt. Soc. Am. A*, vol. 2, pp. 1160-1169, 1985.

Douglas, R.J. and Martin, K.A.C., "A Functional Microcircuit for Cat Visual Cortex", *J. Physiol.*, vol. 440, pp. 735-769, 1992.

Hubel, D.H. and Wiesel, T.N., "Receptive Fields, Binocular Interaction, and Functional Architecture in the Cat's Visual Cortex", *J. Physiol.*, vol. 160, pp. 106-154, 1962.

Jones, J. and Palmer, L., "The Two-Dimensional Spatial Structure of Simple Receptive Fields in Cat Striate Cortex", *J. Neurophysiol.*, vol. 58, pp. 1187-1211, 1987a.

Jones, J. and Palmer, L., "An Evaluation of the Two-Dimensional Gabor Filter Model of Simple Receptive Fields in Cat Striate Cortex", *J. Neurophysiol.*, vol. 58, pp. 1233-1258, 1987b.

Jones, J., Stepnoski, A. and Palmer, L., "The Two-Dimensional Spectral Structure of Simple Receptive Fields in Cat Striate Cortex", *J. Neurophysiol.*, vol. 58, pp. 1212-1232, 1987.

Julesz, B., "Textons Gradients: the Theory Revisited", *Biol. Cybern.*, vol. 54, 245-251, 1986.

Koch, C., "Seeing Chips: Analog VLSI Circuits for Computer Vision", *Neural Comp.*, vol. 1, pp. 184-200, 1989.

Kulikowski, J.J. and Bishop, P.O., "Linear Analysis of the Responses of Simple Cells in the cat Visual Cortex", *Exp.Brain Res.* vol. 44, pp. 386-400, 1981.

Lazaro, J., Ryckebusch, R., Mahowald, M., and Mead, C.A., "Winner-take all networks of O(N) complexity", in *Advances in Neural Information Processing Systems*, vol. 1 Los Altos, CA: Morgan Kaufmann pp. 703-711, 1989

Malik, J. and Perona, P., "Preattentive Texture Discrimination with Early Vision Mechanisms", *J. Opt. Soc. Am.*, vol. 7, pp. 923-932, 1990.

Mead, C.A., *Analog VLSI and Neural Systems*, MA: Addison-Wesley, Reading, 1989.

Mead, C.A. "Neuromorphic Electronic Systems", *Proceedings of IEEE*, vol. 78, 1629-1636, 1990.

Mitchison, G., "Axonal Trees and Cortical Architecture", *Trends in Neurosci.*, vol. 15, pp. 122-126, 1992.

Morrone, M.C., Burr, D.C. and Maffei, L., "Functional Implications of Cross-Orientation Inhibition of Cortical Visual Cells.I. Neurophysiological Evidence", *Proc. R. Soc. Ser. B*, vol. 215, pp. 335-354, 1982.

Mullikin, W.H., Jones, J.P. and Palmer, L.A., "Periodic Simple Cells in Cat Striate Cortex", *J. Neurophysiol.*, vol. 52, pp. 372-387, 1984.

Poggio, T., Torre, V. and Koch, C., "Computational vision and regularization theory", *Nature*, vol. 317, pp. 314-319, 1985.

Raffo, L., "Classificazioni di Tessiture con Reti Neurali nello Spazio di Gabor", in *Proc. Symposium Image Processing:Application and Trends Genova-Italy*, pp. 159-167, 1991.

Schwartz, D.B., Howard, R.E. and Hubbard, W.E., "A programmable analog neural networks chip", *IEEE Journal of Solid State Circuits*, vol. 24, pp. 313-319, 1989.

Seitz, C., "VLSI Concurrent Architectures", *IEEE Trans. on Comp.* December 1984.

Sillito, A.M., "Inhibitory Mechanisms Underlying the Directional Selectivity of Simple, Complex and Hypercomplex Cells in the Cat's Visual Cortex", *J. Physiol.*, vol. 271, pp. 699-720, 1977.

Turner, M.R., "Texture Discrimination by Gabor Functions", *Biol. Cybern.* vol. 55, pp. 71-82, 1986.

Valle, M., Caviglia, D.D. and Bisio, G.M., "An Experimental Analog VLSI Neural Chip with On-Chip Back-Propagation Learning", in *Proc. ESSCIRC'92 Copenaghen*, September 1992.

Vittoz, E.A., "Analog VLSI Implementation of Neural Networks", in *Proc. Int. Symp. Circuit and Systems ISCAS '90, New Orleans*, pp. 2524-2527, 1990.

Webster, M.A. and De Valois, R.L., "Relationships Between Spatial-Frequency and Orientation Tuning of Striate-Cortex Cells", *J. Opt. Soc. Am. A*, vol. 2, pp. 1124-1132, 1985.

Worgotter, F., Niebur, E. and Koch, C., "Isotropic Connections Generate Functional Asymmetrical Behavior in Visual Cortical Cells", *J. Neurophysiol.*, vol. 66, pp. 444-459, 1991.

A VLSI PIPELINED NEUROEMULATOR

José G. Delgado-Frias, Stamatis Vassiliadis, Gerald G. Pechanek, Wei Lin, Steven M. Barber and Hui Ding

INTRODUCTION

Applications and interest on artificial neural networks (ANN) have been increasing in recent years. Applications include pattern matching, associative memory, image processing and word recognition (Simpson 1992). ANNs is a novel computing paradigm in which an artificial neuron produces an output that depends on the inputs (from other neurons), the strength or weights associated with the inputs, and an activation function.

In this paper, the computational task of the artificial neuron is given by (Rumelhart, McClelland, and the PDP Group 1986)

$$Y_i(t+1) = F\left(\sum_{j=1}^{N} W_{ij} Y_j(t)\right) \qquad (1)$$

where N is the number of neurons in the neural network and $F(Z_i)$ is the neuron activation function that is usually a non-linear sigmoid function (Hopfield 1984).

In the case of completely connected network, the total computational task requires: N^2 multiplications, N product summations, N activation functions, and NXN communications. Additionally, we are interested on the incorporation of learning algorithms to neuroemulators. In particular, in this paper we study the implementation of the backpropagation in the sequential pipelined neuroemulator (SPIN) proposed by Vassiliadis, Pechanek and Delgado-Frias (1991).

In the following sections we first describe the SPIN structure and report an evaluation of several neuroemulators (Vassiliadis, Pechanek and Delgado-Frias 1991). An implementation of the backpropagation algorithm on the SPIN structure is thoroughly described. Finally a performance evaluation of the backpropagation algorithm on SPIN is provided.

SPIN DESCRIPTION

All the Y_j's are involved in the multiply accumulate function for each neuron as indicated in equation 1. This also indicates that each neuron is connected through a connection weight Wij to every other neuron in the network; i.e. a 1-to-N communication is necessary. The design of the interconnection strategy is a major consideration since it is required to establish N(1-to-N) communication paths. A direct connection between neurons would result in N^2 buses. As the number of neurons increases, directly connection N neuron becomes extremely costly and may be unrealizable. SPIN (Vassiliadis, Pechanek, and Delgado-Frias 1991) as well as other architectures (Kung and Hwang 1989, Blayo and

Hurat 1989, Weinfeld 1989, Pechanek, Vassiliadis, and Delgado-Frias 1992) offer alternatives to the interconnection problem.

SPIN is based on the observation that the neuron output equations as generated from (1) are of the following form (Vassiliadis, Pechanek and Delgado-Frias 1991):

$$
\begin{aligned}
Y_1 &= F(W_{11}Y_1 + W_{12}Y_2 + \cdots + W_{1N}Y_N) \\
Y_2 &= F(W_{21}Y_1 + W_{22}Y_2 + \cdots + W_{2N}Y_N) \\
&\vdots \\
Y_i &= F(W_{i1}Y_1 + W_{i2}Y_2 + \cdots + W_{iN}Y_N) \\
&\vdots \\
Y_N &= F(W_{N1}Y_1 + W_{N2}Y_2 + \cdots + W_{NN}Y_N)
\end{aligned}
\tag{2}
$$

This suggests that for each neuron i there exists a set if N weight multiplications using the same Y values. In order to compute a new output value of a neuron i, it would be necessary to have the Y input values (Y_1 to Y_N) and their associated weights (W_{i1} to W_{iN}) separately available. In addition, N multipliers and weight storage are required. The N products can be formed in parallel in one multiplier delay time which is denoted as δ_M. The N products can be added together forming a final summation (Z_i) which is passed through an activation function $F(Z_i)$ to produce a neuron output. To update another neuron for instance j, the Y input values (Y_1 to Y_N) remain as constant terms in the multipliers while a new set of weights (W_{j1} to W_{jN}) is read. Then the products and $F(Z_j)$ are handled in similar fashion as above. This process continues until all the Y values are updated which is the end of an *update cycle*. The update cycle requires to compute N new Y values.

The **sequential pipelined neuroemulator** (SPIN), shown in Figure 1, incorporates the previous observation and is implemented as a single pipelined physical neuron. Figure 1 shows the hardware organization of SPIN.

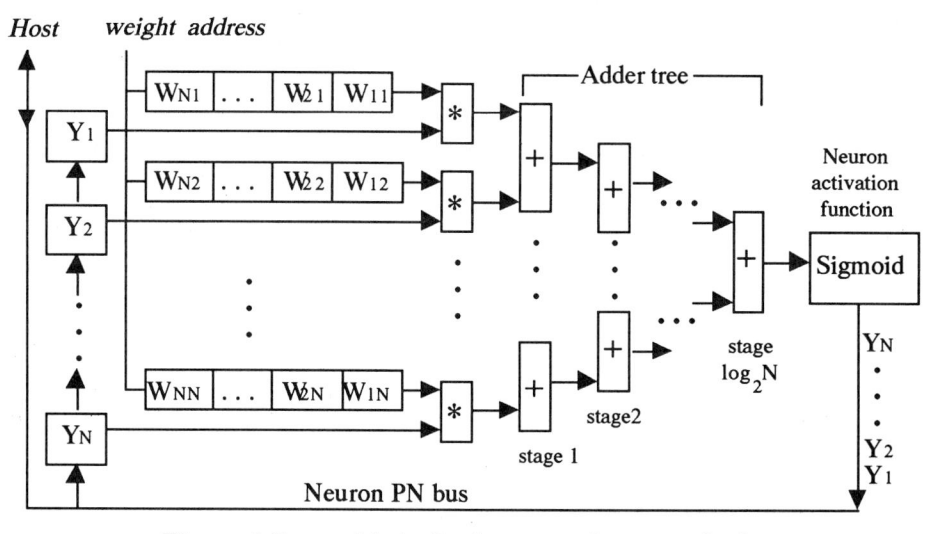

Figure 1 Sequential pipelined neuroemulator organization

SPIN contains N processing elements (denoted as Y_1 to Y_N) that hold the values of Y and have limited computational capabilities, N weight storage units with N weights each, N multipliers, an adder tree containing $\log_2 N$ adder stages, a single sigmoid generator, and a bus that connects the sigmoid generator, Y_N, and the Host computer.

SPIN contains a Host interface that provides the initial Y and W values. Once all the N Y values and N^2 W values are loaded, SPIN updates neuron values in a pipelined fashion. The N multipliers are used to compute N W·Y products in parallel. The results are sent to the adder tree that computes the accumulation. The activation function is considered to be a sigmoid (Hopfield 1984).

SPIN PERFORMANCE ON A FULLY CONNECTED NETWORK

In order to properly compare different special purpose neurocomputer architectures, we have established a set of common assumptions. The list of assumptions begins with the equation (1) for a completely connected neural network with a non linear activation function. All the designs utilize L bits of precision for both the neural output values and the connection weights. Full overlapping of operations with a bit-serial format is assumed in an attempt to show maximum performance.

The equation for the period of generating new neuron values is the performance comparison mechanism. Since all the architectures considered in this paper are evaluated based on the recursive equation (1), the computation of $Y_i(t+1)$ cannot begin before the previous N Y(t) values have been calculated and received at the input logic. The delay variables are denoted as δ_{NAME} which represents the delay through the named element. The common variables are: δ_M equal to a multiplier delay, δ_A equal to an add delay, and δ_{SIG} equal to a sigmoid generator delay. To guarantee meeting the maximum rate of incoming summation values, the sigmoid generator should have the following restriction $\delta_{SIG} \leq \delta_M$. A high performance hardwired sigmoid generator can be designed (Zhang, *et.al.* 1992) at the expense of extra hardware. Alternatively, a single table lookup sigmoid generator could be used (Treleaven 90).

An N neuron emulation is achieved on SPIN with a period specified by

$$N\delta_M + \delta_A \cdot (\log_2 N) + \delta_{SIG} + \delta_{STORE} + \delta_{BUS} \quad (3)$$

where δ_{STORE} and δ_{BUS} are the delay of the PE to store and shift values and the bus delay to transfer data from the sigmoid generator to the PE.

The systolic ring neural network has N processing elements, each PE contains a multiplier, an adder, an N-weight memory, and a non-linear activation function generator (Kung and Hwang 1989). The systolic operation can be executed in a pipelined fashion for which we have estimated the period delay would be

$$N\delta_M + \delta_A + \delta_{SIG} + \delta_{BUS} \quad (4)$$

The recurrent systolic array (Blayo and Hurat 1989) is made up of N^2 processor cells, each consisting of a simple adder/subtracter, accumulation, and controls that are connected on a square systolic array with nearest neighbor connections. There are N binary threshold (+1, -1) neuron activation functions connected at the edge of the array. We estimated for a neural network of N neurons, that the recurrent systolic array would hgave the following delay:

$$\delta_M + 2\delta_A + \delta_{SIG} + (2(N-1)+1+L)\delta_{BUS-C} \quad (5)$$

where δ_{BUS-C} is the neuron output value transmission delay between cells.

The neuromimetic architecture (Duranton, Gobert and Mauduit 1989) reformats equation (1) into a sum of bit products as:

$$Y = F\left(\sum_{j=1}^{N} W_{ij} Y_j^1 + 2\sum_{j=1}^{N} W_{ij} Y_j^2 + \cdots + 2^{L-1} \sum_{j=1}^{N} W_{ij} Y_j^{L-1}\right) \quad (6)$$

where the L bit Y values are indicated in the super-scripts. The reformatted equation (6) is implemented with N-L bit ANDs, an adder tree for the summation of Lbits, an iterative adder, an accumulator storage element, one neuron activation function generator and one output bus. Using a full adder (Vassiliadis 1989, Quach and Flynn 1992) the update equation of the neuromimetic architecture is:

$$\delta_{AND} + (NL + \log_2 N) 4\delta_A + \delta_{SIG} + \delta_{BUS} + \delta_{ACUM} \quad (7)$$

The amount of hardware that is required for each emulator has been compared. This comparison is in terms of weight store (the amount of memory required for storing the weights), the number of multipliers, adders and sigmoid generators that are required to compute the neurons, and the number of buses that are required to communicate the neuron values. Table 1 shows this comparison. All the neuroemulators require the same size of weight store. There are significant differences between SPIN and the Recurrent Systolic in their structures. These hardware comparisons do not include hardware for learning algorithms.

Table 1 Hardware requirements for the neuroemulators

ARCHITECTURE	WEIGHT STORE	MULTIPLIERS	BIT-SERIAL ADDERS	SIGMOID GENERATORS	BUS
SPIN	N^2	N	N-1	1	1
Systolic ring	N^2	N	N	N	1
Recurrent systolic	N^2	N^2	N^2	N	N
Neuromimetic	N^2	N·L ANDs	N·L parallel adds	1	2

The performance comparison of these four neuroemulators is shown in Figure 2. For this comparison it was assumed that the length of the word is eight bits (L=8) and one full bit adder will take a cycle to operate. This graph shows that the performances of both Systolic Ring and SPIN are comparable. The recurrent systolic shows a better performance at extra expense as shown above. The neuromimetic architecture show a slightly lower performance.

The implementation of the non-linear function generator may require a large amount of hardware (Zhang, *et.al.* 1992) if high precision and/or a large number of sigmoid generators is required. SPIN needs only one sigmoid generator; this in turn would allow to implement a high precision sigmoid generator if it is required at low cost. For learning algorithms having a pipeline approach provides some advantages as explained in the following section.

The SPIN architecture has been implemented in VLSI CMOS technology. A two's complement 16-bit multiplier similar to Gnanasekaran's (1983) bit-serial multiplier has been designed. The adder tree which is formed of bit-serial adders performs the addition of the weight-input products. The sigmoid function utilizes a lookup table.

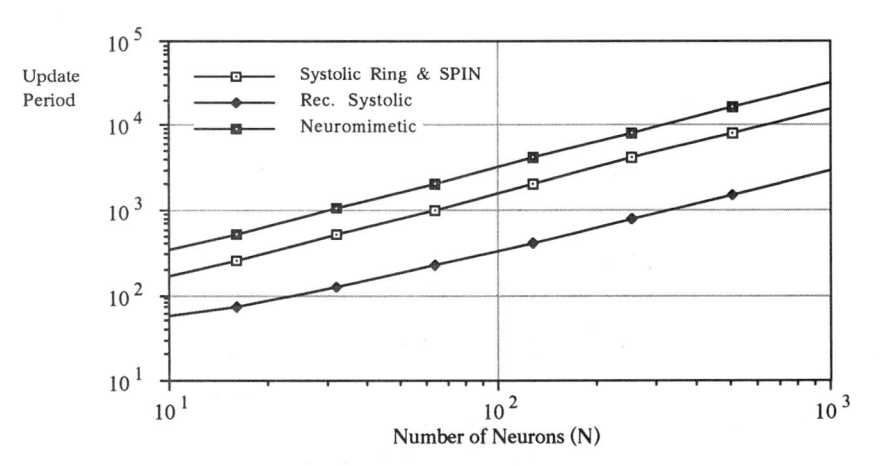

Figure 2 Period update comparison with L=8 bits.

BACKPROPAGATION LEARNING ALGORITHM IN SPIN

Backpropagation is one of the most widely used learning algorithms in multi-layered neural networks (Widrow 1990). Figure 3 shows a multi-layered neural network. There two weight matrices: weight matrix (1) specifies the strengths between the input layer (shown as layer 0) and the hidden layer (shown as layer 1) and weight matrix (2) specifies the strengths between the hidden and output (shown as layer 2) layers. At the learning stage the network is presented with a target pattern.

Using a sigmoid as activation function, the output error value ($\delta_i^{(2)}$) is computed as:

$$\delta_i^{(2)} = (T_i - Y_i^{(2)}) Y_i^{(2)} (1 - Y_i^{(2)}) \tag{8}$$

where T_i is the target value for neuron i and $Y_i^{(2)}$ is the output value of the same neuron. The error value in the hidden layer is computed as follows:

$$\delta_i^{(1)} = \left(\sum_{j=1}^{N^{(2)}} W_{ji}^{(2)} \delta_j \right) Y_i^{(1)} (1 - Y_i^{(1)}) \tag{9}$$

where $N^{(2)}$ is the number of neuron at the output layer.
The weight adjustments are performed as follows:

$$\Delta W_{ij}^{(2)} = \eta \, \delta_i^{(2)} Y_j^{(1)} \quad \text{for weight matrix (2)} \tag{10.a}$$

$$\Delta W_{ij}^{(1)} = \eta \, \delta_i^{(1)} Y_j^{(0)} \quad \text{for weight matrix (1)} \tag{10.b}$$

where η is the learning rate. The new weight value is computed by adding to the old weight value and the weight adjustment as shown in equation (11).

$$W_{ij}^{new} = W_{ij}^{old} + \Delta W_{ij} \tag{11}$$

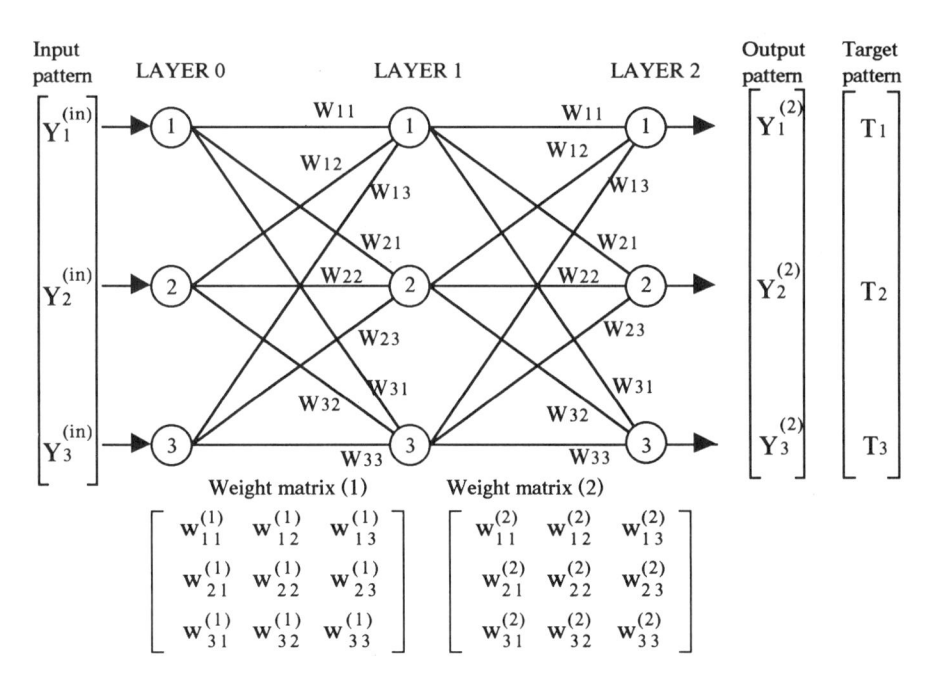

Figure 3 Three-layered backpropagation neural network and the weight matrices

In order to accommodate the requirements imposed by equations (8) to (10), the original SPIN structure has been modified as shown in Figure 4. The SPIN structure does not operate as a full pipelined structure when the learning algorithm is run; this is done in order to avoid conflicts in the bus and PEs.

Two register files are used to store the two weight matrices of a three-layer neural network. The register file for weight matrix (2) should be able to store a copy of the original matrix as well as to provide access to the transposed matrix. It is assumed that the maximum number of neuron per layer is N. Thus, these files have the same weight store capacity (N^2). If smaller number of neurons per layer are used in a given network, the corresponding weight strengths could be set to zero and SPIN would update only the existing neurons. If more neuron per layer than N are required the SPIN structure should go into virtual mode (Vassiliadis, Pechanek and Delgado-Frias 1991); this mode is not included in this paper.

Additional hardware is needed to compute the error; this hardware is located after the sigmoid generator. In addition, N adders are required in order to modified N weights in parallel. The processing elements (PEs) need to store information such as the three neuron outputs ($Y_i^{(0)}$, $Y_i^{(1)}$ and $Y_i^{(2)}$) and the error value of the output layer δ. A simplified processing element is shown in Figure 5. It should be noticed that a broadcast line has been added; this line is required to broadcast values that all the weights should operate with.

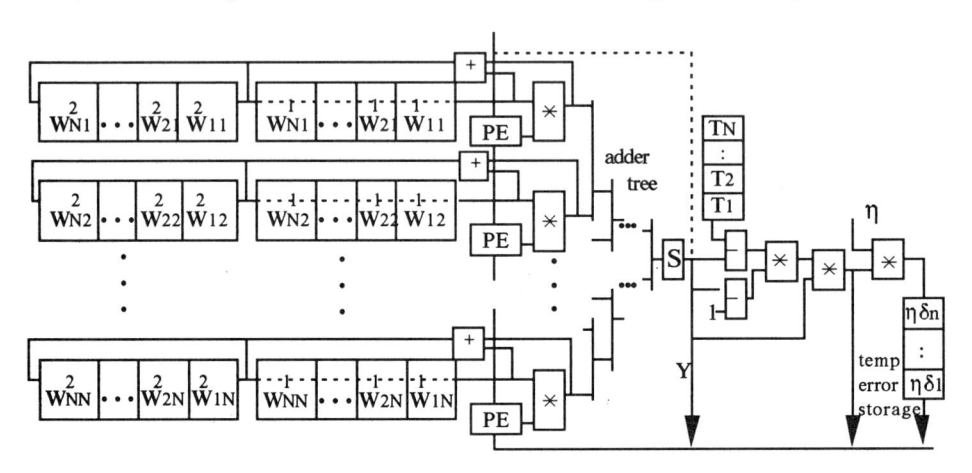

Figure 4 Multi-layered SPIN organization

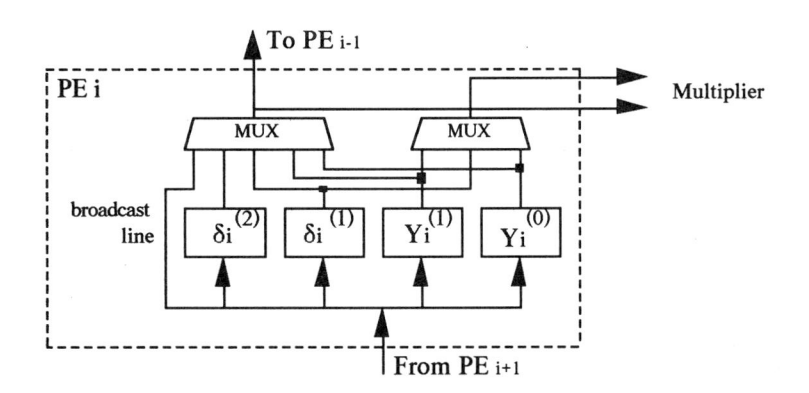

Figure 5 Multi-layered SPIN's processing element

Learning algorithm at the output layer

It is assumed that the output values of the input and hidden layers have been already computed and stored in the PE registers. At the output layer equations (1), (8), (10.a) and (11) have to be implemented. To accomplish this, the following steps are performed.

- The output values ($Y_i^{(2)}$) are computed. These values are passed on as input to the hardware that implements eq. (8); this hardware is located after the sigmoid generator (see Figure 4).

- The output error values $\delta_i^{(2)}$ (eq. 8) are computed by means of the hardware provided after the sigmoid generator. Using the neuron PN bus, each value is sent to PE_N where is stored.
- The products of the output error values and the learning rate (η) are obtained and stored in a temporary error storage.
- The weight adjustments (eq. 10.a) can be calculated by using the N multipliers with the broadcast value $\eta\,\delta_i^{(2)}$ and the hidden layer outputs $Y_j^{(1)}$ that are available at each PE.
- The new weight values are obtained (eq. 11) and stored in the proper location in the weight matrix (2).

The last two steps need to be done after the computation of the error values in the hidden layer ($\delta_i^{(1)}$). The old values of the weight matrix (2) are used in the computation of such values; thus, these values have to be preserved. This however has no major impact on the performance of the learning algorithm.

Figure 6 shows how the new weights for matrix (2) are computed; in particular, it illustrates how SPIN works when implementing the equations (8) and (10.a) with i=1. The thicker lines indicate the active modules in the structure. The update delay due to the output layer is (assuming N neurons in each layer):

$$(2N+3)\delta_M + (\log_2 N+2)\delta_A + \delta_{SIG} + 2\delta_{STORE} + \delta_{BUS} \quad (12)$$

where δ_{STORE} refers to the time that takes to store the error values into the temporary register and the new weights into the file. Sending $\eta \cdot \delta_i$ to the PEs takes a delay of δ_{BUS}. Once the PEs have received the products, the PEs send them to the multipliers along with the output values (to implement equation 10.a).

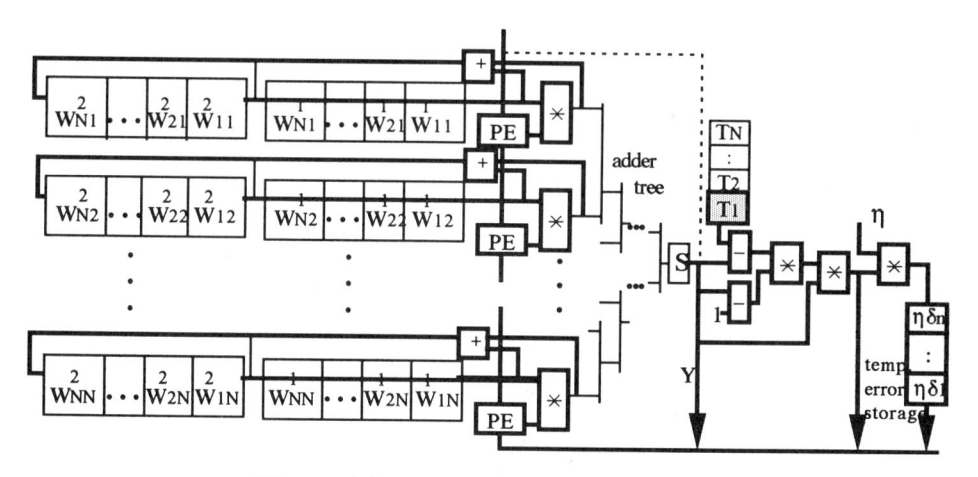

Figure 6 Learning update at the output layer

Learning algorithm at the hidden layer

For the learning algorithm, the hidden layer has slightly different requirements than the output layer. Equations (9), (10.b) and (11) are implemented by following the steps described below.

- The N products $W_{ji}^{(2)}\,\delta_j^{(2)}$ are computed in parallel. The output error values from the output layer are stored in the PEs. For these computations the transposed weight matrix (2) is used (see Figure 7); a custom hardware design is required to hold the old value of the weights and to provide the transposed matrix as output.
- The adder tree is used to perform the summation of the N products of eq. (9).
- The sigmoid function unit is bypassed; it is not required for any of the computations.

- The output error values $\delta_i^{(1)}$ (eq. 9) are computed by means of the hardware provided after the sigmoid generator (see Figure 7). Using the neuron PN bus, this value could be sent to PE_N if more than one hidden layer is used; this case, however, is not considered in this paper.
- The product of the output error values and the learning rate (η) are obtained and sent to a temporary register file. This process is done in a pipeline fashion.
- The weight adjustments (eq. 10.b) can be calculated by using the N multipliers with the broadcast value $\eta \, \delta_i^{(1)}$ and the hidden layer outputs $Y_j^{(0)}$ that are available at each PE.
- The new weight values are obtained (eq. 11) and stored in the proper locations in the weight matrix (1).

Figure 7 shows the case when i=1 in the equations (9), (10.b) and (11). The update delay for the hidden layer, assuming that there are N neurons in each layer, is:

$$(2N+3)\delta_M + (\log_2 N+2)\delta_A + 2\delta_{STORE} + \delta_{BUS} \quad (13)$$

Using a sequential processor to update the output and hidden layer weights would take:

$$N[(2N+3)\delta_M + 2N\delta_A + \delta_{SIG} + N\delta_{SEQ}] \quad \text{output layer} \quad (14.a)$$
$$N[(2N+3)\delta_M + 2N\delta_A + N\delta_{SEQ}] \quad \text{hidden layer} \quad (14.b)$$

Where δ_{SEQ} is the time require to control a sequential program; this includes loops, conditions, and data transfer.

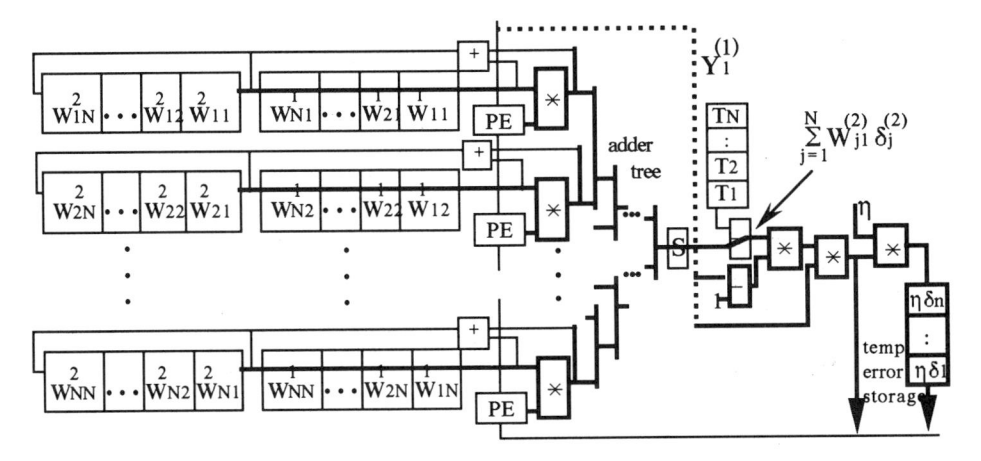

Figure 7. Accumulation for equation (9) and errot coputation for hidden layer.

Equations (12) and (13) are required to compute the total training period. The following assumptions were made: $\delta_M = 2\delta_A$ and $\delta_A = \delta_{SIG} = \delta_{BUS}$. Figure 8 shows a comparison of SPIN and a sequential implementation. In this case it is assumed that all the SPIN operations are bit-serial while the sequential machine is assumed to perform the operations in a word-parallel fashion. It should be noted that as N increases the difference between these two approaches favors SPIN.

CONCLUDING REMARKS

In this paper we have incorporated the backpropagation algorithm into the sequential pipelined neuroemulator (SPIN) proposed by Vassiliadis, Pechanek and Delgado-Frias (1991). SPIN has approximately the same performance as the ring systolic array; SPIN however requires N-1 less sigmoid generators. The recurrent systolic array has higher performance; this is at the expense of an additional N^2-N processing elements and N-1 sigmoid generators.

It has been shown that the backpropagation algorithm can be incorporated with minor modifications to the original SPIN. It should be noted that the additional hardware that is put after the sigmoid generator (in order to compute equation (8)) can be omitted, since the computation is almost sequential. This computation can be done by the host computer. This approach offers advantages such as flexibility for different learning algorithm. Furthermore, an embedded weight modification mechanism has been studied. This approach could allow one to implement a number of learning algorithms. The SPIN machine is currently being built.

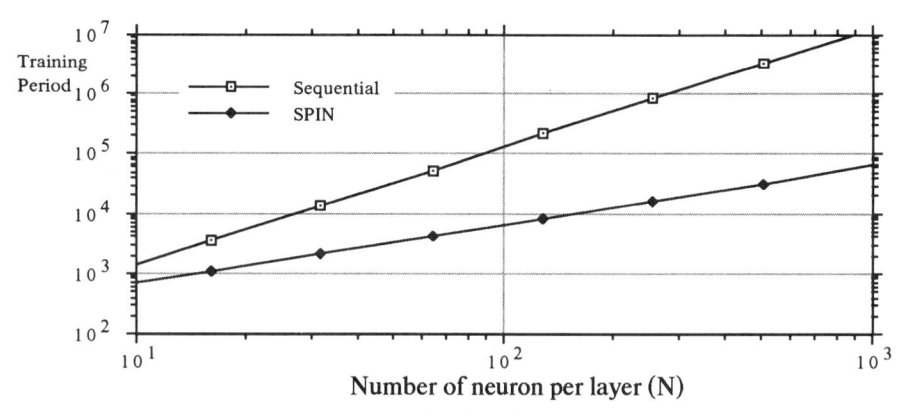

Figure 8 Training period update for SPIN and sequential

REFERENCES

Blayo, F. and Hurat, P., "A VLSI Systolic Array Dedicated to Hopfield Neural Networks." *VLSI for Artificial Intelligence,* J. Delgado-Frias and W. Moore (Eds), pp. 255-264, Kluwer Academic Publishers, 1989.

Duranton, M., Gobert, J. and Mauduit, N., "A Digital VLSI Module for Neural Networks," *Neural Networks from Models to Applications,* L. Personnas and G. Dreyfus (EDS.) Paris: I.D.S.E.T., 1989.

Gnanasekaran, R., "On a Bit-Serial Input and Bit-Serial Output Multiplier," *IEEE Transactions on Computers,* Vol. C-32, no. 9, pp. 878-880, 1983.

Hopfield, J., "Neurons with Graded Response Have Collective Computational Properties Like Those of Two-State Neurons," *Proceedings of the National Academy of Sciences,* pp. 3088-3092, May 1984.

Kung, S. Y. and Hwang, J. N., "A Unified Systolic Architecture for Artificial Neural Networks," *Journal of Parallel and Distributed Computing,* Vol. 6, pp. 358-387, 1989.

Pechanek, G.G., Vassiliadis, S., and Delgado-Frias, J. G., "Digital Neural Emulators Using Tree Accumulation and Communication Structures," *IEEE Transactions on Neural Networks,* Vol. 3, no. 6, pp. 934-950, November 1992.

Quach, N. T. and Flynn, J., "High-Speed Addition in CMOS," *IEEE Transactions on Computers,* Vol 41, no. 12, pp. 1612-1615, December 1992.

Rumelhart, D. E., McClelland, J. L. and the PDP Research Group, *Parallel Distributed Computing, Vol. 1: Foundations.* Cambridge, Mass.: The MIT Press, 1986.

Simpson, P. K., "Foundations of Neural Networks," in *Artificial Neural Networks: Paradigms, Applications and Hardware Implementations.* E. Sánchez-Sinencio and C. Lau (Eds), New York: IEEE Press, pp. 3-24, 1992.

Treleaven, P. and Vellasco, M. "Neural Networks on Silicon," *Wafer Scale Integration, III,* M. Sami and F. Distante (Eds), pp. 1-10, Elsevier Science Publishers, 1990.

Vassiliadis, S., "Recursive Equations for Hardwired Binary Adders," *Int. Journal of Electronics,* vol. 67, no.2, pp. 201-213, 1989.

Vassiliadis, S., Pechanek, G. G. and Delgado-Frias, J. G., "SPIN: A Sequential Pipelined Neurocomputer," *IEEE Int. Conference on Tools for Artificial Intelligence,* pp. 74-81, San Jose, Calif., November 1991.

Weinfeld, M., "A Fully Digital Integrated CMOS Hopfiled Network Including the Learning Algorithm," *VLSI for Artificial Intelligence,* J. Delgado-Frias and W. Moore (Eds), pp. 169-178, Kluwer Academic Publishers, 1989.

Widrow, B. and Lehr, M. A., "30 Years of Adaptive Neural Networks: Perceptron, Madaline, and Backpropagation," Proceedings of the IEEE, vol 78, no.9, pp. 1415-1442, 1990.

Zhang, M., Delgado-Frias, J. G., Vassiliadis, S. and Pechanek, G. G., "Hardwired Sigmoid Generator," *IBM Technical Report TR01.C492,* pp. 1-38, IBM, Endicott, NY, September 1992.

A LOW LATENCY DIGITAL NEURAL NETWORK ARCHITECTURE

William Fornaciari and Fabio Salice

INTRODUCTION: THE PSEUDO-NEURON APPROACH

Neural networks are an effective approach to solve non-standard or non-algorithmic problems such as system control, classification and pattern recognition. These important capabilities, are balanced by the costs both of design and silicon implementation.

The solutions presented in literature for ANNs implementation on silicon adopt two main technological approach: the *analog* or the *digital* ones. Digital solutions allow to combine a number of features such as modularity, flexibility and arbitrary precision (typical of digital system) with the large availability of design tools for VLSI digital architecture. For this we have chosen this class of architecture as the object of our study.

Many digital architecture has been presented in the recent literature (Distante *et al* (1990), S.Y.Kung (1988), Dreyfus *et al* (1990), Soucier and Ouali (1990), Blayo and Hurat (1989)). One of the main problems to be solved when considering digital implementation is the high connectivity requirement. In order to reduce the connectivity, time multiplexing of signals onto a reduced number of interconnection lines must be adopted; as a consequence, a relatively high *latency* must be expected since actual operation parallelism obviously decreases. Due to time multiplexing, *throughput* limitations are also present. For some applications e.g. fast control systems, or for any *hard real time* applications, high *latency time* may not be acceptable.

To overcome these limitations, Distante *et al* (1989), presented a proposal based on decomposition of functionality of a neuron onto a (suitably chosen) number of processing elements, called Pseudo Neurons (PNs). This units can behave either as a neuron or as a part of one neuron (in fig.1 a functional schema of the PN is reported).

The PNs can be assembled together to constitute a neuron (see fig.2). In this way, each PN, must be able to receive input values from both the outside world (or previous neural network *layer*) and the preceding PN. In the second case, the input value is the partial sum of the weighted inputs of the neuron. The PNs according to its position inside the neuron, must also be able either to forward the partial sum computed on its subset of inputs or to evaluate the non-linear output function.

With this approach, neurons with any number of inputs can be obtained by connecting the proper number of PNs. Fornaciari and Salice (1991), has shown that if all PNs are uniform, i.e., they deal with the same number of weights, synchronization is simplified and no *wait cycle* for inter-layer data transfer must be introduced. In fact, even though time multiplexing of signals is adopted, all the building blocks which constitute neurons of different size have the same computational cycle length.

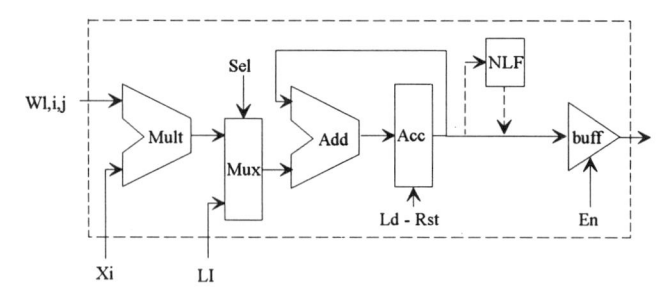

Figure 1 **Blocks-schema of one PN.** Signals inside the box are PN local synchronization while signals external to the box came from both others PNs and PN local memory.

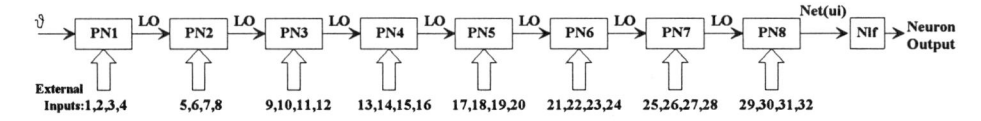

Figure 2 **Internal PNs-based structure.** One single neuron with 16 inputs is considered; the overall computation is splitted over eight PNs organized as a *linear array*.

Some digital architecture for multi-layered, feed-forward neural network try to overcome connectivity and synchronization problems by adopting a placement of PNs in a *linear array* (Distante *et al* (1990), Fornaciari and Salice (1991)); in such solutions *latency* and *throughput* are improved because of pipelining evaluation of the final neuron output.

Before discussing and comparing this architecture and our new one, we introduce the following notation and parameters:

- n number of layers in the neural network
- N_l number of neurons in layer l; in each layer, neurons are numbered from 1 to N_l
- K_l number of PNs per neuron in layer l
- r_l number of weights in each PN in layer l; in our case of fully connection between layers, evaluates $\lceil N_{l-1}/K_l \rceil$
- $w_{l,i,j}$ logical weight value of connection between neuron i of layer l and neuron j of layer $l-1$;
- $\vartheta_{l,i}$ threshold value for neuron i in layer l
- $X_{l,i}$ output value of neuron i in layer l
- PNw number of words stored locally in each PN memory
- $Net(u_i)$ is the weighted neuron sum $Net(u_i) = \sum_{j=1}^{N_{l-1}} W_{l,i,j} X_{l-1,j}$

Linear array architecture are specified in terms of the number of weights for each PN and the number of PNs per neuron; Fornaciari and Salice (1991) have shown that the degree of parallelism is maximized by fixing the number of weights per PN to a constant (this value is referred as r). We adopt the same position so that, in the remaining part of the paper, we will use these following notations $r = r_l$, $K_l(r,l) = K_l$.

General relationship between the architectural parameters will be presented by considering an example in which each neuron has 32 inputs and r is equal 4.

Since we consider a full connection between layers, the number of neuron inputs in layer l is equal to the cardinality of layer $l-1$. As a consequence the number of PNs for each neuron is $K_l = \lceil N_{l-1}/r \rceil = \lceil 32/4 \rceil = 8$

Propagation of the data and computation in the set of PNs is shown in fig.3.

			←		r + 2		→	←			K-1			→			
PN1	9	y	1	2	3	4	9	y	1	2	3	4	9	y	...	...	...
PN2	8	x	y	5	6	7	8	x	y	5	6	7	8	x	y	...	...
PN3	11	12	x	y	9	10	11	12	x	y	9	10	11	12	x	y	...
PN4	14	15	16	x	y	13	14	15	16	x	y	13	14	15	16	x	y
PN5	17	18	19	20	x	y	17	18	19	20	x	y	17	18	19	20	x
PN6	...	21	22	23	24	x	y	21	22	23	24	x	y	21	22	23	24
PN7	...	...	25	26	27	28	x	y	25	26	27	28	x	y	25	26	27
PN8	...	...	...	29	30	31	32	x	Fire	29	30	31	32	x	Fire	29	30
Time	-1	0	1	2	3	4	5	6	7	8	9	10	11	12	13	14	15

Figure 3 Neuron weights temporal distribution. On the X axis different time steps are reported while the Y axis reports the PNs identifiers. The row length differs from r because is necessary to spend two time steps to forward the PN output value (Firing or Linear Output) before to start a new computational cycle. Bolded lines shown the temporal evolution of one computational cycle.

Each cell of fig. 3 contains and identifier representing a weight inside the PN memory, i.e. the identifier q represents the logical weight $w_{l,i,q}$, while x, y and *Fire* are not important for computation. The time steps corresponding to x and y are used only for synchronization since they allow propagation of the partial sum through PNs, while the time step marked *Fire* allows the forwarding of the neuron output to the following neurons. The input of the partial sum from the previous PN take place during the time step x whilst the output of the new partial sum is performed during the y time step.

Referring to the example reported in fig.3, the delay between the process of the first weight pertaining to first PN and the firing, namely the neuron *latency*, is 13 clock cycles. In general, Distante et al (1990) have been shown that *latency* can be expressed by eq.(1); moreover, since for full connection between layers $r = \lceil N_{l-1}/K_l \rceil$ eq.(1) becomes eq.(2).

$$\text{latency}_{\text{linear array}} = ((r + 2) + (K_i - 1)) = PNw + (K_i - 1) \quad (1)$$

$$\text{latency}_{\text{linear array}} = ((\lceil N_{l-1}/K_l \rceil + 2) + (K_i - 1)) \quad (2)$$

These expressions will be used in the following sections to compare this structure with the architecture we propose.

THE PROPOSED ARCHITECTURE

In this section we introduce a novel neuron architecture, still based on the PN approach, but characterized by a much lower *latency* than the *linear array* approach.

Latency is minimized when parallelism of PNs operations (i.e. the partial sums) is maximized. Each PN is able only to add two terms at a time: one term can be either a product between neuron input and correspondent weight or a linear output (partial summation) from another PN. To improve parallelism it is possible to organize the set of PNs in pairs working together as reported in fig.4.

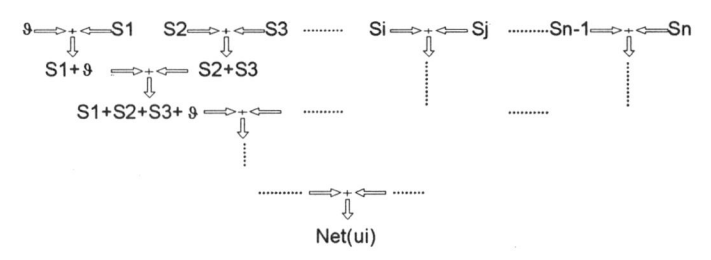

Figure 4 Sequence of operation to obtain *Net(u$_i$)* for the *tree* structure. S$_i$ is the computation performed by PN$_i$.

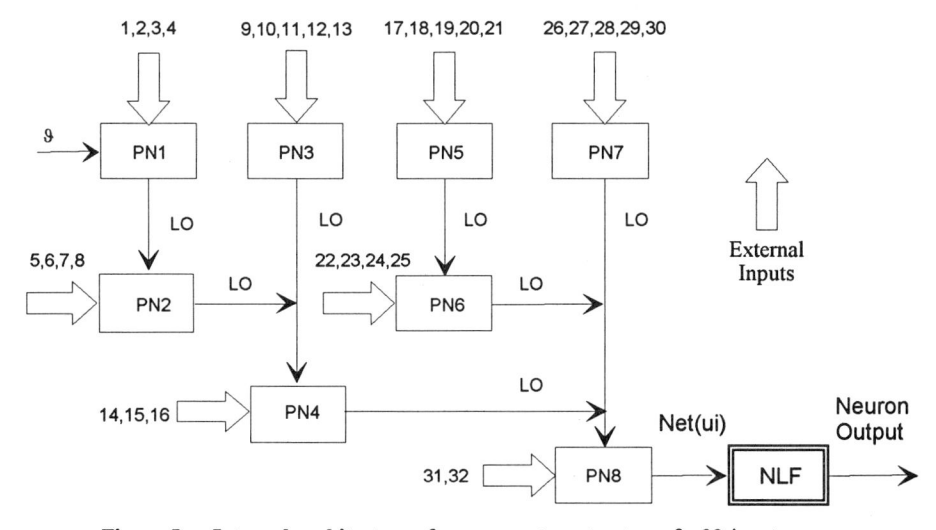

Figure 5 Internal architecture of a neuron: *tree* structure of a 32-inputs neuron.

The internal architecture of each neuron is organized as a multi-layered structure, each internal layer consisting of PNs connected as in fig.5. Correspondingly, the overall neuron computation is mapped onto a sequence of steps: each step is associated to a different internal layer of the neuron structure.

The operation performed by the *tree* structure can be formally described as:

$$Net(ui) = \left(...(((\vartheta + S_1) + (S_2 + S_3)) + ((S_4 + S_5) + (S_6 + S_7))) \,...\, + (S_{n-1} + S_n)...\right) \quad (3)$$

where the most internal operations are pairwise sums of partial summations, followed by pairwise sums of such first results, and so on until the final sum (the *root* of the *tree*) is reached. To perform this operation, the internal structure of the PN as well as its interconnections must be modified; each PN must be able to calculate its Linear Output as the summation of a number of products $w_{l,i,j} \cdot X_{l-1,i}$ and of a number of Linear Outputs produced by PNs belonging to the previous internal layers of the neuron. In fact, the number of PNs forwarding their linear inputs to the considered PN depends on the position of the PN itself within the *tree*.

The organization of PNs in the *tree* is based on the following definitions:

Definition: a block consists of a pair of sub-blocks: *right* and *left*;
- the output of the *left* sub-block is connected to the input of the *right* sub-block
- the input and output of the block is the same as the *right* sub-block;

The algorithm to build the *tree* structure is therefore:

Initialization: All PNs are disposed orderly from left to right. Every single PN is a block

REPEAT

Action: from left to right, using adjacent blocks if it is possible, build a new *super* block according to the *Definition*

UNTIL (all the PNs are enclosed into one single block)

Since the number of blocks at a particular level is half of the blocks at the previous level, the number of levels of the whole structure for K PNs in each neuron, by assuming that K is a two's power, is $n\_levels = \lfloor \log_2(K) \rfloor$

In order to compare *tree* and *liner array* structures by means of *latency*, we use the *tree* structure to build the same neuron of the example in fig.3. Fig.6 reports the new temporal diagram.

			← PNw →					← $\lceil \log_2(K) \rceil$ →							
PN1	9	y	1	2	3	4	9	y	1	2	3	4	9	y	...	...	...
PN2	8	x	y	5	6	7	8	x	y	5	6	7	8	x	y	...	...
PN3	13	y	9	10	11	12	13	y	9	10	11	12	13	y	9	10	...
PN4	16	x	x	y	14	15	16	x	x	y	14	15	16	x	x	y	14
PN5	21	y	17	18	19	20	21	y	17	18	19	20	21	y	17	18	19
PN6	...	x	y	22	23	24	25	x	y	22	23	24	25	x	y	22	23
PN7	...	...	26	27	28	29	30	y	26	27	28	29	30	y	26	27	28
PN8	...	...	...	x	Fire	31	32	x	x	x	Fire	31	32	x	x	x	Fire
Time	-1	0	1	2	3	4	5	6	7	8	9	10	11	12	13	14	15

Figure 6 Temporal distribution of weights and linear outputs for one neuron with an internal *tree* structure. PNs belonging to the same layer forward their linear output at the same time step. Some PN weights positions are used to collapse partial results from previous layers.

As it can be observed, the *latency* is reduced to 9 clock cycles instead of the 13 ones necessary in a *linear array*. Since for the *tree* structure the synaptic weights $w_{l,i,j}$ have not an uniform distribution over all the PNs, it is not possible to use a common value r as the number of weights for each PNs (e.g. compare PN4 in fig.4 and fig.5). The general expression of *latency* for a *tree* structure based neuron, is given by using only *PNw*:

$$\text{latency tree} = PNw + \lceil \log_2(K_l) \rceil \qquad (4)$$

As depicted in fig.6, the *latency* of one neuron is given by two terms: the first one is the time necessary to pass through the first PN (i.e., *PNw* time steps) plus a number of time steps that is equal to the number of levels composing the *tree* structure plus *1* if *K* is not two's power (since the output of the sub-part of the neuron composed by a number two's

85

power of PNs, cannot be the neuron output but have to became LI for the last PN of the remaining neuron part); the latter term thus evaluates $\lceil \log_2(K) \rceil$.

Actually, if multi-layer neural network are envisioned, it is necessary to subtract 1 to this value because of two subsequent neurons (i.e. belonging to two adjacent layers) have one position overlapped (the time step position marked *Fire* is overlapped to the first position of the PN_1 belonging to the following neuron).

THE LINEAR ARRAY VS THE TREE STRUCTURE

Aim of this section is to present a comparative analysis between the *linear* structure and our *tree* architecture. First we show that both structures (using the same number of PNs) are functionally equivalent; then, a comparison in term of speed is provided.

Functional equivalence of the *linear array* and the *tree* structure

In this paragraph we argue that a neuron of *I* inputs can be realized using the same number and the same kind (i.e. the same memory size *PNw*) of PNs for both the considered structures, namely *linear array* and *tree*. In other words for both structures is employed the same hardware resources.

As discussed in the previous paragraph, the memory position are not used only for synaptic weights storage; therefore, according to their usage, we have grouped all the memory words of the local memory of the PN_i in the following classes:

- LW_i which represents the synaptic weights treated by the considered PN;
- NUW_i which represents the weights positions inside the PN_i local memory not used to store synaptic weights. During the time steps reserved for them, the acquisition of the partial sums (Linear Inputs) computed by the previous PNs take place (in fig.3 and fig.6 these positions are marked *x*); the threshold ϑ is treated as a Linear Input.
- One memory position is used to allow either the forwarding of the computed Linear Output between PNs or the propagation of the firing value among neurons (marked respectively *y* or *Fire* in fig.3 and fig.6).

According to this notations it is possible the following formulation for *PNw* (as stated *PNw* is a constant for the whole neural network):

$$PNw = LW_i + NUW_i + 1 \qquad (5)$$

Consider now a neuron composed of *K* PNs: the number *I* of inputs supported by the whole neuron is equal to the sums of all LW_i related to each PN (eq.(6)). If the *linear array* is adopted, *I* can be written as in eq.(7), since there is exactly one LI per PN (for PN_1 this represents the threshold acquisition).

$$I = \sum_{i=1}^{K} LW_i = \sum_{i=1}^{K} (PNw - NUW_i - 1) = \sum_{i=1}^{K} (PNw - 1) - \sum_{i=1}^{K} (NUW_i) \qquad (6)$$

$$\forall i \quad NUW_i = 1 \implies NUW_{linear\ array} = \sum_{i=1}^{K} (NUW_i) = K \qquad (7)$$

$$I_{linear\ array} = (PNw - 2)K \qquad (8)$$

On the other hand, since LWi is constant in each PN, we can derive:

$$\forall i \quad LW_i = r \quad \Rightarrow \quad I_{linear\ array} = \sum_{i=1}^{K} LWi = rK \quad (9)$$

$$I_{linear\ array} = (PNw - 2)K = rK \quad (10)$$

To analyze of the *tree* structure, we consider the case at first of K equal to a power of 2. The number of neuron inputs may be computed from eq.(6) Under this assumptions on K is possible to prove the following lemma:

Lemma 1: *Let PNw be the memory size (constant) of each PNs and K the number of PNs used to build a neuron (where K is power of 2). The same number I of inputs can be considered both for a neuron based upon a linear array and for a neuron with a tree structure.*

Proof: Referring to fig.8, a neuron based on a *tree* structure is build by considering internal hierarchical levels of PNs. The number of LI belonging to a generic level j is:

$$number\ of\ LIs\ in\ level\ j = 2^{n\_levels-j} \quad (11)$$

where $n\_levels = \log_2(K)$ is the number of levels in the *tree* structure.

The number of LIs in the whole structure consists in the sum of all the LIs presents in each layer of the neuron (eq.(12)); from eq.(12) we can derive eq.(13) by means of the arithmetical progression law:

$$total\ number\ of\ LI\ per\ neuron = \sum_{j=1}^{n\_levels} 2^{n\_levels-j} = \sum_{i=0}^{n\_levels-1} 2^i \quad with\ i = n\_level-1 \quad (12)$$

$$total\ number\ of\ LI\ per\ neuron = ((1-2^{n\_levels})/(1-2)) = 2^{n\_levels} - 1 = K-1 \quad (13)$$

The number of memory words which are not used to store weights, for all the neuron is equal to the total number of LI plus 1; the terms $+1$ arises from the use of one time interval (usually reserved for one memory word of PN_1) to input the threshold value 9.

$$NUW\ tree = \sum_{i=1}^{K} NUW_i = total\ number\ of\ LI\ per\ neuron + 1 = (K-1) + 1 = K \quad (14)$$

By substituting eq.(6), the number of memory words for synaptic weights related to neuron inputs is:

$$Itree = \left(\sum_{i=1}^{K}(PNw - 1)\right) - NUW\ tree = K(PNw - 1) - K = K(PNw - 2) \quad (15)$$

This is the same result founded in eq.(8) for the *linear array* structure.

Using *lemma 1*, it is possible to prove the following theorem which generalize the result achieved by removing the assumptions on the K values.

Theorem 1: *Let PNw be the memory size of each PNs and I the number of neuron inputs A tree structure dealing with I inputs can be realized using the same number K of PNs necessary for the linear array structure.*

Proof: It can always applied the following decomposition rule:

Rule 1: $$K = K^P + K^R \qquad (16)$$

where K^P is the maximum power of two such that $1 < K^P \leq K$ (obviously K^R is not a power of two). Referring to K_1^P as the greatest number power of two such that $K > K_1^P$ and to K_1^R as the rest such that $K = K_1^P + K_1^P$, we can apply the *Rule 1* to K_1^R

$$K = K_1^P + K_1^R = K_1^P + (K_2^P + K_2^R) \qquad \text{with } K_2^P < K_1^R < K_1^P \qquad (17)$$

by applying iteratively the *Rule 1* to the terms K_i^R we decompose K in the following monothonic decreasing succession of $m + R$ terms:

$$K = K_1^P + K_2^P + K_3^P + \ldots + K_i^P + \ldots + K_m^P + R = R + \sum_{i=1}^{m}(K_i^P) \qquad (18)$$

where each terms K_i^P is power of two, $K_i^P > K_{i+1}^P$, and R is either 0 or 1.

By exploiting this decomposition of K we can partition the *tree* structure. It is so possible to identify a set of m sub-structures S_i which correspond to the m terms K_i^P; each of them is composed by a number of PNs equal to a power of two (see fig.8) and, possibly, a block S_R (if exists) is composed by a single PN. It is now possible to computes the number of LI per neuron by considering the neuron subdivided in two components:

C1: the LIs inside each substructures S_i;
C2: the LIs which are necessary to link all the S_i to the last one (either S_R or S_m);

The evaluation of such components may be performed as follows:

C1. Eq.(11) can be applied to each terms of the succession K_i^P for *Lemma 1* and eq.(18)

$$C1 = \sum_{i=1}^{m}(K_i^P - 1) = \sum_{i=1}^{m}(K_i^P) - m = (K - R) - m \qquad (19)$$

C2. The number of sub-blocks which constitute the neuron are $m+R$, since it is m if $R=0$, and $m+1$ if $R=1$. The number of links needed to join all this blocks to the last one is:

- if $R=1$, there are m sub-blocks S_i connected to one final element S_R, so that m links are needed;
- if $R=0$, the structure is composed by m blocks S_i; the blocks $S_1 \ldots S_{m-1}$ are connected to the last one S_m via $m-1$ links;

Therefore, it is $C2 = m-1+R$. The total number of LIs per neuron is:

$$\text{total number of LIs per neuron}^{tree} = C1 + C2 = (K-R) - m + (m-1+R) = K-1 \qquad (20)$$

The number of memory words unused to store weights value for a *tree* structure is thus:

$$NUW_{tree} = \sum_{i=1}^{K} NUW_i = \text{total number of LIs per neuron tree} + 1 = (K-1) + 1 = K \qquad (21)$$

where the term $+1$, is due to the time step for the threshold as discussed above for eq.(14).

By substituting eq.(21) in eq.(6), we can compute the number of synaptic weights treated by the neuron based on the *tree* structure:

$$Itree = \sum_{i=1}^{K}(PNw - 1) - \sum_{i=1}^{K} NUW_i = K(PNw - 1) - K = K(PNw - 2) \tag{22}$$

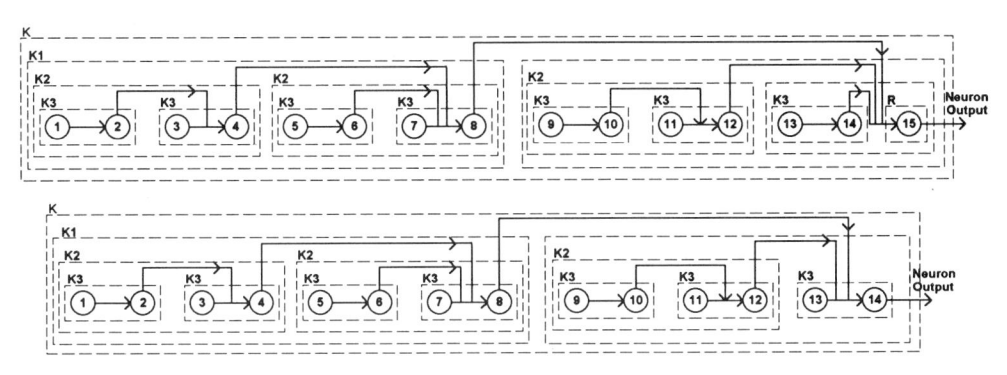

Figure 7 Division of a *tree* structure in sub-blocks power of 2: K=15, K1=8, K2=4, K3=2, R=1, m=3 (upper); K=14, K1=8, K2=4, K3=2, R=0, m=3 (lower).

This is equal to the eq.(8), that has been found for a *linear array* structure.

The main topological difference between *linear array* and the *tree* structure is the different weights distribution among PNs: despite of the *linear array*, in the *tree* structure it is not homogeneous. As a consequence, r in the *tree* structure has not the same value for all PNs (note that in the *linear array* it has a fixed value).

According to the proved functional equivalence, the comparison between equation (8) and (22) allows to define a fictitious parameter $r_{eq} = PNw-2$.

This parameter represents the number of synaptic weights which are treated by each PN if, instead of a *tree* structure, a *linear array* architecture, dealing with the same number of inputs were adopted; it can be used to obtain the same expression for both the formulas (8) and (22): $I\ tree = I\ linear\ array = r_{eq}\ K$.

Comparison

Both the *linear* and *tree* structures, require the same hardware resources (i.e. the same number of PNs with the same local memory size). Moreover, they have the same throughput since, after the initial *latency*, they perform the same computation in the same time.

Using eq(1) and eq.(4) reporting *latencies* for the *linear* array and our *tree* structure, respectively $O(K)$ and $O(\log_2(K))$, we define the following latency gain that is quite-linearly dependent on K, (fig.9).

Latency gain = latency linear array - latency tree $= (K_i-1) - \lceil \log_2(K_i) \rceil$

The actual gain starts from K=3; this happens because both structures for K < 4 corresponds to the same connection of PNs.

Fig.8 shows the latency gain with respect to the latency of a linear architecture for different value of the local memory size. The main evidence is that latency gain is improved if low local memory size (PNw) are considered. On the other hand this induces a high value of K (for the same number of neuron inputs), i.e. the area occupied is expensive.

If no other constraints are defined, we obtain the value of K and PNw from considering an area Vs *latency* optimization; this approach may be driven by:

$$cost\ function = Area^{\alpha} * Latency^{\beta} \qquad \text{with } \alpha = \beta-1$$

Obviously α and β must be chosen according to the relative importance between silicon costs and speed requirements for the specific application.

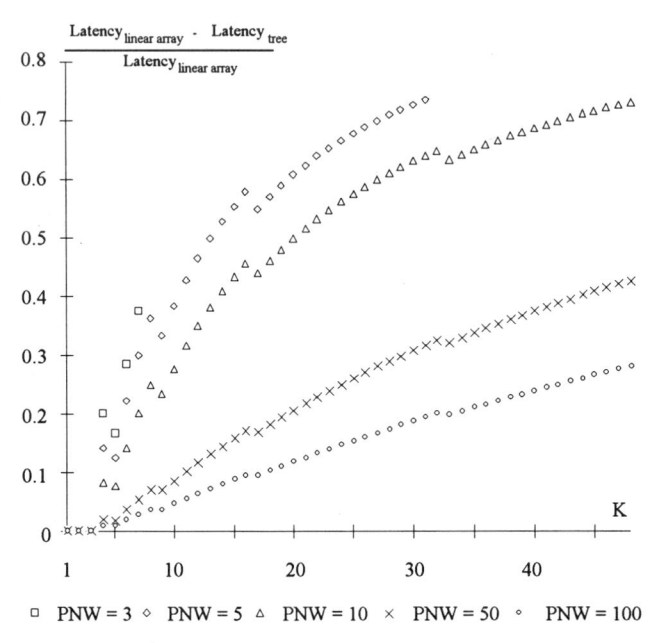

Figure 8 Relative *latency* gain with respect to the *linear array* for several architectural parameters, *PNw* and *K*.

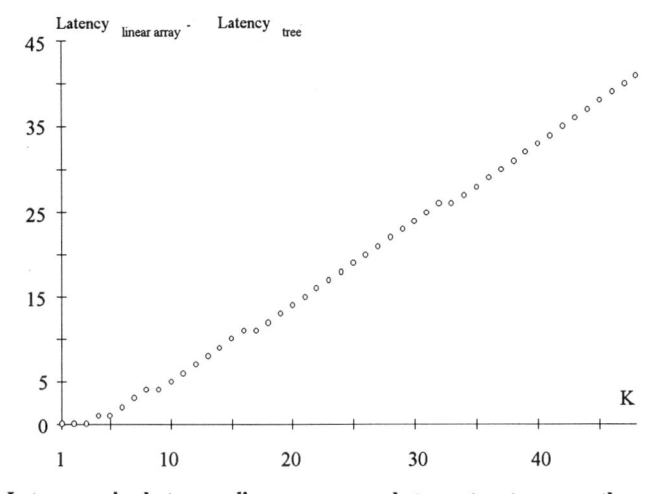

Figure 9 *Latency* gain between *linear array* and *tree* structure vs. the architectural parameter *K*.

REFERENCES

Alla P.Y., Dreyfus G., Gascuel J.D., Johnnet A., Personnaz L., Roman J., Weinfeld M.; "Silicon integration of learning algorithm and other auto-adaptative properties in a digital feedback neural network", *Proc. IEEE-ITG Workshop on Microelectronics for Neural Networks*, Dortmund, Germany, 1990.

Blayo F., Hurat P.; "A reconfigurable WSI neural network", *Proc. IEEE Int'l Conf. on WSI*, San Francisco, Jan. 1989.

Distante F., Sami M.G., Stefanelli R., Storti Gajani G.; "A compact and fast silicon implementation for layered neural nets", in *VLSI for Artificial Intelligence and Neural Networks*, José G. Delgado Frias and William R. More Eds., Plenum Press, New York, 1991.

Distante F., Sami M.G., Stefanelli R., Storti Gajani G.; "Multistage Interleaved Architectures for Implementation of Neural Networks", *ICS89*, Santa Clara CA, 1989.

Fornaciari W., Salice F., Storti Gajani G.; "A formal methodology for automatic synthesis of neural networks". *IEEE-IFIP MicroNeuro-91*, Munich, Oct. 1991.

Kung S.Y.; "Parallel architectures for artificial neural nets", *Proc. Int'l Conf. on Systolic Arrays*, pp. 163-174, San Diego, CA, USA, May 1988.

Ouali J., Saucier G.; "Fast generation of neuro ASICs", *Proc. INNC 90*, Paris, 1990.

MANTRA: A MULTI-MODEL NEURAL-NETWORK COMPUTER

Marc A. Viredaz, Christian Lehmann, François Blayo and Paolo Ienne

INTRODUCTION

Though artificial neural networks have been studied for five decades, they have experienced, in the past ten years, a very rapidly growing interest. Most of the applications in this domain are however either simulated on conventional machines or implemented on some specialized hardware dedicated to a given model. At the present day, there is no platform which is, at the same time, versatile enough to implement any neural-network model and learning rule, and fast enough to be used on large problems. This fact prompted many researchers to work toward the design of a *generic neural-network computer* (Treleaven *et al* 1989). As an intermediate step in this quest, a *multi-model* hardware implementation, whose architecture is influenced by biological considerations, is presented in this paper.

One of the authors proposed in his Ph.D. thesis (Blayo 1990) to base such a machine on a 2-D systolic array. He observed that several widely used neural models could be decomposed into a few operations, and showed how these operations could be mapped on a systolic structure. On the basis of these considerations, a few versions of a VLSI digital building block called *Generic Elements for Neuro-Emulators using Systolization* or *GENES* have been implemented (Lehmann and Blayo 1991).

The first prototype, called *GENES HN8* (for *Hopfield Network 8 bits*) has already been used on an evaluation board described here. A more general chip, referred to as *GENES IV*, has been sent to foundry and will be used in the *MANTRA* neural-network computer to implement the following models:

- Mono-layer networks: Perceptron (Rosenblatt 1958), ADALINE (Widrow and Hoff 1960), and delta rule.
- Multi-layer feed-forward networks: back-propagation rule (Le Cun 1985).
- Fully-connected recurrent networks (Hopfield 1982).
- Self-organizing feature maps (Kohonen 1982).

Special care has been taken to sustain a utilization rate of 100% and to keep the communication and computation local, two conditions which are necessary for a truly scalable architecture.

THE GENES STRUCTURE

In the above-cited thesis (Blayo 1990), it has been shown that the following four operations are sufficient to implement most neural models: (1) the matrix-vector multiplication, (2) the vector outer product (yielding a matrix), (3) the addition and subtraction of vectors and matrixes, and (4) the application of a usually non-linear activation function (if learning is not required, operations (1) and (4) are enough). Although this fact is easily verified for the Perceptron, the ADALINE, the delta rule, and the Hopfield model, other algorithms — such as the back-propagation rule or the Kohonen model — deserve further discussion.

Concerning the Kohonen model, only a version of this algorithm can be implemented using these four operations. It uses the scalar product as a distance:

$$y_i(0) = \mathbf{W}_i \cdot \vec{x} \tag{1}$$

where $\mathbf{W}_i$ represents the ith line of the weight matrix $\mathbf{W}$, and $\vec{x}$ the input vector. The resulting output vector $\vec{y}(0)$ is then fed to a recurrent network of fixed weight $\mathbf{W}^{\text{rec}}$ in order to determine the winner (the most active neuron) and its neighborhood:

$$\vec{y}(\tau) = \sigma(\mathbf{W}^{\text{rec}} \cdot \vec{y}(\tau - 1)) \quad \text{for: } \tau = 1, 2, 3, \ldots \tag{2}$$

With appropriate recurrent weights $\mathbf{W}^{\text{rec}}$, this network will eventually stabilize, producing the vector $\vec{y}^{\text{stable}}$ used to update the synaptic weights:

$$\mathbf{W}_i := \mathbf{W}_i + \alpha \cdot y_i^{\text{stable}} \cdot (\vec{x}^{\text{T}} - \mathbf{W}_i) \tag{3}$$

This model is the closest to biological ones. Unfortunately the scalar-product distance may require the normalization of the inputs and/or weights to produce correct results.

To avoid these difficulties, the more conventional Kohonen algorithm is often used. The latter uses the Euclidean or Manhattan distance:

$$y_i = \sqrt{\sum_{j=1}^{N}(x_j - W_{i,j})^2} \tag{4}$$

$$y_i = \sum_{j=1}^{N}|x_j - W_{i,j}| \tag{5}$$

The winner is then identified by finding the neuron i_{win} such that $y_{i_{\text{win}}} = \min(y_i)$, and the synaptic weights are updated as follows:

$$\mathbf{W}_i := \begin{cases} \mathbf{W}_i + \alpha \cdot (\vec{x}^{\text{T}} - \mathbf{W}_i) & \text{if: } v \in N_{i_{\text{win}}} \\ \mathbf{W}_i & \text{if: } v \notin N_{i_{\text{win}}} \end{cases} \tag{6}$$

Where $N_{i_{\text{win}}}$ is the neighborhood of neuron i_{win}. If, as t is often the case, α is chosen as a decreasing to zero function of the distance between the neurons i and i_{win}, the updating can be done on all the neurons.

The identification of the smallest and largest element of a vector allows any combination of distance and winner determination (recurrent network or maximum/minimum).

The mapping of these operations onto the GENES structure, a square 2-D array of simple *processing elements (PEs)*, is described here. Each PE implements a synaptic connection, storing the value of the corresponding weight. They are only connected to their four neighbors, hence keeping the communication local (systolic paradigm). All

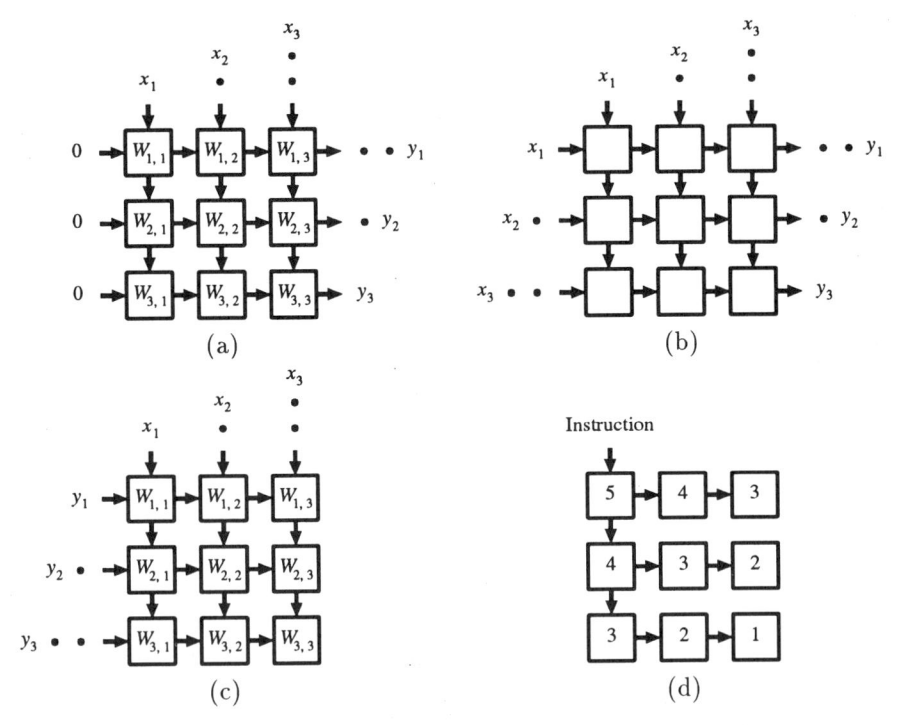

Figure 1 Systolic flows of data and instructions in the GENES structure. (a) Matrix-vector multiplication or Euclidean distance computation. (b) Identification of the smallest/largest element of a vector. (c) Synaptic weight updating using the Hebbian or Kohonen learning rule. (d) Instruction flow. These are represented by numbers in order of dispatching (1 is the oldest instruction, and 5 the newest one).

the operations are executed by the array itself, except the application of the activation function which is left to an external circuit connected to the outputs of GENES.

The matrix-vector multiplication is implemented in GENES using the "classical" systolic algorithm shown in figure 1 (a). Each cell contains one element $W_{i,j}$ of the matrix $\mathbf{W}$. The input vector $\vec{x}$ is "diagonally" entered from the top and flows downwards. The horizontally-flowing data — the partial sums — are initially set to zero. Each PE adds the product of the top input and the stored weight to the left input, and transmits the result to the right. This operation globally yields the output vector $\vec{y} = \mathbf{W} \cdot \vec{x}$.

The implementation of the Euclidean or Manhattan distance (equations (4) and (5)) on GENES can be done using the same systolic flow. Each PE then adds to the partial sum, either the square (Euclidean) or the absolute value (Manhattan) of the difference between the top input and the synaptic weight. The described scheme actually results in the square of the Euclidean distance. The Manhattan distance is usually justified by a simpler computation on a conventional machine. Since the Euclidean distance was found easier to implement on GENES, the Manhattan one was abandoned.

The search of the smallest (largest) element of a vector can easily be implemented on the GENES structure, as shown by figure 1 (b). For this operation, the synaptic weights stored in the cells are not used. The same vector $\vec{x}$ is injected on the top and left sides. While the vertical data flow unchanged downwards, a horizontal value entering a cell is compared to the vertical one. If the former is strictly larger (smaller), it is replaced by the largest (smallest) representable value. Otherwise it is simply propagated to the

95

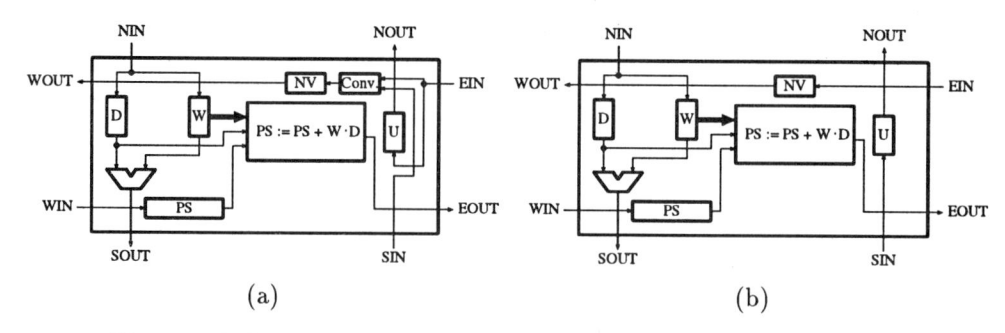

Figure 2 Architecture of GENES HN8. (a) Diagonal cells. (b) Non-diagonal cells.

right. In this way, each element y_i of the output vector $\vec{y}$ is equal to the corresponding input x_i only if this is the smallest (largest) element of $\vec{x}$. The output vector can then be applied to a certain activation function to produce a vector with a one at the winner's place and zeros everywhere else. This vector, multiplied by an *ad hoc* matrix, produces the neighborhood.

The synaptic weight updating can be done using either a Hebbian learning rule (Perceptron, ADALINE, delta rule, and back-propagation rule) or the Kohonen learning rule. Figure 1 (c) show the data flow for both cases. Two vectors $\vec{x}$ and $\vec{y}$ are injected from the top and left sides, flowing unmodified respectively towards the bottom and to the right. At their intersection, the synaptic weight stored in the PE is updated as shown by equation (7) in the Hebbian mode and by equation (8) in the Kohonen mode.

$$W_{i,j} := W_{i,j} + y_i \cdot x_j \tag{7}$$
$$W_{i,j} := W_{i,j} + y_i \cdot (x_j - W_{i,j}) \tag{8}$$

Equation (7) corresponds to the addition of the outer product $\vec{y} \cdot \vec{x}^T$ to the matrix $\mathbf{W}$.

In the learning phase of the back-propagation rule, some matrix-vector multiplications must be executed using the transposes of the weight matrixes. In GENES, instead of implementing the matrix transposition as an operation *per se*, any operation can be executed either on the matrix stored in the array or on its transpose. This is easily implemented by allowing the horizontal and vertical flows of data to be exchanged.

THE GENES HN8 EVALUATION BOARD

The first member of the GENES building block family, GENES HN8 (Lehmann and Blayo 1991), was built as a full-custom chip using a 2μm CMOS technology. This chip implements only the matrix-vector multiplication and therefore has no learning capabilities. It contains an array of 2×2 cells. Figure 2 (a) shows a diagonal cell, and figure 2 (b) a non-diagonal one. Diagonal cells can be configured as non-diagonal ones on a chip level, hence making the device fully cascadable. This circuit accepts 8-bit data and synaptic weights (registers D and W) as inputs, and computes 24-bit partial sums (register PS). The computation and communication being serial, a cell needs 24 clock cycles to compute a partial sum and to add it to the previous one. A new vector can therefore enter the array every 24 clock cycles. Finally, the registers U and NV are used to feed the computed data back.

The size of the PS register allows the use of 256×256 cells — corresponding to a fully-connected network of 256 neurons with 256 inputs — with no risk of overflow. Greater networks may still be used thanks to a very simple overflow detection mechanism which freezes the partial sum if it leaves the interval $[-2^{22}, 2^{22} - 1]$.

A first evaluation system has been built around 64 GENES HN8 chips arranged as an array of 16×16 cells. A 16-bit look-up table is used for the activation function. This board, interfaced to a SUN SPARCstation, proved the usefulness of a GENES-based system as a neural accelerator.

GENES IV: THE MULTI-MODEL GENES

The latest version of the circuit, GENES IV, implements matrix-vector multiplication, Euclidean distance computation, identification of the smallest/largest element of a vector, and Hebbian or Kohonen learning rules. It also includes several additional features necessary to sustain a utilization rate of 100 % and to handle overflows.

The strength of the GENES architecture, like any systolic structure, comes from its large number of simple PEs computing in parallel. Since the data flow systolically in the structure, GENES behaves as a pipeline. The utilization rate, and hence the performance, may therefore decrease dramatically if the pipeline is not kept full. It would therefore be disastrous to implement GENES as a conventional SIMD machine, because all the PEs would execute the same operation at the same time, and the pipeline would have to be emptied before each change in the operating mode, and refilled afterwards. It was instead decided to enter an instruction with each vector, and have the former following the latter. Figure 1 (d) shows the way instructions flow in the structure. This concept corresponds to an *SIMD architecture with systolic flow of instructions*.

The resulting architecture is shown in figure 3. A VLSI chip, containing 2×2 PEs, has been designed using standard cells and a CMOS 1μm technology (die size: $6.3 \times 6.1\,\text{mm}^2$). In the evaluation phase, it accepts 16-bit data and synaptic weights (D, W_0, and W_1 registers) as inputs, and computes 39-bit partial sums or Euclidean distances (PS register). In the learning phase, the weight registers are extended with 16 additional bits. These bits can be viewed as placed after the binary point, and are invisible in the evaluation phase. This added precision is necessary to reduce the effect of quantization in the learning process.

The systolic algorithms used with GENES require the data to be entered and retrieved "diagonally," that is, all the elements of a vector should be input into and output from the array at consecutive cycles. To avoid this burden, GENES IV allows the input/output to take place on the diagonal. In this way, all the elements of a vector can be entered at the same time. They then flow through the U or L register before being fed into the D and PS registers respectively. Similarly, the output data can be fed into the U or L register and retrieved on the diagonal.

GENES IV handles positive and negative overflows by setting the result respectively to the largest or smallest representable number. A sticky bit — appended to the PS, W_0 and W_1 registers, hence making their respective lengths equal to 40 bits and 33 bits — is set whenever an overflow occurs. This allows either to discard a result if it has overflown, or to consider this mechanism as a kind of "saturation effect." The latter is very well suited to update the synaptic weights. In fact, it is possible that a synaptic weight overflows while it is still far away from its convergence point. By saturating it and allowing the learning process to continue, it will probably converge.

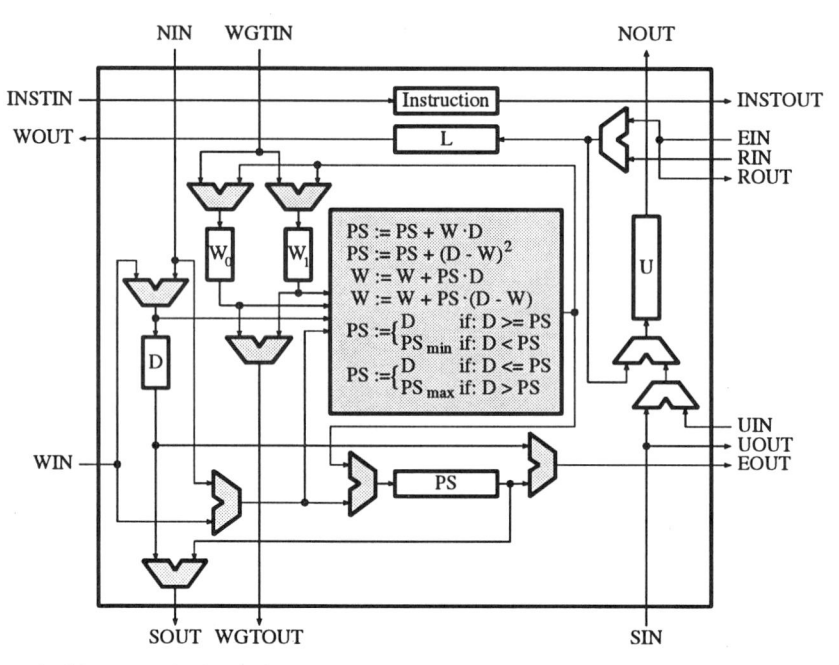

Figure 3 Architecture of a GENES IV diagonal cell. Non-diagonal cells have neither the input/output signals: UIN, LIN, UOUT, and LOUT, nor the corresponding three multiplexers. Shaded areas represent the units controlled by the instruction word, other units are globally controlled by external signals.

Another important feature of GENES IV is the ability to work with virtual neural networks larger than the physically implemented structure. This is done by dividing the network into sub-networks, and by time-sharing the structure between the sub-networks. This mechanism, very similar to that of virtual memory on a conventional computer, is required to allow the user to implement any arbitrarily large network, on a given physical machine. Using this feature, a sub-network can be swapped to or from memory in the background, while another sub-network is being used for the computation, hence maintaining a utilization rate of 100 %. The same mechanism is also used to implement multi-layer networks.

Finally, since the computation and communication are serial, a PE requires 40 clock cycles to complete one instruction. Thanks to the pipelined nature of the structure, a new vector can be entered every 40 clock cycles.

There are two traditional metrics to measure the computational power of neural-networks computers: (1) the number of *connections per seconds* or *CPS* (evaluation phase) and (2) the number of *connection updates per seconds* or *CUPS* (learning phase). The expected performance of GENES, with respect to these metrics, is given by:

$$P = \frac{N_n^2 \cdot f}{N_{op} \cdot m} \cdot U \qquad (9)$$

where N_n^2 is the number of PEs, f is the clock frequency, N_{op} is the number of operation necessary to compute a connection (evaluation phase) or a connection update (learning phase), $m = 40$ is the number of clock cycles per operation, and U is the utilization

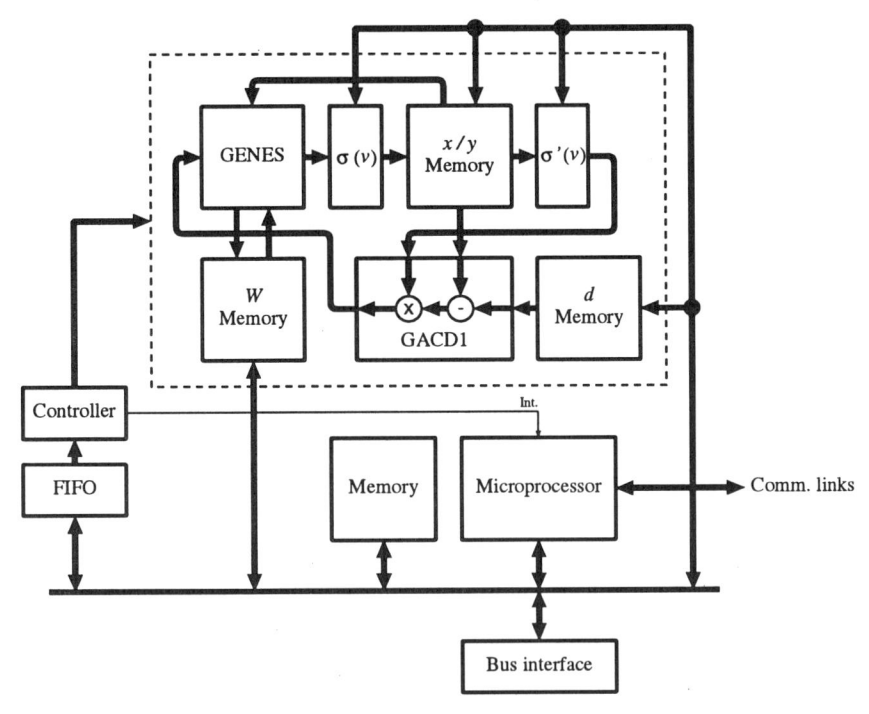

Figure 4 General architecture of the MANTRA machine.

rate. Using a 40 × 40 array with a frequency of $f = 10\,\text{MHz}$, the peak performance ($U = 100\,\%$) of GENES IV is illustrated by table 1.

Table 1 GENES IV peak performance.

Model	Evaluation phase N_{op}	CPS	Learning phase N_{op}	CUPS
Perceptron, ADALINE, Delta rule	1	$400 \cdot 10^6$	2	$200 \cdot 10^6$
Back-propagation rule	1	$400 \cdot 10^6$	3	$133 \cdot 10^6$
Kohonen with minimum/maximum	2	$200 \cdot 10^6$	4	$100 \cdot 10^6$

THE MANTRA MACHINE

The MANTRA machine, whose general architecture is shown in figure 4, is a neural-network computer based on a GENES IV array of 40 × 40 PEs. A commercial microprocessor (the TMS320C40 from Texas Instruments) was chosen to act as the sequencer of the GENES SIMD structure. Its tasks will be to configure the system, to dispatch instructions, and to handle the inputs and outputs. It will also manage the communications with a host computer and between different MANTRA machines connected through the microprocessor's own dedicated communication links.

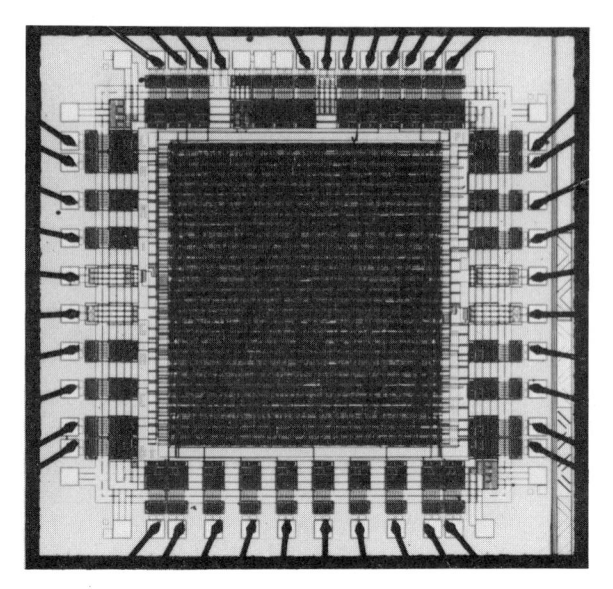

Figure 5 The GACD1 chip (die size: $3.2 \times 3.2\,\text{mm}^2$).

Two look-up tables, symbolized by $\sigma(v)$ and $\sigma'(v)$ on figure 4, are used to compute the evaluation function and its derivative.

Another VLSI chips, called *GACD1*, has been built to compute the error in the delta and back-propagation rules. The purpose of building this chip was double, first, the design of a discrete version of a rather complex unit could be avoided, and second, it provided a first experience with the technology and standard-cell library later used for GENES IV. This chip has already been manufactured and successfully tested. Its die is shown in figure 5.

The MANTRA machine also features the possibility to hook a DSP-based engine that can be used, for instance, to display snapshots of the computation in a graphical form. Finally, a standard bus interface is implemented, allowing the machine to be plugged into a commercial workstation offering a user-friendly programming environment. One of the first applications of the MANTRA machine will be static security assessment in power distribution networks (Niebur and Germond 1991).

RELATED WORKS

The project presented here is a follow-on of the work done by F. Blayo and P. Hurat (1989) on the APLYSIE chip, a systolic array dedicated to the Hopfield model.

U. Ramacher *et al* (1991) used a similar systematic description of several neural algorithms based on elementary operations, to propose a neural-network computer. Their description is based on the arithmetical concept of the Hamiltonian. The proposed hardware is therefore rather different from GENES.

J.-H. Chung, H. Yoon and S. R. Maeng (1992) proposed a systolic array dedicated to the back-propagation rule, which is, in some respects, similar to GENES. However, while MANTRA is a versatile platform that can be used for several algorithms and any network topology — the same hardware is time-shared between the different layers — they proposed a highly specialized implementation with the network topology (number of layers and number of neurons on each layer) being hard-wired. Moreover, in GENES the same hardware is time-shared between the evaluation and the learning phases, while in their array different units working in parallel are used for these two phases.

J. D. Gascuel *et al* (1992) proposed a systolic array dedicated to Hopfield networks and based on an associative chip. Their solution is however a linear structure, with each PE implementing a neuron and all its synapses. This architecture results in an algorithm mapping which is rather different from that of GENES. This chip is currently used as an accelerator in a software simulator.

PROJECT STATUS

As of July 1992, the first phase of the project — the GENES HN8 accelerator board — is completed. The presented evaluation board is fully working and reliable. The second phase — the GENES IV chip and the MANTRA machine — was started at the beginning of 1992. The auxiliary VLSI chip GACD1 has been manufactured and successfully tested. The GENES IV chip has been designed and sent to foundry, the dies should be available by the end of September 1992. The discrete part of the MANTRA machine has been designed, and the boards are almost ready to be sent for manufacturing. We hope to have the hardware working by the beginning of 1993.

CONCLUSION

The GENES architecture, presented here, is a good candidate for a generic neural-network computer. Its strength comes from its simple processing elements and its very good scalability, making it well suited for WSI. However, its pipelined nature imposes the use of epoch updating in the learning algorithms that can be implemented. In order to sustain a 100 % utilization rate, the number of prototype vectors presented per epoch must be at least the double of the physically implemented number of neurons. This makes a GENES-based computer well tailored to sufficiently large problems.

Acknowledgements

The authors would like to thank Prof. J.-D. Nicoud, the laboratory director, without whom this project would not have been possible. Special thanks go to R. Beuchat for his help concerning the hardware development and to C. Marguerat for his advise and help with the TMS320C40. The authors greatly acknowledge the support of the *Fond National Suisse de la Recherche Scientifique* through the Esprit-BRA Nerves project N° 3049, and the Swiss Federal Institute of Technology through the MANTRA project.

REFERENCES

Blayo, F. and Hurat, P., "A VLSI Systolic Array Dedicated to Hopfield Neural Network," in *VLSI for Artificial Intelligence*, J. G. Delgado-Frias and W. R. Moore (ed.), Norwell, MA: Kluwer Academic Publishers, pp. 255–264, 1989.

Blayo, F., *Une Implantation Systolique des Algorithmes Connexionnistes*, Ph.D. thesis N° 904, EPFL, Lausanne, Switzerland, 1990.

Chung, J.-H., Yoon, H. and Maeng, S. R., "A Systolic Array Exploiting the Inherent Parallelisms of Artificial Neural Networks," *Microprocessing and Microprogramming*, vol. 33, no. 3, pp. 145–159, May 1992.

Gascuel, J.-D., Delaunay, E., Montoliu, L., Moobed, B. and Weinfeld, M., "A Software Reconfigurable Multi-Networks Simulator Using a Custom Associative Chip," in *Proc. of the Int. Joint Conference on Neural Networks*, pp. II-13–II-18, June 1992.

Hopfield, J. J., "Neural Networks and Physical Systems with Emergent Collective Computational Abilities," *Proc. of the National Academy of Sciences*, vol. 79, pp. 2254–2258, April 1982.

Kohonen, T., "Analysis of a Simple Self-Organizing Process," *Biological cybernetics*, vol. 44, pp. 135–140, 1982.

Le Cun, Y., "A Learning Scheme for Asymmetric Threshold Network," in *Proc. of Cognitiva 85*, June 1985.

Lehmann, C. and Blayo, F., "A VLSI Implementation of a Generic Systolic Synaptic Building Block for Neural Networks," in *VLSI for Artificial Intelligence and Neural Networks*, J. G. Delgado-Frias and W. R. Moore (ed.), New-York, NY: Plenum Press, pp. 325–334, 1991.

Niebur, D. and Germond, A. J., "Power System Static Security Assessment Using the Kohonen Neural Network Classifier," in *Proc. of the IEEE Power Industry Computer Application Conference*, pp. 270–277, May 1991.

Ramacher, U., Raab, W., Anlauf, J., Hachmann, U. and Wesseling, M., "SYNAPSE-X: A General-Purpose Neurocomputer," in *Proc. of the 2nd Int. Conference on Microelectronics for Neural Networks*, pp. 401–409, October 1991.

Rosenblatt, F., "The Perceptron: A Probabilistic Model for Information Storage and Organization in the Brain," *Psychological Review*, vol. 65, pp. 386–408, 1958.

Treleaven, P., Pacheco, M. and Vellasco, M., "VLSI architectures for neural networks," *IEEE Micro*, vol. 9, pp. 8–27, 1989.

Widrow, B. and Hoff, M. E., "Adaptive Switching Circuits," in *IRE-WESCON Convention Record*, pp. 96–104, 1960.

SPERT: A NEURO-MICROPROCESSOR

Krste Asanović, James Beck, Brian E. D. Kingsbury, Phil Kohn, Nelson Morgan and John Wawrzynek

INTRODUCTION

High-speed special-purpose digital systems for artificial neural network (ANN) calculations have been proposed by a number of researchers [Hammerstrom, 1990, Ramacher et al., 1991]. In 1989, our group designed and built a programmable machine called the Ring Array Processor (RAP) for general backpropagation training of layered feed-forward networks [Morgan et al., 1992]. This used multiple TMS320C30 DSP chips [Tex, 1988] connected by a ring of programmable gate arrays. We are currently developing a single chip CMOS microprocessor, called the Synthetic PERceptron Testbed (SPERT), to make fast and flexible neurocomputing more cost-effective.

SPERT is fully programmable, executes a wide range of connectionist computations efficiently, and is designed for fast access to relatively large off-chip memories. Special emphasis is placed on support for variants of the backpropagation training algorithm for multi-layer feedforward networks[Rumelhart et al., 1986, Werbos, 1974]. This class of networks is of interest in the speech recognition task that is the focus of our applications work [Morgan and Bourlard, 1990]. At the design clock rate of 50 MHz, detailed performance analysis indicates that this chip will sustain over 300×10^6 connections per second (CPS) when performing classification, and around 100×10^6 connection updates per second (CUPS) when performing backpropagation training.

SPERT's performance is comparable to that of a 40 processor RAP machine. Such a RAP system occupies ten 9U VME boards, while in contrast the initial SPERT system design will be a double SBus card that fits inside a workstation. SPERT achieves this dramatic improvement in cost/performance through a combination of using reduced precision fixed point arithmetic, providing high on- and off-chip operand and instruction bandwidth, and employing a highly parallel, pipelined architecture.

SPERT fulfills an intermediate role in our neural network system development plans. The RAP gave us an early speed-up over existing workstations so that we could develop our speech algorithms using large databases (over 1,000,000 patterns) and large networks (100,000 to 400,000 connections). The use of commercial chips for this system permitted a short (one year) development time. SPERT is our first full microprocessor, and is the result of several years of work on our VLSI design path. SPERT tests this path as a first step towards our next full multiprocessor machine (tentatively named CNS-1), which will use similar technology. A consistent set of software abstractions will assist in porting code between these platforms.

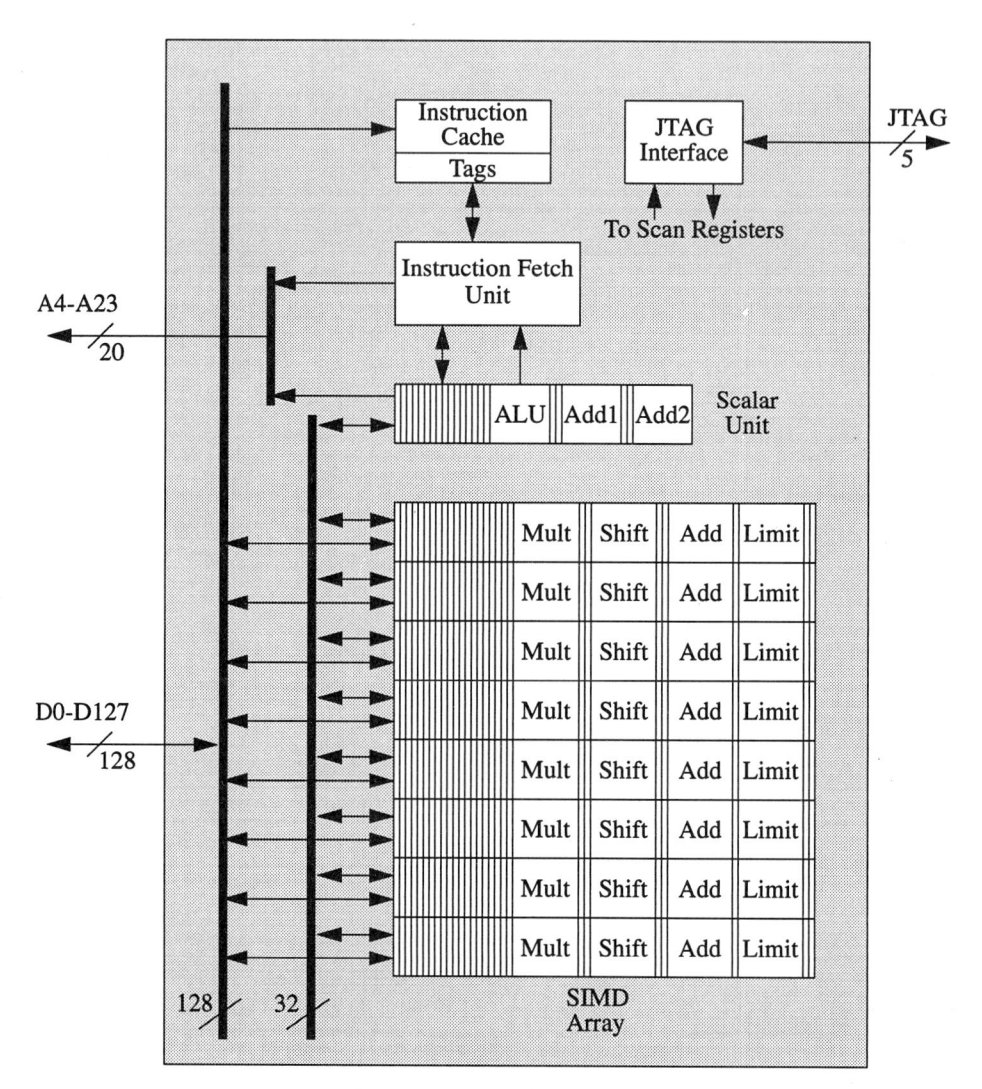

Figure 1. SPERT Structure.

SPERT ARCHITECTURE

The overall structure of the SPERT microprocessor is shown in Figure 1. SPERT includes an instruction fetch unit with an instruction cache, a scalar 32b integer datapath, a small SIMD array containing eight 32b fixed point datapaths, a 128b wide external memory interface, and a JTAG interface and control unit.

SPERT is intended to function as an attached processor for a conventional workstation host. An industry standard JTAG serial bus is used as the host–SPERT interface. Chip and board testing, bootstrap, data I/O, synchronization, and program debugging are all performed over the JTAG link. The host can read or write 128b data words from SPERT's memory using JTAG, and the shifting of memory data through the JTAG scan ring can be overlapped with SPERT computations. These extra JTAG functions have been carefully designed so as not to compromise JTAG's use as a test access port.

SPERT is a VLIW architecture, with a single 128b instruction format. The execution pipeline has 7 stages, and one instruction can be completed every cycle. SPERT has a 1KB on-chip instruction cache holding 64 instructions. Typical SPERT applications are dominated by tight inner loops, and the small instruction cache is very effective at freeing the external memory port for data accesses.

The scalar unit is similar to the integer datapath of a RISC processor. It is used for general scalar computation and to support the SIMD array by providing address generation and loop control. It contains a triple-ported 16×32b general purpose register file, a fully bypassed ALU, and two address generators.

The SIMD array contains 8 fixed-point datapaths, similar to those found in DSP chips. Each SIMD datapath contains a triple-ported 16×32b general purpose register file, a 24b$\times 8$b multiplier, a 32b saturating adder, a 32b shifter, and a 32b limiter. In addition each SIMD datapath contains a number of datapath registers. These are located on the inputs of the functional units, and allow results to be moved between functional units without requiring extra read and write ports on the general purpose register file. All of the functional units can be active simultaneously, delivering up to 400×10^6 fixed-point multiply-accumulates per second at the maximum design clock rate of 50 MHz.

The external memory interface supports up to 16 MB of external memory over a 128b data bus. Either 8×8-bit words, 8×16-bit words, or 4×32-bit words can be transferred in a single cycle. At the maximum 50 MHz clock rate, memory bandwidth is 800 MB/s. The external memory will typically be used to hold weight values and program code. Input and output training patterns will be supplied by the host from a large database held on disk. The serial JTAG link provides a significantly greater bandwidth for this purpose than the maximum transfer rate of a SCSI-I disk.

SPERT SBUS BOARD

We are designing a two-slot SBus board around the SPERT chip. In its standard form, the board will contain a SPERT chip clocked at 33 MHz, 8 MB of 20 ns SRAM, and the SBus to JTAG interface. This will be placed inside a SPARC desktop workstation to form a powerful and flexible, low cost neural network accelerator. The same board will be used to test SPERT chips after fabrication. A 50 MHz version with faster but smaller memory will also be built.

SPERT SOFTWARE

In many ways, SPERT looks like a RISC microprocessor to the user. C/C++ is used as our standard programming language, and we are porting the GNU gcc compiler to SPERT. An assembler is also available.

The application developer's interface to SPERT will be a set of C++ library classes. These class libraries will encapsulate carefully hand-crafted assembly language routines for common matrix-vector operations, and for common neural network algorithms. These predefined objects can be composed to construct arbitrary network topologies. New networks and training algorithms can first be derived from existing classes, then later refined into optimized assembly language versions if required. Further SPERT libraries provide standard I/O facilities by forwarding I/O requests to a SPERT-server process running on the host workstation.

This programming environment is modelled on that used successfully by the RAP [Kohn, 1991], and we intend to port much of the higher level simulator software from the RAP.

STATUS

SPERT is being implemented in a $1.2\,\mu m$ CMOS process using MOSIS scaled CMOS design rules, with a worst case target clock rate of 50MHz. The design uses a combination of full-custom cells for the datapaths and pads, with standard cells for control and other random logic.

We have recently successfully fabricated a test chip containing a single, complete SIMD datapath together with test logic. The datapath is still under test at this time but appears fully functional at over 50MHz. The SBus board has been designed and fabricated and is being used to test the datapath chip.

The SPERT assembler and instruction set simulator have been completed.

SUMMARY

We have presented an overview of the SPERT project. For those connectionist applications that do not require high precision arithmetic, SPERT represents a very high performance, low cost acceleration tool. Detailed performance analysis predicts a sustained performance of over 300×10^6 CPS for forward propagation, and around 100×10^6 CUPS for error backpropagation in layered feedforward networks. A consistent set of software abstractions unify SPERT with our other systems, including the earlier RAP, and a larger multiprocessor we are currently designing.

REFERENCES

[Hammerstrom, 1990] Hammerstrom, D. A VLSI architecture for High-Performance, Low-Cost, On-Chip Learning. In *Proc. International Joint Conference on Neural Networks*, pages II–537–543, 1990.

[Kohn, 1991] Kohn, P. CLONES: Connectionist Layered Object-oriented NEtwork Simulator. Technical Report TR-91-073, International Computer Science Institute, 1991.

[Morgan and Bourlard, 1990] Morgan, N. and Bourlard, H. Continuous speech recognition using Multilayer Perceptrons with Hidden Markov models. In *Proc. IEEE Intl. Conf. on Acoustics, Speech, & Signal Processing*, pages 413–416, Albuquerque, New Mexico, USA, 1990.

[Morgan et al., 1992] Morgan, N., Beck, J., Kohn, P., Bilmes, J., Allman, E., and Beer, J. The Ring Array Processor (RAP): A Multiprocessing Peripheral for Connectionist Applications. *Journal of Parallel and Distributed Computing*, 14:248–259, April 1992.

[Ramacher et al., 1991] Ramacher, U., Beichter, J., Raab, W., Anlauf, J., Bruls, N., Hachmann, M., and Wesseling, M. Design of a 1st Generation Neurocomputer. In *VLSI Design of Neural Networks*. Kluwer Academic, 1991.

[Rumelhart et al., 1986] Rumelhart, D., Hinton, G., and Williams, R. Learning Internal Representations by Error Propagation. In *Parallel Distributed Processing. Exploration of the Microstructure of Cognition*, volume 1. MIT Press, 1986.

[Tex, 1988] Texas Instruments, Houston, Texas, USA. *Third-Generation TMS320 User's Guide*, 1988.

[Werbos, 1974] Werbos, P. *Beyond Regression: New Tools for Prediction and Analysis in the Behavioral Sciences*. PhD thesis, Dept. of Applied Mathematics, Harvard University, 1974.

DESIGN OF NEURAL SELF-ORGANIZATION CHIPS FOR SEMANTIC APPLICATIONS

Tadashi Ae, Reiji Aibara and Kazumasa Kioi

INTRODUCTION

For Kohonen's LVQ algorithm, we need the operation as follows;

$$\min.\{|X_1 - X|, |X_2 - X|, \ldots, |X_k - X|\} \qquad (1),$$

where $X_1, \ldots, X_k$ are the reference vectors, X is the input vector, and $|A - B|$ is the norm between A and B (Kohonen 1989).

The hardware Kohonen net is directly realized by the systolic processing circuit of $k \times r$ processing units (where r is the dimension of the vector) to compute the norm $|X_i - X|(i = 1, 2, ..., k)$, and the minimum norm detector of k units.

In this paper, first, we focus on the case that $X_i(i = 1, 2, \ldots, k)$ or X is given by the subset of fundamental elements, which may be distinct to each other. It is redundant to execute directly Eq.(1) for "Set Norm", i.e., the norm between two sets, using the circuit described as the above, since many elements of two vectors are not used for norm calculation. Then, we use the norm processor for the reference vector each. The first part of our system consists of a set of fundamental Kohonen nets, each of which corresponds to a single chip with 64 norm processors. In Eq.(1), k is supposed to be 64.

Next, we prepare one more fundamental Kohonen net which is similar to the above, and it follows the automatic key-word generator. Last, we introduce the cooperation among the fundamental Kohonen nets. The cooperation is also divided into two parts; one is the cooperation among fundamental Kohonen nets for the quantitative input, and the other is the cooperation between the quantitative input and the qualitative one. The former is similar to the *join* operation among the relational databases. The latter is not so stronger cooperation than the former, and the qualitative input plays a role of complementary classification to the quantitative input. We combine these fundamental Kohonen nets by the multiport memory. The multiport memory is designed by the optically-coupled three-dimensional common memory (Ae 1988, Koyanagi 1990), which is a major project of Research Center of Integrated Systems of Hiroshima University.

The system is mainly designed for the database applications, but is also applicable for many other purposes.

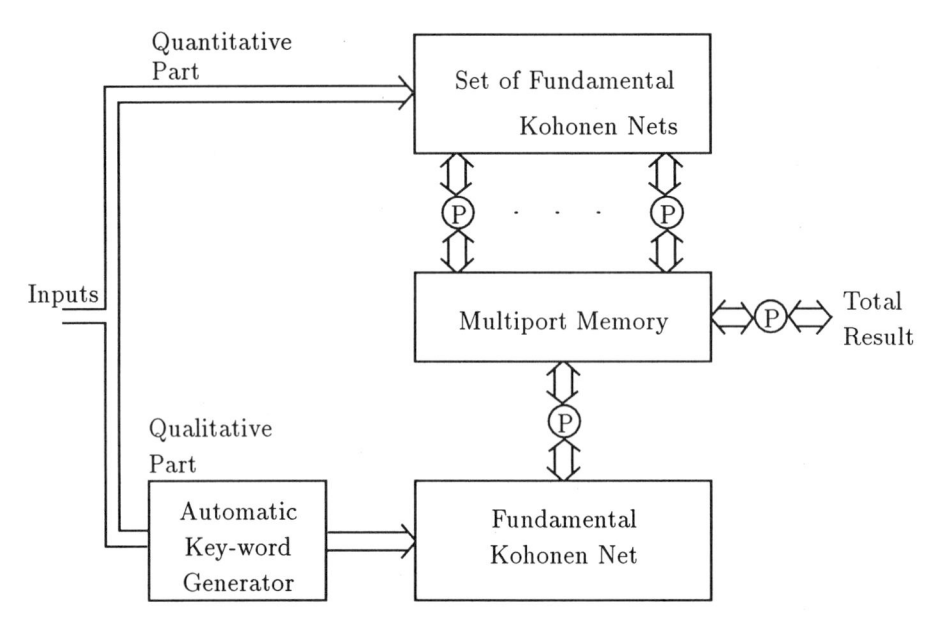

Figure 1 Total System
(where the input includes two parts;
quantitative and *qualitative*.)

SYSTEM OVERVIEW

To demonstrate the cooperating Kohonen net described as the above, we design the system and the chip set as follows;

1 Total System

The system is designed for recognition of the *database*, each data of which consist of the quantitative part and the qualitative part. Then, the system as in Figure 1 is divided into three parts as follows;

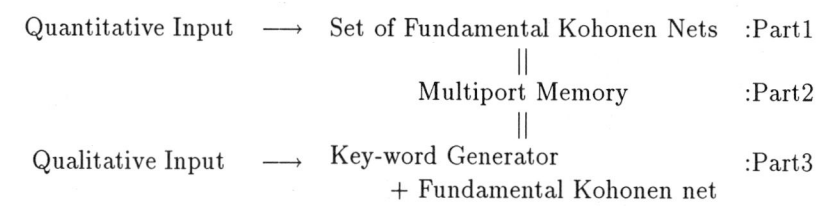

```
Quantitative Input  ⟶  Set of Fundamental Kohonen Nets   :Part1
                                    ‖
                        Multiport Memory                  :Part2
                                    ‖
Qualitative Input   ⟶  Key-word Generator                :Part3
                        + Fundamental Kohonen net
```

The input object includes the two types, i.e., *quantitative* and *qualitative*. The quantitative input is a vector whose element consists of the term identifier (id) and the value (e.g., (price, 10,000)). The number of fundamental elements of vector, i.e., the domain, is supposed to be 128, but that of each vector is limited to 16, where the vector may have each own subset of elements. In general, the quantitative input consists of more than one databases, and then the operation among databases is needed. To apply the LVQ algorithm to each database, k, *the number of reference vectors is supposed to be at most 64 each.*

The qualitative input is a sentence to explain the object (in short, *abstract*). The length of an abstract is supposed to be 200 words. In order to transform the abstract into the vector (actually, the key-word vector), the automatic key-word generator (in short, KWG) is used, which consists of CAMs (content-addressable memories). The KWG is a translator using the process in the MBR (memory-based reasoning) in the artificial intelligence techniques (Stanfill et al. 1986). Every word generated by the KWG has a coefficient that corresponds inversely to the number of appearances in the abstract, and the key-word vector includes the largest sixteen words. The coefficient of each word equals to its value, and then, 1 is the largest (only once appearance) and 1/2 is the second largest (twice appearances) and so on. For this vector, the fundamental Kohonen net works like the previous one for the (translated) quantitative input.

Totally, Part1 and Part3 cooperate with each other through Part2, i.e., the multiport memory, which is the same one used within Part1.

After training by the LVQ algorithm, the system can recognize (i.e., classify) the input object automatically and reply for user's queries.

2 Chip set

Part1 consists of more than one fundamental Kohonen nets, each of which is a chip described as follows;

a) 64 norm processors (corresponding to the reference vectors) to compute Eq.(1) for vectors representing the set each, which uses at most 16 elements out of 128 domain elements, and
b) Winner-take-all circuit. After each norm processor calculates its own 1-norm for $|X_i - X|(i = 1, 2, \ldots, 64)$, the *minimun detector* searches the minimum among all norms. This "winner-take-all" circuit is the binary tree of six stages for 64 candidates (the results of 64 norm processors).

Part2 is the multiport memory which has two kinds of cooperation;

i) Cooperation among quantitative inputs. More than one fundamental Kohonen nets output the result each. The multiport memory is used in this part for a kind of the relational operations (e.g., *join*) among these results.
ii) Cooperation between the quantitative part and the qualitative part.

Part3 is a chip for constructing the semantic map, which is given by the subsidiary input, that is , the qualitative input. Part3 consists of two types of circuits; one is the KWG, and the other is the fundamental kohonen net. The former is mainly constructed by the CAMs, and the latter is essentially the same one as in Part1. The interaction between Part1 and Part3 is not so strong, but the output of Part3 is unique and complementary, since the qualitative input is given by the abstract (i.e., the sentence of 200 words).

FUNDAMENTAL KOHONEN NET

In our system the Kohonen net plays a role of fundamental operation based on the LVQ algorithm. When applying the LVQ algorithm to the semantic applications such as the database, the behavior is a little different from the pattern recognition and the signal processing, since the object to be quantized is originally virtual and then the quantization itself does not need to be so strict. Then, we can reduce the learning time steps to at most 1,000 from 100,000 recommended for the pattern recognition (Kohonen 1990).

Instead, the object for quantization is a little cumbersome. For instance, we show a list of used cars;

USED CARS 1 of five elements : (name, year, price, purpose, imported of not).
 (Rover, '90, $10,000, leisure, imported)
 (Audi, '88, $20,000, leisure, imported)
 (Peugeot, '90, $15,000, business, imported)
 (Civic, '90, $10,000, business, domestic).

For the databases, we need the category for each reference data (vector) to apply Kohonen's LVQ algorithm. Then, we add the category to the reference vector and it is easy for the case of USED CARS. For the case of easy categorization, the learning time steps are extremely reduced.

In the case of more than one lists (databases), however, the terms (elements) in each list may be different. For instance, we have another list as follows;

USED CARS 2 of six elements : (name, year, price, owner or not, fuel, max.speed)
 (Fiat, '88, $10,000, rent-a-car, gasoline, 200km)
 (Pajero, '90, $20,000, owner, diesel, 180km).

This list differs in several terms (elements) from the previous list (USED CARS 1). In our system, the number of domain elements is 128, but each data is supposed to use at most 16 out of 128.

The fundamental Kohonen net is shown as in Figure 2, which consists of 64 norm processors and the minimum norm detector of six-stage binary tree. It is an appropriate size for the applications and also for a single chip realization. Every norm processor is shown as in Figure 3, where only terms (elements) used in each data are the objects for operation. For the input data $X = (x_1, \ldots, x_p)$ and a reference vector $X_i = (x_{i1}, \ldots, x_{iq})$, the norm processor executes sequentially the step;

$\{Norm\ between\ X_i\ and\ X\}$
 repeat { *initially, $j = 1$ and* **partial norm** $= 0$}
  obtain $d_j = |x_{ij} - x_j|$ in the operation unit;
  add d_j to **partial norm** of up to $(j-1)$-th terms in the adder and register;
 until $j = 128$ {*calculation of d_j is omitted when both elements do not exist*}.

The input data includes at most 16 blocks (i.e., 16 terms), each of which consists of 32 bits (term id : 16 bits, and the value : 16 bits). For a block (e.g., (price, 10,000)), id (e.g., price) is sent to the decoder, and the value in the same term (i.e., price) of the reference vector is selected. The control in the operation unit can choose the norm. The sequential step is flexible for database applications, since the choice of terms is free and the order of elements in the vector is also free. For the reference vectors, however, the parallel execution is achieved by 64 norm processors and also by the minimum norm detector of six-stage binary tree ("winner-take-all" circuit). For a data of 16 blocks, the total execution time is estimated to be less than 1,600ns for the gate-level circuit design.

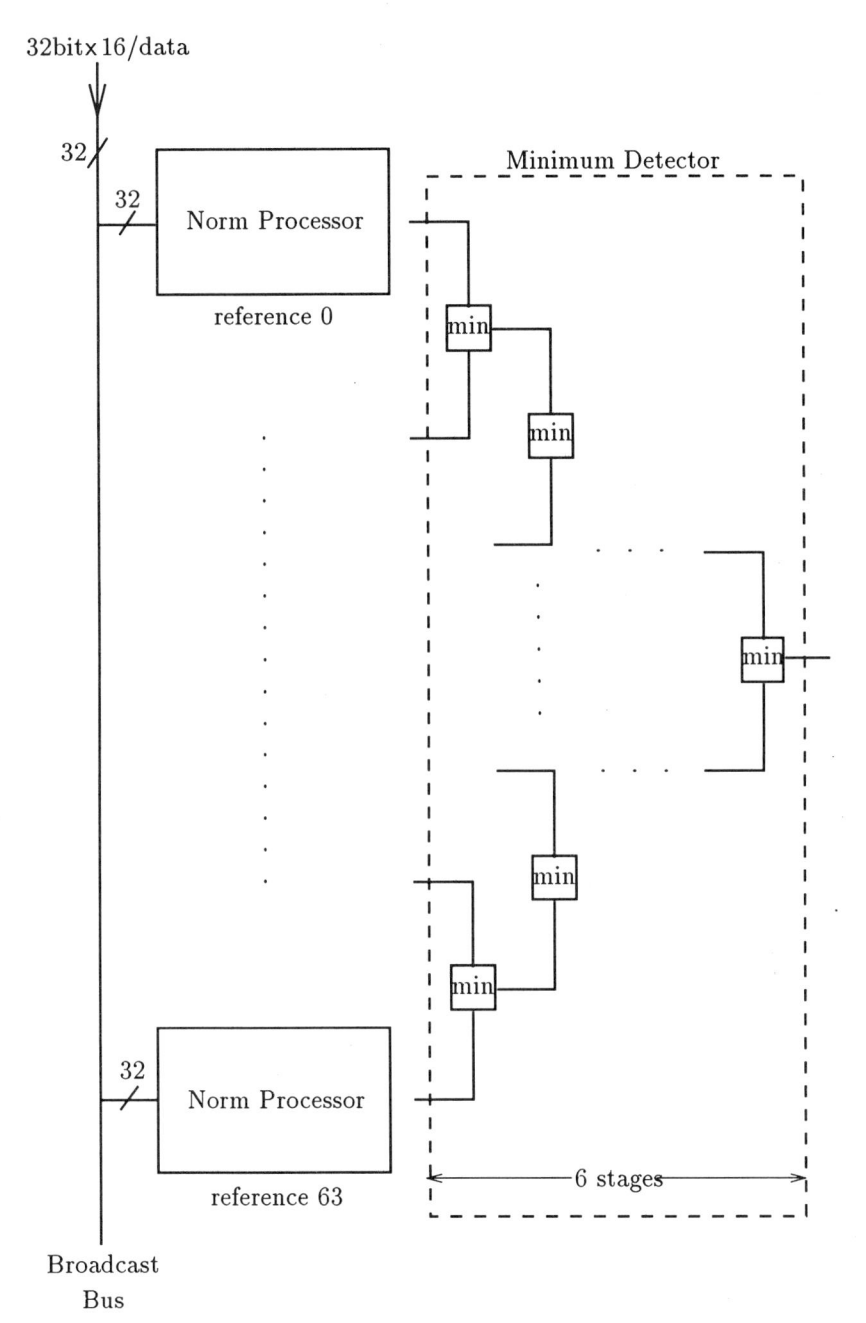

Figure 2 Fundamental Kohonen Net

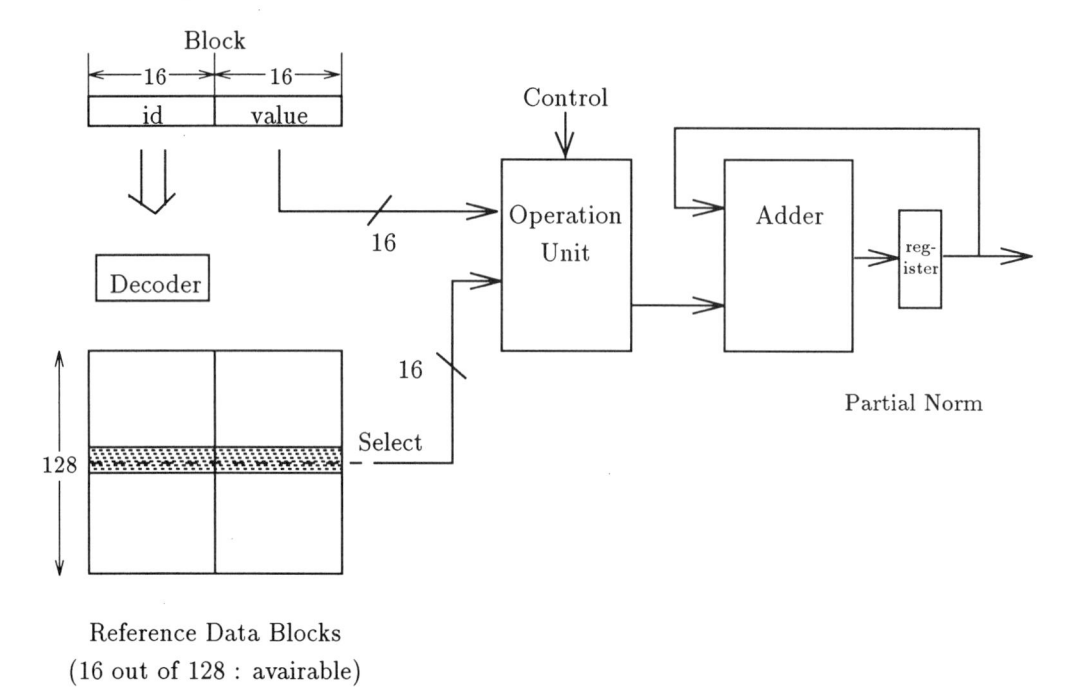

Figure 3 Norm Processor

KEY-WORD GENERATOR (KWG)

The qualitative part of input is supposed to be an abstract of about 200 words. To transform it to a vector, we use the technique which is the same as the process in the MBR (Stanfill 1986). Every word generated by the KWG has a coefficient that corresponds inversely to the number of appearances in the abstract. Although the Connection Machine is used in the MBR, it is realized by the CAM and the slightly-modified CAM (Extended CAM) as in Figure 4. The behavior of this automatic key-word generator is simple, and the details are omitted. The fundamental Kohonen net follows the KWG, and executes the LVQ algorithm for the coefficient vector obtained in the KWG.

COOPERATIVE KOHONEN NET

The cooperation among more than one Kohonen nets is achieved through the multiport memory (Koyanagi et al. 1990), whose details are not described here. First, we do not regard the direct product itself, that is, the combination of more than one lists (e.g., USED CARS 1 and USED CARS 2) as the cooperation. Then, we focus on the real cooperation which becomes a proper subset of the direct product, such as the *join* operation. The join is classified into many types, but we focus on only Ichikawa's join (Ichikawa et al. 1986). The join operation between two relational databases of n elements each takes $O(n^2)$ steps, and therefore, the hardware acceleration is needed for

Input Data

| vector |
| abstract |

vector → Quantitative Part

abstract → (Qualitative Part) → 200 Word Stream → CAM

CAM

Preset Words (1K Words)

word | flag

if word.flag = "on"

then trap word.

↓ Untrapped Word Stream

Extended CAM

w_0 | c_0
w_1 | c_1
w_2 | c_2
⋮ | ⋮
w_n | c_n

Initialize : all $c_i \leftarrow 0$

if $c_i = 0$

  then $c_i \leftarrow 1$

  else $c_i \leftarrow c_i + 1$.

↓

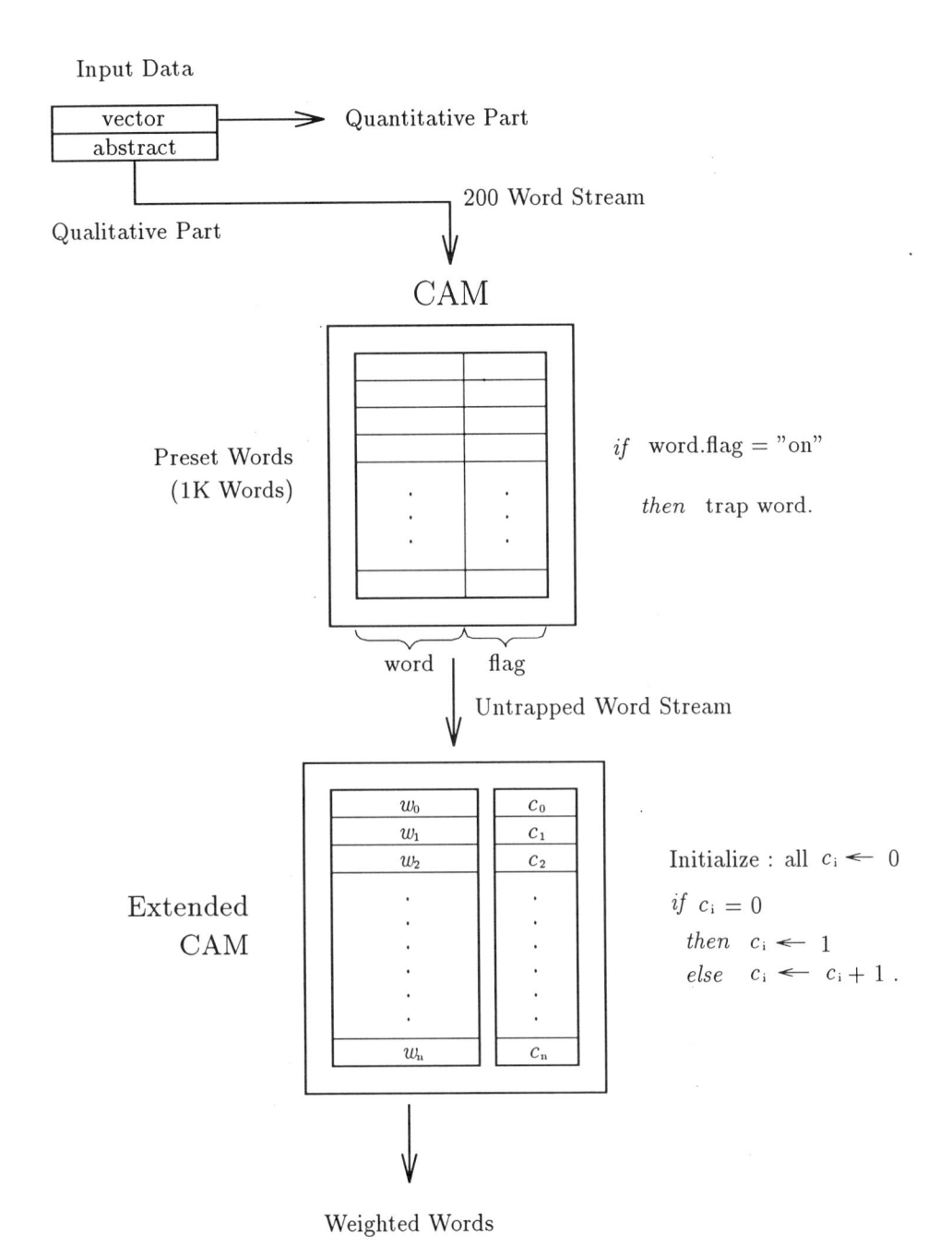

Weighted Words

Figure 4 Automatic Key-Word Generator (KWG)

the large-scale databases. Then, the systolic chip for the join operation is proposed, but the conventional processors (i.e., Ps in Figure 1) associated with each database execute the *join* among databases through the multiport memory in Figure 1. Of course, the system in Figure 1 cannot achieve much parallelism, but it is not needed, since the system executes the **similarity join** ; an extension of Ichikawa's join, instead of the true join. The similarity join is the join between the sets of reference data (vectors), supposed to be 64 in the system, and is unchanged for n. Mathematically, k is the number of equivalence class (based on "similarity") and is extremely reduced from n, the number of data (vectors) of the original set.

To demonstrate the similarity join, we introduce a new list as follows;

CUSTOMERS
(Customer1, >'90, <$10,000, business, domestic)
(Customer2, >'87, <$18,000, leisure,   imported).

Clearly, *Customer1 is just recommended to buy Civic* in USED CARS 1, since the join (then, also the similarity join) holds between two lists. No other true join hold between two lists, but another similarity join may hold. Really, *Audi and Customer2* corresponds to the case of **similarity join** between USED CARS 1 and CUSTOMERS. Each list described as the above should be viewed as the list of reference data (vectors).

The cooperation among more than one databases is mainly applied for the **SBR** (similarity-based reasoning), although it can be used for the similarity-based inference. The reasoning is quite natural for the neural networks The MBR is also a natural generalization of neural networks, but needs the huge system. On the other hand, the SBR is realized in the small system using several chips as in this paper. The SBR in the neural network is born with the combination of Kohonen net and Ichikawa's database.

REMARKS

We extend the self-organizing system for *semantic applications*, adding Ichikawa's database to the Kohonen net. The three circuits, i.e., *the Kohonen net, the multiport memory and the CAM* play an important role of realizing the system. In this paper, we focus on the Kohonen net and the cooperation among more than one Kohonen nets. The multiport memory and the CAM are also the projects of realizing the 3D optically-coupled IC and of developing the parallel processor (Ae et al. 1992, Koyanagi et al. 1990).

References

Ae,T.,"3D Microprocessors and Devices",Digest 5th International Workshop on Future Electron Devices,pp.55-60,1988.

Ae,T.,Fujita,S.,Yamanaka,T.and Aibara,R.,"Hypercube is better than De Bruijn for Connectionist",5th ISMM International Conference on Parallel and Distributed Computing and Systems,Pittsburgh,pp.370-372,1992.

Ichikawa,T.and Hirakawa,M.,"ARES:A Relational Database with the Capability of Performing Flexible Interpretation of Queries",IEEE Trans.Software Engineering,vol.SE-12,no.5,pp.624-634,1986.

Kohonen,T., Self-Organization and Associative Memory (3rd Ed.),Springer-Verlag, 1989.

Kohonen,T.,"The Self-Organizing Map",Proc.IEEE,vol.78,no.9,pp.1464-1480,1990.

Koyanagi,M.,Takata,H.,Mori,H.and Iba,J.,"Design of 4-kbit×4-Layer Optically Coupled Three-Dimensional Common Memory for Parallel Processor System",IEEE J.Solid-State Circuit,vol.25,no.1,pp.109-116,1990.

Stanfill,C. and Waltz,D., "Toward Memory-Based Reasoning",C.ACM,vol.29,no.12, pp.1213-1228,1986.

VLSI IMPLEMENTATION OF A DIGITAL NEURAL NETWORK WITH REWARD-PENALTY LEARNING

Terence Hui, Paul Morgan, Hamid Bolouri and Kevin Gurney

INTRODUCTION

The use of networks of logical processing elements to effect pattern matching and pattern recognition has been the subject of much research since the birth of digital microelectronic systems. In 1959, Bledsoe and Browning (59) proposed the N-Tuple method of pattern recognition. The technique offered a relatively low-cost approximation to Bayesian classification and formed the basis for many pattern recognition studies in the sixties (see for instance, Steck 62, Ullmann 69). Later, Aleksander and colleagues exploited the potential of RAM technology to develop hardware realisations of the the N-Tuple method which allowed real time learning and classification (the WISARD system, Aleksander 84).

In the WISARD system, the RAM address lines are randomly connected to elements of the input pattern space. Each possible pattern presented at the address terminals of the 1-bit wide RAM forms an address to one of 2^n sites within the node. Training involves writing a '1' to addressed sites in the RAM. Networks of many such nodes are then used to determine the degree of similarity between any 'unseen' pattern and the patterns in the training set.

In 1987, Kan and Aleksander proposed the concept of Probabilistic Logic Neurons (PLNs) in which the RAM site values could be '0','1' or Undefined (output '0' or '1' with 50% probability). They showed that PLN networks offer greater learning and generalisation potential compared to WISARD-like networks of deterministic boolean nodes.

In 1988, Gorse and Taylor proposed a probabilistic RAM-based model of a neuron (called a pRAM) which they demonstrated closely matched models of biological neurons. The memory words in the pRAM are assumed to be many bits wide, such that the stored values may be thought of as real numbers (Clarkson, Gorse and Taylor (89) suggest a minimum of 16 bits). The output of a pRAM neuron is a logic '1' with a probability dependent on the value of the memory site addressed. The function relating the memory site values to the probability of firing a '1' is termed the activation-output function and is often taken to be the sigmoid function.

Gurney (89) exploited the representation of a boolean logic node as a population of values at the vertices of a hypercube to develop a number of training algorithms for networks of probabilistic RAM neurons. He also proposed that for most real applications the number of bits per RAM word need not exceed 8. These findings have also been confirmed by learning algorithms developed by Gorse and Taylor (90) and Myers (90).

When the number of inputs to a RAM-based node becomes large, the size of the RAM required to store all possible combinations of input bits becomes prohibitive. Gurney (89, 92) made the observation that for large numbers of inputs, training results in sparsely populated RAMs. He therefore suggested that the memory requirements for large-input nodes may be reduced by dividing the node input space into a number of smaller sub-spaces each addressing a separate subunit. Such nodes are termed Multi Cube Units (MCUs). Figure 1 shows an example of a network of MCUs organised as a multi layer feed forward network with three hidden layer neurons and one output neuron. Each MCU comprises a number of independent subunit RAMs which receive only a portion of the total unit input. The multi bit value output by each subunit is averaged over all the subunits in the MCU and a single '1' or '0' is output with a probability equal to the sigmoid of the average site value.

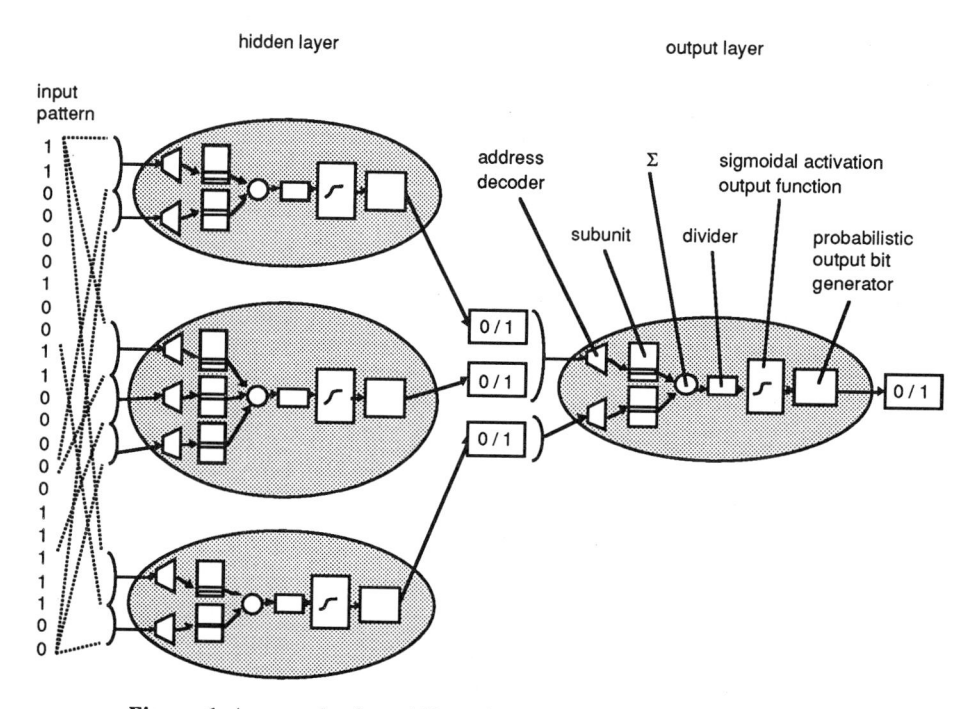

Figure 1 An example of a multi layer feed forward network of Multi Cube Units

For the sake of brevity, we refer to the above architecture as HyperNet (short for probabilistic hypercube-based Artificial Neural Network). In the sections that follow, we first give a brief description of the training algorithms used in the VLSI implementation of HyperNet, and then proceed to describe the chip and system designs.

REWARD-PENALTY LEARNING IN HyperNets

The algorithm used is an adaptation of the method proposed by Barto and Jordan (87) and is described in detail in (Gurney 89, 92) where proof of convergence is also provided. Here, we only give a brief description in order to enable a better understanding of the chip.

The basic idea is to make use of a single globally broadcast performance measure (the one bit reward/penalty or reinforcement signal 'r') to update the internal parameters (site values) of each node. This has obvious advantages in terms of ease of implementation over comparable gradient descent techniques such as Error Back Propagation.

When the network performs well, it is "rewarded" (r=1) by changing its parameters so as to increase the probability of repeating this successful outcome in future. When the network response is unsatisfactory, it is "penalised" (r=0): the addressed site values in the network are modified so as to increase the probability of generating the opposite output values in future. The change in site values may then be described by the following formulae:

$$\delta S_\mu^j = \begin{cases} \alpha (y^j - \sigma(S_\mu^j)) & \text{if } r = 1 \\ \lambda\alpha (1 - y^j - \sigma(S_\mu^j)) & \text{if } r = 0 \end{cases}$$

where S is the site value, j is the neuron index, μ indicates the site address, y is the neuron output, and $\sigma(.)$ represents the activation output function (usually taken to be a sigmoid). The two parameters α and λ control the reward and penalty rates respectively. Ferguson et al (92a,b) discuss the significance of these parameters in more detail.

In the case of the chip implementation, we make use of a target output pattern in a supervised training situation to calculate the overall output error value which is then used to produce a reinforcement signal 'r'.

$$r = \begin{cases} 1 & \text{with probability } (1 - e) \\ 0 & \text{otherwise} \end{cases}$$

It is perfectly possible to train all layers of a multi layered HyperNet network using the Reward-Penalty scheme described above. However, since the chip implementation uses the output error value to compute the reinforcement signal, it is relatively inexpensive to train the output layer neurons individually using an adaptation of Widrow-Hoff's (60) delta training rule, as follows.

$$\delta S_\mu^j = \alpha \rho \frac{2}{N_v} \sigma'(S_\mu^j)(y_t^j - \sigma(S_\mu^j))$$

Which simplifies to:

$$\delta S_\mu^j = \alpha \frac{2}{N_v} \sigma(S_\mu^j)(1 - \sigma(S_\mu^j))(y_t^j - \sigma(S_\mu^j))$$

In the above, ρ is the sigmoid "steepness" parameter, N_v is the number of output (visible) nodes, $\sigma'(.)$ is the derivative of the sigmoid function, and y_t is the target output value for the neuron in question. All of the above computations are performed using RAM look up tables in the chip. This means that the rules and their various parameters may be changed arbitrarily by the user. Thus, the use of delta rule training for output layer neurons speeds up the learning process without incurring any notable overheads. Figure 2 below, shows a schematic representation of the learning scheme implemented in the HyperNet chip.

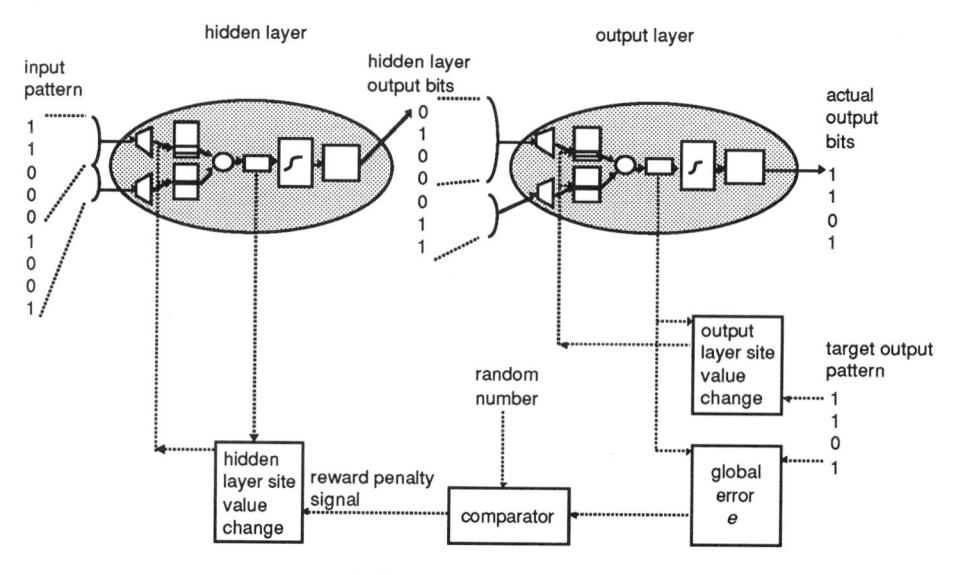

Figure 2 Schematic view of Reward-Penalty learning in HyperNets

THE CHIP

The following sections describe the design and implementation of HyperNet as a VLSI IC. The chip was designed for the ES2 1.5 micron double metal CMOS process using the ChipCrafter Silicon compiler from Cascade Design Automation. It comprises approximately 16K gate equivalents in an area of 62mm^2, operates with a 10MHz input clock, and has a power dissipation of about 160mW in operation. Packaging is provided by a standard 68-pin plastic leadless chip carrier. Figure 3 shows a plot of the HyperNet chip layout. The three large boxes implement the look up tables used for the learning algorithms.

Overview

The HyperNet chip has been designed to be used in fully connected multi layer feed forward neural networks. The number of layers in the neural network and the number of neurons per layer are determined by the number of chips used in a system. A single chip implements the necessary logic for up to 2048 MCU neurons each with up to 32K bits input. The limiting factor is the size of the activation memory which is implemented as an external RAM directly coupled to each chip. All or some of the neurons on each chip can be used as all or part of a layer in a network. Neurons within a chip are evaluated sequentially so that processing hardware can be shared between neurons. On the other hand, chips which are part of the same neural layer operate in parallel. Therefore, the user can trade off system cost and complexity (viz. number of chips used) against processing speed (viz. the proportion of neurons processed in parallel).

The number of subunits per neuron, the learning rules and the learning parameters α, ρ and λ are all user programmable. All parameters are loaded one byte at a time through a host IBM-compatible PC-AT. Testability is provided through a scan path. For the chip described here, all neurons in a layer receive the same input pattern (i.e a fully connected network). The design of a second VLSI device in which the mapping of the inputs to individual neurons can be arbitrarily defined is currently in progress. Figure 4, shows an overview of the HyperNet system. As shown, apart from the HyperNet chips and their associated Site Store RAMs, only a minimum of logic to support the input, output, and target output pattern buffers is required.

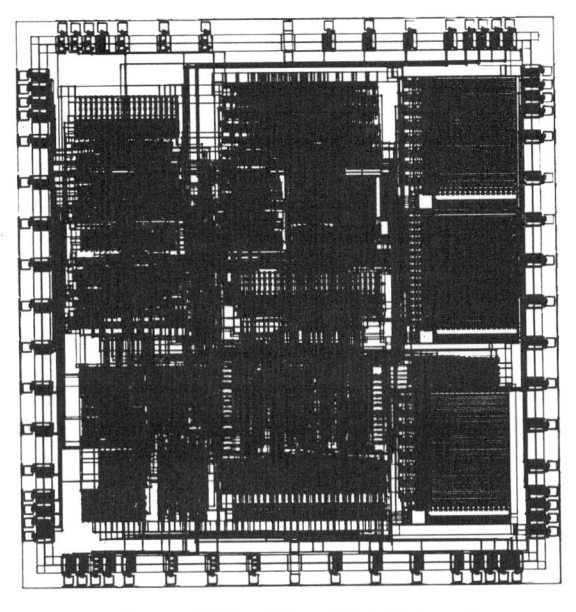

Figure 3 Plot of HyperNet chip layout

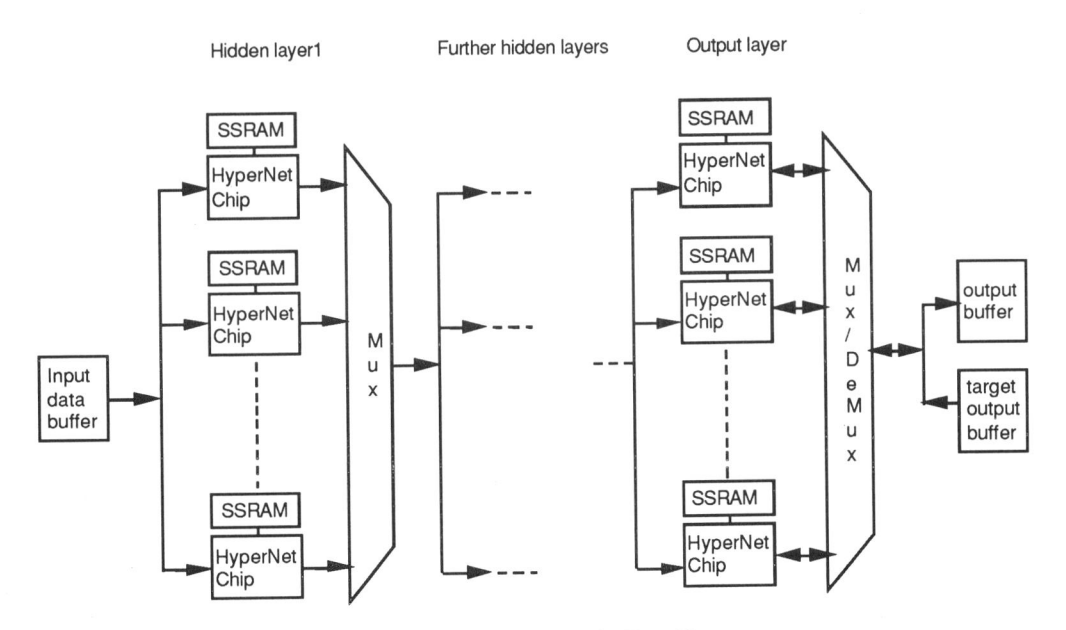

Figure 4 Schematic overview of a HyperNet system

Modes of operation

The chip has seven modes of operation, as follows.

The *Write* mode is executed to configure the network. A host computer is required to download information. The number of neurons used on the chip, the number of subunits per MCU neuron, the number of input bits per subunit, and the various look up table contents (including the learning rules and parameters) are all loaded onto the chip using this mode. Additional parameters such as whether the chip forms (part of) a hidden or output layer, and the number of training/evaluation cycles to be performed are also written in this mode. Furthermore, site values stored from a previous run can be downloaded into the Site Store RAMs associated with each chip. This means that the state of a network can be saved (using the *Read* mode below) and recreated at a later date using the *Write* mode.

The *Read* mode allows all network information (see above) to be read back from the chips to the host PC. This mode can be used by the host PC to examine or save a network.

In *Run* mode, a feed forward pass of the HyperNet network is performed. The inputs to each neuron are evaluated sequentially one byte at a time. The activation value corresponding to the addressed site in each subunit is read from the external Site Store RAM and the resultant probabilistic bit is stored in the output buffer.

The *learn* mode implements the Reward-Penalty and Delta Rule learning algorithms described earlier. The input and target output pattern buffers present the training pairs to the network one byte at a time. All other processing is carried out entirely by the HyperNet chips.

The *Test* mode involves three operations. Firstly, a set of test patterns are written to the chip. Secondly, a test cycle is entered to exercise the combinational circuits in the chip. Finally, the results of the test cycle are scanned out one byte at a time.

Finally, the *Halt* mode stops all chip operations while retaining all network information. When followed by a *Read* mode, this mode can be used to inspect or save the status of a network during learning.

Configurability

The chip is designed so that the number of neurons per chip, the number of inputs per neuron, and the number of subunits per MCU are all programmable. To ease implementation, all parameters are only programmable in multiples of 2. Furthermore, in order to limit the size of the external Site Store RAM (to 128M words in this case), some combinations of configuration parameters are not allowed. Table 1 below summarises the available configuration options.

THE DEMONSTRATOR SYSTEM

To demonstrate the feasibility of building large scale neural networks using the HyperNet chips, a system comprising five HyperNet chips has been designed. Figure 5 shows a schematic view of the HyperNet demonstrator system.

The system implements a three layer feed forward network. Four of the chips are used in the hidden layer and one in the output layer. To reduce RAM costs, only 256 neurons from each chip are utilised. Furthermore, the maximum number of inputs per neuron is limited to 256 (32, 8-bit subunits). Thus, each chip requires a Site Store RAM of $256 \times 32 \times 2^8$ or 2M words. The RAM wordlength was chosen to be 4-bits based on analysis using a behavioural/structural

Table 1 HyperNet chip configuration options

no. of subunits	subunit width options				corresponding no. of input bits per neuron			
1	1	2	4	8	1	2	4	8
2	1	2	4	8	2	4	8	16
4	1	2	4	8	4	8	16	32
8	1	2	4	8	8	16	32	64
16	1	2	4	8	16	32	64	128
32	1	2	4	8	32	64	128	256
64	1	2	4	8	64	128	256	512
128	1	2	4	8	128	256	512	1024
256	1	2	4	8	256	512	1024	2048
512	1	2	4	8	512	1024	2048	-
1024	1	2	4	-	1024	2048	4096	-
2048	1	2	4	-	2048	4096	8192	-
4096	1	2	4	-	4096	8192	16384	-
16384	1	2	-	-	16384	32768	-	-
32768	1	-	-	-	32768	-	-	-

simulator written in C. The system can support upto 10240 neurons in total if additional RAM is provided. Larger networks would require the use of more HyperNet chips. The speed of operation of the system depends on the network configuration and the way in which neurons are distributed across chips. Table 2 below gives performance figures for minimum and maximum size networks in the demonstrator system. Since all input/output is performed in 8-bit words, some configuration variations have no effect on the system performance.

The chip and system designs were completed in December 1991. System level simulations (using Mentor Graphics Quicksim) with worst case parameters were completed by the end of Easter 1992. The chip was submitted to Eurochip for manufacture in May 1992. However, the automatically compiled layout contained a number of geometric design rule violations. These are currently being corrected and we hope to resubmit the design for manufacture shortly.

CONCLUSIONS

The paper has described the VLSI implementation of a probabilistic hypercube-based digital neural network named HyperNet. On-chip learning logic, programmable network configuration, and simple inter chip communications, enable the construction of low-cost, high-performance neural networks of thousands of neurons.

Acknowledgements

We are grateful to the Royal Society for funding to manufacture the chip and system.

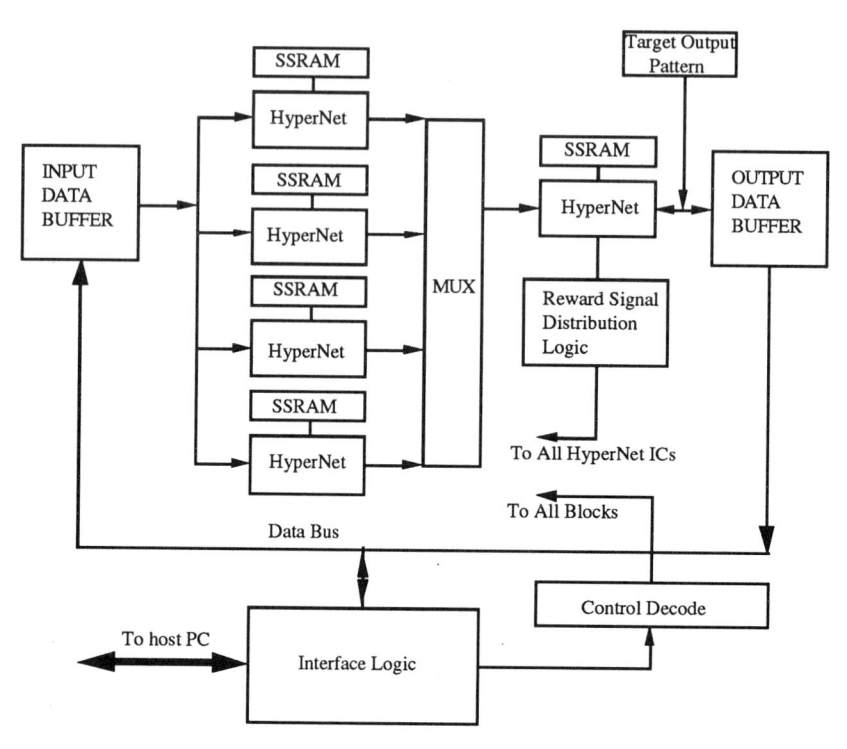

Figure 5 HyperNet demonstrator system

Table 2 Example HyperNet demonstrator system performance figures

no. neurons/layer		no. inputs/neuron	no. subunits	subunit size	system cycle time		
Hidden	1	1,2,4, or 8	1	1,2,4,8	run	= 4.2	μsec
Output	1	1	1	1	learn	= 8	μsec
Hidden	256x4	256	32	8	run	= 16.4	msec
Output	256x1	256	32	8	learn	= 42.6	msec

References

Aleksander I, Thomas W V, Bowden P A, *WISARD, a radical step forward in image recognition,* Sensor Review, July 1984, pp.120-124.

Aleksander I, *Canonical neural nets based on logic nodes,* Proc. 1st IEE Conf. on Artificial Neural Networks, 1989, pp.110-114.

Barto A G, Jordan I, *Gradient following without backpropagation in layered networks,* Proc. 1st IEEE Int. Conf. on Neural Networks, vol. 2, San Diego, 1987, pp.629-636.

Bledsoe W, Browning I, *Pattern recognition and reading by machine,* Proc. Eastern Joint Computer Conference, 1959, pp.225-232.

Clarkson T G, Gorse D, Taylor J G, *Hardware realisable models of neural processing,* Proc. 1st IEE Int. Conf. on Artificial Neural Networks, 1989, pp.242-246.

Ferguson A, Bolouri H, Gurney K, *The significance of learning parameters in the probabilistic Hypercube-based (HyperNet) neural network,* Document no. ERDC/1992/0033, Electronics Research and Development Centre, University of Hertfordshire, AL10 9AB, UK.

Ferguson A, Bolouri H, Gurney K, *The effects of learning parameter variations on the performance of HyperNets,* Document no. ERDC/1992/0039, Electronics Research and Development Centre, University of Hertfordshire, AL10 9AB, UK.

Gorse D, Taylor J G, *On the equivalence and properties of noisy neural and probabilistic RAM nets,* Physics letters A, vol. 131, no. 6, August 1988, pp.326-332.

Gorse D, Taylor J G, *Training strategies for probabilistic RAMs,* in Parallel Processing in Neural Systems and Computers, Eds. Eckmiller R, Hartmann G, and Hauske G, Elsevier, 1990, pp.161-164.

Gurney K, *Learning in networks of structured hypercubes,* PhD Thesis, Department of Electrical Engineering and Electronics, Brunel University, Uxbridge, Middlesex, May 1989.

Gurney K, *Training hardware realisable Sigma-Pi units,* Neural Networks, vol. 5, no.2, 1992, pp.289-303.

Kan W-K, Aleksander I, *A probabilistic logic neuron network for associative learning,* Proc. 1st IEEE Int. Conf. on Neural Networks, San Diego, 1987, vol. 2, pp.541-548.

Myers C, *Learning with delayed reinforcement in an exploratory probabilistic logic neural network,* PhD Thesis, Department of Electrical Engineering, Imperial College, 1990.

Steck G P, *Stochastic model for the Browning-Bledsoe pattern recognition scheme,* IRE Transactions on Electronic Computers, vol. EC-11, April 1962, pp.274-282.

Ullmann J R, *Experiments with the N-Tuple method of pattern recognition,* IEEE Transactions on Computers, December 1969.

Widrow B, Hoff M, *Adaptive switching circuits,* 1960 IRE WESCON Convention Records, part 4, pp.23-24.

ASYNCHRONOUS VLSI DESIGN FOR NEURAL SYSTEM IMPLEMENTATION

John F. Hurdle, Erik L. Brunvand and Lüli Josephson

INTRODUCTION

Synchronous VLSI design has proven to be enormously useful. Clocked circuits comprise practically every interesting piece of electronic hardware in use today. They are supported by a large and practical set of computer-aided design (CAD) tools, from high-level specification languages to automated synthesis systems to placement and routing aids. So, given the success of synchronous design, why consider an *asynchronous* approach? What advantages might it offer the neural hardware designer? This paper offers answers to these questions by arguing that asynchronous systems match neurocomputing in a natural and complementary way. In particular, we argue that an asynchronous paradigm holds great promise in solving several thorny design issues facing neural hardware researchers: scaling of neural circuits, composing neural modules, circuit robustness and process tolerance, and performance.

To frame this argument we begin in the next section with a brief overview of synchronous approaches to neural VLSI circuits. Treleaven *et al.* (1989) and Zornetzer *et al.* (1990) offer readable summaries of the progress made in electronic implementations of neural systems. A more contemporary summary is needed but in the meantime a useful resource may be found in the *IEEE Transactions on Neural Networks* yearly issue devoted to neural hardware. The papers in the last two special issues have been more or less evenly divided between digital and analog implementations, including several hybrid systems. This suggests a taxonomy by which work in the field may be classified. Research either falls into the digital, the analog, or the hybrid class. The Treleaven article suggests another classification. It categorizes by the processing engine used in the design and relies on the traditional computer taxonomy "special-purpose" and "general-purpose" (Treleaven *et al.* 1989). An example of the former would be any design that directly implements a specific neural model. The latter would include designs predicated on virtual processing nodes which support arbitrary interconnection schemes.

For the neural hardware designer, though, implementation issues are the central focus. We concentrate in this paper on an important alternative axis of classification for digital systems: synchronous versus asynchronous circuit implementations. The latter has received scant attention in the neural literature. One of the goals of our work is to help address this imbalance. The rationale for this new perspective is provided by exploring the shortcomings of synchronous design for neural hardware and then contrasting these to the advantages offered by asynchronous design. With this emphasis on circuit control and timing, it may seem that this new axis of classification bears only on digital design, but as shown below asynchronous design is suitable for hybrid neural circuits as well. In the third section of the paper we review the basics of asynchronous circuit theory and discuss how asynchrony suits neurocomputation.

The final section presents a practical example, the design of an asynchronous version of Albus' (Albus 1975) Cerebellar Motor Articulation Controller (CMAC) model. This design is presented in some detail to underscore the utility of the principles outlined in the first part of the paper. Field programmable gate arrays (FPGAs) were chosen for the implementation, a decision based on economics and timeliness, and the circuit is still undergoing design and testing. The circuit is entirely digital. However, nothing in this design approach limits us to all-digital circuits or to FPGAs. In fact, one of the unique benefits of this approach is the ability to switch the underlying technology or to mix one technology with another in the same neural system. These ideas are explored in more detail in the third section.

TRADITIONAL SYNCHRONOUS NEURAL VLSI

A complete neural system properly consists of four models: a *synapse* model, a *neuron* model, an *architectural* model, and a *learning* model. Each offers unique circuit challenges. It is instructive to see how these model classes have typically been built, so we offer a brief overview in the next subsection. In the concluding subsection we underscore some pertinent weaknesses in the traditional synchronous approach, especially with regard to the building of flexible, large-scale neurocomputers.

A Synopsis of Traditional Design

In both neural software and hardware the *synapse* is the easiest piece to model, usually as a programmable weight. In hardware specifically, weights have been implemented as fixed or variable resistors (Moon et al. 1992 is typical), switched capacitors (e.g., Redman-White *et al.* 1989), values in random access memory (RAM) or shift registers (e.g., Wike *et al.* 1990), or as combinational logic (CL) blocks in several pulse-encoded and boolean networks (e.g., Murray *et al.* 1991). Except in fully analog systems, synchronous methods are used to "connect" the proper weight to the proper neuron at the proper time. Even in hybrid systems where the neuron is analog the weights are usually digital. As the degree of connectivity approaches biologically-inspired ranges measured in the hundreds and thousands, the distribution of weight values promises to be a serious factor in performance. So the synchronous methods used to control the distribution of these weights warrant equally serious review.

Neuron models are more complex. The classic formulation is built on the inner-product calculation,

$$\sum x_j * w_{ij}, \text{ for all neurons } j \text{ connected to neuron } i \text{ through weight } w_{ij}$$

This sum is passed to an activation function, f, and the result is the output of that neuron. The activation function typically is a simple threshold, linear, semi-linear, logistic, transcendental (e.g., tanh), exponential, or something more exotic (e.g., probability density functions like Gaussians). These are harder to implement in hardware than weights. One approach is to use the current-mode subthreshold conductance characteristics of CMOS to emulate selected activation functions (e.g., Andreou *et al.* 1991). Another is to use dedicated digital adders, multipliers, pulse counters, or other CL blocks (e.g., Murray *et al.* 1991). Kung (Kung and Hwang 1989) suggests using general purpose central processing units (CPU) arranged in systolic arrays for the direct computation of nodal functions. For some time Hecht-Nielsen (Hecht-Nielsen 1986) has advocated virtual nodes implemented as accelerated special-purpose nodal computation engines. Analog and hybrid circuit designers suggest a variety of analog elements to perform node calculations. The work of Card and Moore (1990) is representative.

Neuron implementation has received the lion's share of attention in the neural hardware literature. This is understandable since the neuron is the computational locus in neural systems.

Synapses, modelled as weights, are straightforward to build so research tends to focus on the more challenging neuron model. Designers of neural chips have tended to focus on the "microarchitecture" of neural systems, that is, the weights and neurons. Neural circuits based on integrated circuit technology have been around for at least 20 years (Nakano 1972). In that time much of the research effort spent on neural hardware has been spent on microarchitectural issues. This focus may have been at the cost of progress in the areas of "macroarchitectural" issues, especially the area of wiring together neurons (what we call here the *architecture* model) and implementing hardware-realizable *learning* models. A small but important research niche in neural hardware is occupied by design studies of sensory systems and biological neural ganglia. These systems are, by definition, special purpose and the boundaries between microarchitecture and macroarchitecture blur. Some especially interesting examples of sensory and ganglia-level modelling may be found in Mead (1989) and Card and Moore (1990).

To neural researchers who focus exclusively on software simulations or neural network theory, a learning model implies an architectural model. Such workers pay scant attention to the architecture because it is implicit in the learning model. A Hopfield learning model is built on a single layer of fully interconnected neurons (except w_{ij} where i = j). Standard backward error propagation models (BP) are built on feedforward, multiple, fully-interconnected layers of neurons. Hardware designers, though, have learned that there *is* a significant distinction between the abstract architecture of a model and the physical architecture when implemented in VLSI. It suddenly becomes important to view learning and architecture as distinct things because each has a profound effect on the way neural chips are built. This is especially true when such chips are designed to be cascaded, or *scaled*, into larger ensembles to form true neurocomputers.

Important progress has been made with neural hardware systems which directly implement a software model. Practically all of the major neural models have been attempted in hardware. The Hopfield model is popular and well studied (e.g., Wike *et al.* 1990), as is BP (e.g., Caviglia *et al.* 1990). Associative memories are also common (e.g., Rückert 1990). Even the challenging adaptive resonance theory (ART) models of Grossberg and Carpenter has been attempted in hardware (Rao *et al.* 1990). A notable exception which has received little attention by hardware designers is Albus' CMAC model. Since CMAC has virtues in real-time, nonlinear control, it makes sense to consider VLSI versions for real-world applications. We offer a preliminary design in the last section.

Limitations of Synchronous Design

Synchronous design relies on a global clock. Since the circuit must adhere to the clock's schedule this imposes an artificial sense of *timing* on a neural circuit. All circuit components have to wait for the clock and the clock has to wait for the slowest component, since the overall circuit can only run as fast as the slowest component on the critical path. The timing characteristics of many neural parts, such as arithmetic parts like adders, are inherently *data dependent*. Some data take longer to process than others. The addition of binary "0000 + 0000", being carryless, completes faster than binary "1111 + 0001", being laden with carries at each binary stage. One way to allow for differential, data-dependent execution times requires stalling the circuit at various points to allow parts to complete their *worst-case* execution (e.g., the addition of "1111+ 0001"). To maintain data safety circuit timing must be adjusted to ensure proper processing of the legal worst-case data, even if the worst case is a rare case. Alternatively, extra circuitry can be added to speed up worst-case processing. For example, carry lookahead (CLA) can be used to accelerate an adder.

Stalling a circuit carries a time penalty and adding extra circuitry carries a space penalty. The space penalty of CLA, for example, is $O(n*log(n))$ compared to $O(n)$ for a ripple adder, where "n" is the number bit stages. A 16-bit CLA-based adder would be on the order of *four* times larger than the equivalent (and slower) 16-bit ripple adder. Since the speed of 16-bit ripple adder is $O(n)$ and that of a 16-bit CLA adder is $O(log(n))$, the space-time product for both adders is the same to within a constant. This relationship holds in general for 16-bit and 32-bit integer arithmetic circuits since speed is always bounded from below by an O(log(n)) term and yet space typically grows at O(n*log(n)). When the number of bits reaches 64 the log(n) terms become significant and the extra hardware of circuits like CLA become cost effective.

For circuits in the 16-32 bit range, though, there is an effective alternative to these traditional synchronous approaches, as we explore in more detail in the next section. The alternative is to replace the *timing* of circuit parts with the proper *sequencing* of control and data signals by using handshaking. Such circuits are *asynchronous* because under such a scheme circuit modules tell each other when to proceed without reference to a clock. They preserve a correct sequence of execution without making assumptions about how long it takes any given part to complete its job.

Clocked circuits tend to require fewer components than functionally-equivalent asynchronous circuits which must support the overhead of the handshaking (see the next section). Because of the availability of good CAD tools, clocked systems are often easier to design than asynchronous circuits. These are two excellent reasons why synchronous systems remain so popular. However, vexing problems appear as synchronous circuits become faster and silicon area grows larger. Clock skew is a good example. Acting as *the* master control element, the clock has to appear uniform across the chip and between chips (or between modules in multichip module designs). The delay in clock propagation and clock skew become barriers for large, fast circuits. Inevitably designers will want to build large and fast neurocomputers, so they will eventually have to face the issue of clock distribution.

A related problem lies in the power consumed by the clock. Clocked circuits are obliged to maintain the clock even when idling. When a CPU turns over control to a floating point unit (FPU) for a calculation, the CPU is still being clocked and is still dissipating the power to drive that clock. In the specific example of the CPU, utilization can often be increased (e.g., by overlapping instructions), but in general there will always be parts in any large system which are idle. This is especially true for systems where there is a mismatch between data input rates and processing time. A familiar example is a human typing text into a computer. It is easy to imagine neurocomputer applications where some real-world process is providing data in a slow or "bursty" fashion, or where input data first triggers the use of one neural unit and then of another (e.g., limb coordination). Data dependencies focus processing onto local circuit regions while idling other regions. Maintaining a clock in those idle regions wastes power and this waste only gets worse as circuit and system size increase.

The issues outlined above point up weaknesses in clocked protocols as they pertain to *scaling* neural circuits. A related issue is the *composability* of neural circuits. Composabilty refers to the ability to substitute one functionally equivalent module for another. The new module might use a new, faster technology (e.g., gallium arsenide); it might incorporate a new algorithm; or it might replace a digital module with an analog or hybrid one. Replacing a digital adder with an analog summing amplifier would be an example. It makes sense to consider substituting new modules to increase performance, reduce power consumption, decrease circuit area, or to improve accuracy/precision.

It is hard to compose synchronous circuits because of the need to consider the timing behavior of individual circuit components. It is not enough that a new component be the functional equivalent of the old one. The new component must also match the timing behavior of the old one or else the entire circuit needs to be tuned to accommodate the new module. In cases where clock distribution or skew has already led to a tight design, the new timing requirements might simply be impossible to meet.

The limitations of synchronous design for neural hardware may be summarized as a mismatch between the needs of neural designers and the realities of synchronous design. First, neural hardware systems will inevitably get larger, far larger than we have yet built. Nature, having provided the inspiration for neural systems, opted for massive networks of neurons interconnected on truly enormous scales. So scalable architectures constitute one design need. However, synchronous designs often do not scale well and clock distribution is already proving to be difficult in high performance systems. Second, to exploit VLSI process improvements and theoretical improvements to synaptic, neuronal, architectural, and learning models, designers need to be able to compose modules in a circuit. Synchronous systems tend not to compose well because changing a component usually means changing its timing behavior and synchronous systems are inherently sensitive to such changes.

ASYNCHRONOUS NEUROCOMPUTING

Asynchronous systems process without a global clock. One way to implement asynchrony uses modules which signal one another with requests to begin processing and acknowledgments that processing is complete. Such systems are called *self-timed*. We will use the terms "asynchronous" and "self-timed" interchangeably for the remainder of the paper. Research into self-timed systems has been active for several decades (Unger 1969). Seitz (1980) provides a

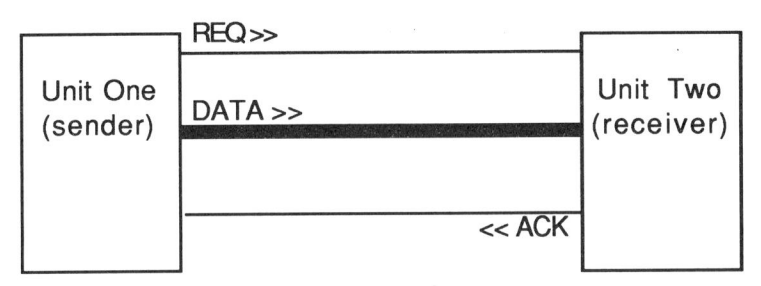

Figure 1 A Simple Self-timed Handshake

concise and readable contemporary summary of self-timed concepts and nomenclature. Sutherland (1989) presents an excellent introduction to the practical use of these techniques. In the next subsection we review a few basic concepts to assist readers new to self-timed design. We conclude the section with an explanation of how self-timed systems help to remedy the mismatch between neural design needs and traditional synchronous design methods.

Basic Self-Timed Circuit Principles

Self-timed circuit elements communicate with each other using handshakes. **Figure 1** shows the simplest possible handshaking scheme, where two control signals are used. By sending a request (REQ) one unit alerts a second that data is ready. By sending an acknowledgement (ACK) the second alerts the first that it is done with the data. A signalling convention must be chosen for the two signalling wires. A *two-phase* (also known as two-cycle) protocol uses transition signalling, where events are coded as a *transition* of any kind (as opposed to a level) on a wire. A *four-phase* (or four-cycle) protocol requires that both the REQ and the ACK return to a known level to completely encode an event. The distinction between the two protocols is very similar to non-return-to-zero (NRZ) and return-to-zero (RZ) encoding found in data communications or recording systems. If a circuit will work correctly irrespective of the delay in its parts as well as irrespective of the delays in its wires, the circuit is called *delay-insensitive* (DI). This is an important notion because strong theoretical results can be proven about such systems, especially in the areas of correct synthesis and verification of DI circuits (Brunvand and Sproull 1989).

On the other hand, it is very hard to build practical DI circuits in VLSI. The control signals and the data run on separate wires, yet they are logically bound together. By the time the second unit of **Figure 1** "sees" the REQ control signal, the data should have arrived and settled. As long as the signals travel on separate wires, it is hard to guarantee at the circuit level the orderly arrival of data and control signals without making assumptions about wire propagation delays. So one style of implementation is to encode data and control on the same wires, typically using a dual-rail scheme where each bit is encoded on two wires. Only certain combinations of the two bits correspond to valid data states. The receiver can figure out when all the data has arrived, bit by bit, and can assume an implicit REQ when this has happened. The significant increase in wires makes this approach expensive for VLSI designs. So other, non-DI, models have evolved where safety properties are weaker. One example is called the *speed independent* model, where the circuit continues to work properly independent of delays inside parts but where delays on wires are explicitly handled by the designer. The easiest way to do this is to impose *bundling constraints* on the data: care is taken in the design of the circuit to ensure that data arrives before control signals do.

We use two-phase transition signalling and the bundled data approach in the design of the CMAC circuit in the next section. The technique is both practical and powerful. For example, our lab has designed a 16-bit reduced instruction set CPU (RISC) using this design methodology (Brunvand 1992).

Advantages of Asynchronous Design

Virtually all of the advantages of asynchronous systems derive from the fact that there is no system clock. Ideally, self-timed systems make no assumptions about time at all. Instead, control is distributed throughout the entire circuit and is implemented by handshaking between parts. Since parts are not tied artificially to time as measured by fixed clock pulses, traditional terms and measures do not apply to these circuits. For example, we do not ask how many clock cycles an operation takes. We are assured that the operation will complete as quickly as it can given the circuit and the data involved. Parts take it upon themselves to notify other parts that data are ready or that an operation is complete. This leads to *average-case* (i.e., data dependent) performance.

Asynchronous systems also do not have the design burden of propagating a clean and uniform clock. This saves valuable design time and eliminates the power consumption of the clock. These two properties mean that asynchronous systems are more easily *scaled* than synchronous ones. Since component timing is not crucial in self-timed systems, correct behavior is ensured when neural modules are interconnected and the system scaled up. Each module need only meet the demands of the handshaking interface to properly participate in the scaling process. Self-timed modules consume the bulk of their power only when processing since there is no clock to consume when they are idle. So there is a simple relationship between power consumed and system size. The average power consumed will follow the demands imposed by the data, just as performance does. As we noted above, scaling promises to be a crucial issue in the design of larger, more powerful neural systems because such systems gain their power directly from their massiveness.

Gaining *composability* and *replaceability* are other important benefits of self-timed design. If all of the circuits and subsystems are self-timed then it is possible to build systems by assembling components. Rather than having to tune an entire ensemble of components to meet a complex and interdependent set of individual timing needs, self-timed components compose with regard only to their functionality. Replaceability means that it is possible to improve the performance or functionality of a system by incrementally improving and replacing subsystems without retiming the whole system.

Naturally, self-timed design has its own set of potential drawbacks. For example, compared to synchronous design there is a significant dearth of computer-aided design tools. This is changing rapidly as new tools are being introduced (Brunvand and Sproull 1989). Additionally, asynchronous circuits do tend to be larger than equivalent synchronous circuits because of the extra elements required by the REQ, ACK, and various rendezvous and control parts. The question "At what level of granularity should handshaking be used?" is still open. The granularity of handshaking determines in large part the degree of extra-circuit overhead in asynchronous designs. The RISC CPU mentioned above was designed at a fairly coarse level of granularity, as is the CMAC model presented below. In both designs the handshaking elements constitute only a modest percentage of the over all circuit.

A SELF-TIMED NEURAL CIRCUIT EXAMPLE: CMAC

Albus first described the Cerebellar Motor Articulation Controller (CMAC) in the mid-1970s (Albus 1975). His goal was to approximate the macroarchitecture of the cerebellum. The cerebellum translates movement orders from higher brain centers into control signals for muscles, taking into account sensory information such as where the limbs are, what loads they are operating under, and so forth. It functions as a real time, non-linear controller.

The CMAC Model

A typical CMAC architecture, in highly schematic form, is shown in **Figure 2**. An input vector (binary or real valued) is mapped into regions which partition the weight space into nondisjoint, overlapping weight sets. This mapping is often termed a "neighborhood function." In software simulation this is often a hash function. The weights in the set are summed to form the output of the unit. The learning model for CMAC uses a supervised technique for adjusting the weights. An input vector is chosen from the training set, applied to the unit (the weights of which are initially zero), and the actual output compared to the desired output for that input vector. The weights are then adjusted using the formula

$$w_i\text{-new} = w_i\text{-old} + \Delta w^*$$

where $\Delta w^* =$ (desired - actual)/ #weight selection units. Only those weights which contribute to the sum are updated. Successive samples of input-output pairs are processed by CMAC until the overall error drops to an acceptable level. The number of weights participating in a specific mapping operation is much smaller than with fully interconnected networks.

The computational complexity of any learning model where error is shared, sample by sample, over a set of weights increases dramatically as the number of weights involved in a

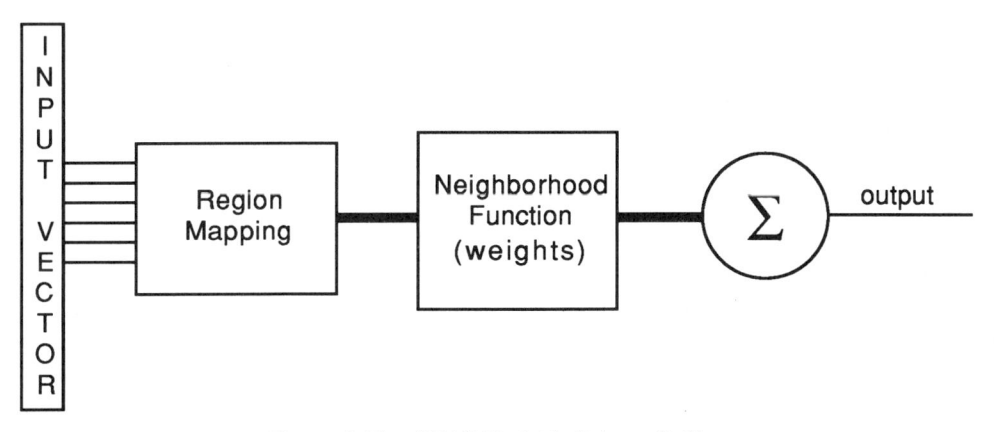

Figure 2 The CMAC Model in Schematic Form

learning step increases. This "credit assignment problem" often stalls learning. CMAC's reliance on *locally*-activated neighborhood weight sets speeds convergence during training. This description of CMAC is necessarily brief and does justice to neither the architecture nor the learning algorithm. An informative review of CMAC which demonstrates both its power and simplicity may be found in Cotter and Guillerm (1992).

A Simple, Self-Timed CMAC Implementation

CMAC is an attractive model to implement in VLSI. If we think of CMAC operating in two modes, *training* (learn the input-output mappings) and *functional* (use what has been learned), then we need only an adder in functional mode and an adder and divider in training mode to implement the neuron model. Typically, CMAC only requires one neuron for each output dimension. The synapses are modelled as weights and, although the weight space might be quite large, any given cycle of operation (training or functional) only requires a few weights at a time. This means that the connectivity, or the virtual connectivity, is low.

We have applied the self-timed circuit techniques described in the previous section to the design of an integer arithmetic, 2-D input version of CMAC (which we call ACMAC for "Asynchronous CMAC"). This version is fully-digital, although nothing in the self-timed approach precludes analog elements. Since self-timed systems compose so well, it should prove easy to insert analog components in this or any other asynchronous design when such parts offer advantages over digital elements.

ACMAC is built with speed-independent circuit elements and is itself a speed-independent element. So arbitrary scaling of ACMAC is possible without regard to clocking constraints. We initially thought that scaling ACMAC would entail adding circuitry to improve the accuracy of the output. However, simulations have shown that with this integer-based design accuracy is high with a relatively small neighborhood weight set. However, scaling could still be used to increase the dimensionality of the output vector.

The high-level circuit design is shown in **Figure 3.** Each circuit module uses two-phase (REQ/ACK) transition signalling (*internal* REQ/ACK lines are not shown in **Figure 3** for clarity). Data is bundled with these control lines with the constraint that each circuit module asserts a request ransition only when the data are stable and have reached their destination. This is sufficient for on-chip communication but may be impractical when sending signals chip-to-chip. We plan to investigate the use of dual rail signalling for data transmission in multichip or multimodule wafer scale CMAC implementations.

If the chip is in *training* mode, the TRAIN/FUNCTION bit will be low. In that case, the CMAC training algorithm will be invoked, using the value present on the TARGET bus as the desired output. If TRAIN/FUNCTION is high, the chip is in *function* mode and it places an output value (the sum of the appropriate weights) on the OUTPUT bus. The data on X-Y-DATA, TARGET, and TRAIN/FUNCTION are bundled with an incoming REQ-LATCH transition. The DATA-LATCH unit will acknowledge to the sender as soon as it has the values latched even though the rest of the ACMAC unit is still working. This allows the sender to forward its next set of data while ACMAC is processing, leading to a naturally pipelined operation. Eventually the rest of the unit finishes and signals completion on the internal line labelled DONE. This transition will rendezvous with the next transition on the REQ-LATCH line using a synchronizing circuit called a Müller C-element. Seitz (1980) provides more detail about this and other asynchronous control elements. This rendezvous initiates the next cycle.

The DATA-LATCH is built using self-timed request-acknowledge handshakes and latch elements. Since the two 16-bit buses for X-Y-DATA and TARGET and the one bit TRAIN/FUNCTION flag are latched together under the control of the same request, the ACKs for all three internal latches are gathered into a C-element tree which forms the ACK-LATCH signal for the sender and the internal START signal (not shown in **Figure 3**) for the rest of the unit. The WEIGHT-UNIT decodes the x- and y- coordinates into a memory address. The current prototype uses an external 16-bit x 64Kb static RAM, contained schematically in the WEIGHT-UNIT. In a full custom version the RAM would be internal and perhaps smaller. Because the RAM is currently off-chip, we use an external delay line to model its access time for the rest of the self-timed circuit.

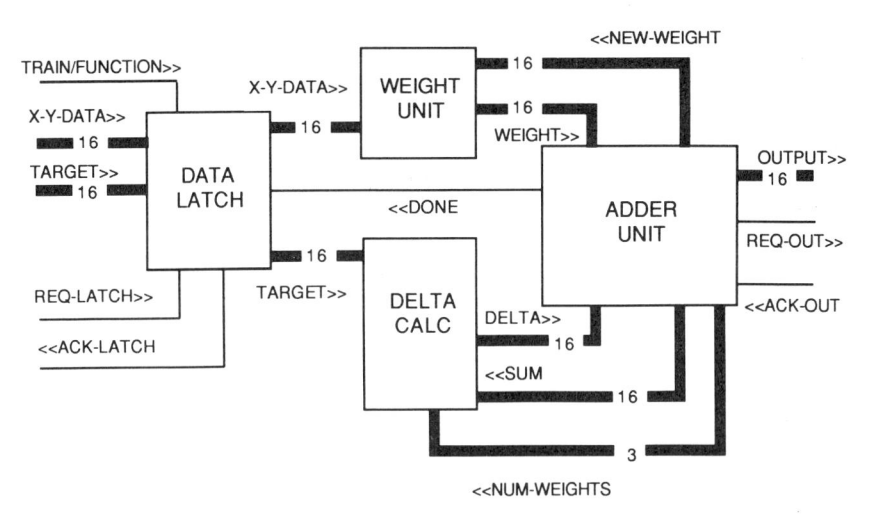

Figure 3 High Level Circuit Design for ACMAC. All buses are bundled with REQ/ACK control lines (not shown for clarity). The number on the bus is its bit width.

The neighborhood function is achieved by having the WEIGHT-UNIT return the data at the address indicated by the x-y input coordinate, as well as its four nearest neighbors. The most significant eight bits of each coordinate are used to partition their respective dimension into 256 equal-sized regions. This provides more than adequate resolution for mapping most functions. The eight bits for y are appended to those for x to form a 16 bit address. If x-y lies on the edge of the memory address space then there may be only three neighbors or in the worst case only two (i.e., x-y lies in a corner). The neighbors must be true neighbors in the weight space, so wrapping around to the opposite edge of memory space to fetch a neighbor is illegal. The weight unit detects such boundary conditions by incrementing and decrementing the x-y values to see if they lie at the extremes of the memory address space. The incrementers and decrementers are self-timed units, generating a completion signal when finished. They show good average-case performance because they do not have to wait for a worst-case increment/decrement.

The addresses of the required weights are sent to memory and the data values returned are forwarded to the ADDER-UNIT asynchronously. If the ACMAC is in training mode, then ADDER-UNIT will be returning updated weights to replace those just sent by the WEIGHT-UNIT. The WEIGHT-UNIT's final task is to collect these and send them back to memory. The ADDER-UNIT accepts weights from the WEIGHT-UNIT, stores them in a small register file, and adds them up in a self-timed carry-completion adder (CCA). In common with the incrementer/decrementer discussed above, the CCA detects when no more carries are propagating and signals completion as soon as the data allow. Once again, average-case behavior predominates in this clockless adder. ADDER-UNIT counts the number of incoming weights until the WEIGHT-UNIT signals completion. If ACMAC is in function mode, it sums the incoming weights and passes the sum to the OUTPUT bus and generates the internal DONE signal. If ACMAC is in training mode, the sum and weight count are sent to the DELTA–CALC unit. That unit will return the Δw^* described above. ADDER-UNIT adds the Δw^* to the register files and returns the updated weights to the WEIGHT-UNIT. When the transfer is complete, the ADDER-UNIT generates the internal DONE signal to allow a new cycle to start.

The DELTA-CALC unit asynchronously gathers the sum and the number of weights (a three bit bus because the number of weights used in a cycle is 3, 4, or 5) from the ADDER-UNIT. It computes the quotient in accordance with the delta rule of the CMAC learning model and returns it to the ADDER-UNIT. The division uses a self-timed "smart" restoring-style division algorithm where subtraction is performed on the remainder register only when necessary. It never physically restores but it otherwise follows the standard restoring algorithm. The quotient is rounded based on the remainder and the remainder is discarded. This divider circuit is a good example of how a designer can exploit data dependencies with self-timed design in a straightforward way. As soon as the circuit detects a negative result from the subtraction it updates the quotient register and immediately moves on to the next cycle without stalling. Being able to exploit the natural flow of an algorithm like this is one of the great benefits of self-timed design.

SUMMARY

Synchronous design, although well understood and well supported by design tools, has serious drawbacks for large-scale neural hardware construction. Neural designers will want to build ever larger hardware platforms but the disadvantages of clock distribution and power consumption in synchronous systems promise to become ever more problematic. Asynchronous design is an alternative paradigm which should allow the graceful scaling and composing of neural components. As an example, we described the design of a simple CMAC architecture based exclusively self-timed parts. Hardware should exploit the inherent data-dependencies found so often in neurocomputing, and we described how various arithmetic parts can be built to do so.

The circuits described above have been coded in a CSP-like language and laid out in a schematic capture program (VIEWLOGIC). Simulation and testing has begun on the system, with the goal of building a prototype using field programmable gate arrays and external RAM. The long-term goal is to build a single chip version with sufficient on chip memory to allow the construction of fast, scalable CMAC hardware.

REFERENCES

Albus, J. A., "A New Approach to Manipulator Control: The Cerebellar Model Articulation Controller", *Trans. of the ASME, J. of Dynamic Sys., Measurement, and Control*, vol. 97 pp. 220-227, 1975.

Andreou, A. G., Boahen, K. A., Pouliquen, P. O., Pavsovic´, A. P. and Stohbehn, K., "Current-Mode Subthreshold MOS Circuits for Analog VLSI Neural Systems", *IEEE Trans. on Neural Networks*, vol. 2 (2), pp. 205-221, 1991.

Brunvand, E. L. and Sproull, R. F., "Translating Concurrent Communicating Programs into Delay-Insensitive Circuits", in *Int. Conf. on Computer-Aided Design, ICCAD 89*, 1989.

Brunvand. E. L., ``Using FPGAs to Prototype a Self-Timed Computer," 2nd International Workshop on Field-Programmable Logic and Applications, Vienna Institute of Technology, August 1992.

Card, H. C. and Moore, W. R., "Silicon Models of Associative Learning in *Aplysia*", *Neural Networks*, vol. 3 (2), pp. 333-346, 1990.

Caviglia, D. D., Valle, M. and Bisio, G. M., "Effects of Weight Discretization on the Back Propagation Learning Method: Algorithm Design and Hardware Realization", in *Int. Joint Conf. on Neural Networks*, pp. 631-637, 1990.

Cotter, N. E. and Guillerm, T. J., "The CMAC and a Theorem of Kolmogorov", *Neural Networks*, vol. 5 pp. 221-228, 1992.

Hecht-Nielsen, R., "Performance Limits of Optical, Electro-Optical, and Electronic Neurocomputers", *Proc. Optical and Hybrid Computing (SPIE)*, vol. 643 pp. 277-306, 1986.

Kung, S. Y. and Hwang, J. N., "Digital VLSI Architecture for Neural Networks", in *IEEE Int. Sym. on Circuits and Systems*, pp. 445-448, 1989.

Mead, C., *Analog VLSI and Neural Systems*. Reading: Addison-Wesley, 1989.

Moon, G., Zaghloul, M. E. and Newcomb, R. W., "VLSI Implementation of Synaptic Weighting and Summing in Pulse Coded Neural-Type Cells", *IEEE Trans. on Neural Networks*, vol. 3 (3), pp. 394-403, 1992.

Murray, A. F., Del Corso, D. and Tarassenko, L., "Pulse-Stream VLSI Neural Networks Mixing Analog and Digital Techniques", *IEEE Trans on Neural Networks*, vol. 2 (2), pp. 193-204, 1991.

Nakano, K., "Associatron-- a Model of Associative Memory", *IEEE Trans. on Sys., Man, and Cybernetics*, vol. 2 pp. 380-388, 1972.

Rao, A., Walker, M. R., Clark, L. T., Akers, L. A. and Grondin, R. O., "VLSI Implementation of Neural Classifiers", *Neural Computation*, vol. 2 (1), pp. 35-43, 1990.

Redman-White, W., Lam, Y. Y. H., Bedworth, M. D. and Bounds, D., "A Limited Connectivity Switched Capacitor Analogue Neural Processing Circuit with Digital Storage of Non-Binary Input Weights", in *First IEE Artificial Neural Networks*, pp. 42-46, 1989.

Rückert, U., "VLSI Implementation of an Associative Memory Based on Distributed Storage of Information", in *Neural Networks,* L. B. Almedia and C. J. Wellekens (ed), Berlin: Springer-Verlag, pp. 267-276, 1990.

Seitz, C., "System Timing", in *Introduction to VLSI Systems (chapter 7),* C. A. Mead and L. Conway (ed), Reading, MA: Addison-Wesley, pp. 218-262, 1980.

Sutherland, I. E., "Micropipelines", *Comm. of the ACM*, vol. 32 (6), pp. 720-738, 1989.

Treleaven, P., Pacheco, M. and Vellasco, M., "VLSI Architectures for Neural Networks", *IEEE Micro*, vol. 9 (6), pp. 8-27, 1989.

Unger, S. H., Asynchronous Sequential Switching Circuits. New York: John Wiley and Sons, 1969.

Wike, W., Van den Bout, D. and Miller, T., "The VLSI Implementation of STONN", in *Int. Joint Conf. on Neural Networks*, pp. 593-598, 1990.

Zornetzer, S. F., Davis, J. L. and Lau, C., "An Introduction to Neural and Electronic Networks", Academic Press, pp. 1990.

VLSI-IMPLEMENTATION OF ASSOCIATIVE MEMORY SYSTEMS FOR NEURAL INFORMATION PROCESSING

Andreas König and Manfred Glesner

INTRODUCTION

Neural Networks and related associative networks or memories have raised a flurry of interest and research activities during the last decades. Currently technology has matured and neural networks and associative memories are employed more and more for demanding applications with real time requirements, e.g., control and automated visual quality inspection. Many neural structures are not easily amenable to formal analysis and efficient VLSI-implementation is hard to accomplish. In contrast, associative memories are more amenable to analysis and permit simple implementation as well as the development of modular architectures that scale up processing power with network size, while providing an easy to implement and inexpensive structure. System implementation providing interfaces and processing structure that allow embedding in a real-time signal processing environment gives access to massive parallelism and learning capability of these structures for real time applications. Waldschmidt (1987) gave a categorization of associative processors and architectures. He classifies an associative processor (ASP) as a SIMD (Single Instruction Multiple Data) consisting in the general case of two units, control unit and associative memory unit. Basically, four different forms of APS architecures can be distinguished: fully parallel ASP, bit-serial ASP, word-serial ASP, and block oriented ASP. The associative memory unit can be realised as a content addressable memory or as a standard RAM in conjunction with an association unit.

The architectures and implementations in the following sections can all be affiliated to the category featuring: Controller and separate scalable associative network, bit-serial processing, and standard RAM with dedicated association unit. The ensuing sections will provide the background of the underlying networks and give a survey on concluded implementation efforts. Furthermore, an augmented architecture will be defined and status of its implementation will be given. Finally, target applications, that gave incentive to the development will be briefly addressed.

THE BASIC BINARY NETWORK

During the last decades authors like Steinbuch, Gabor, Rosenblatt, Kohonen and others have been concerned with associative mappings and memories, their relation to biological evidence and potential application. In general, an associative memory performs a mapping between two finite sets I and O $M : I \rightarrow O$, where I denotes the set of the input stimuli and O the set of the desired output responses. This can also be interpreted as question and answer. The mapping M causes the formation of pairs (I^k, O^k). In the case of arbitrary formation or combination of I^k

and O^k this mapping is called an *heteroassociative mapping*. For $O^k = I^k$ this mapping is called *autoassociative mapping*. In other words, in the latter case an incomplete or corrupted pattern can be restored if an sufficiently large part of the pattern is presented. *Sufficiently large* is of course an imprecise statement, but correct pattern restoration is dependent on all stored associations. This operation is commonly called best match search in terms of pattern recognition and is well suited to implement classification tasks.

In the heteroassociative case, the matrix M that implements the desired mapping can be determined by calculating the inverse or the pseudo inverse of matrix I ($\hat{M} = OI^+$). Limiting considerations to the most simple type of memory with binary elements the matrix M is obtained by the following training scheme:

$$m_{ji} = \bigvee_{k=0}^{N-1} I_i^k \dot{O}_j^k \tag{1}$$

Palm (1980) has shown that the storage capacity or the number of patterns to be stored is maximum for sparse coded input/output patterns.

In the applications investigated in the context of this work the best match search mode of the network is employed, i.e., matrix M is columnwise constituted by a set of reference vectors, which are selected as representatives of the problem instance. No training is employed to generate matrix M, but vectors can of course be selected from an initial larger set using iterative selection procedures (Poechmueller and Glesner 1991). This mode of operation is related to the LVQ algorithm of Kohonen (1989) and the Nestor RCE model.

The best match is determined according to an association criterion, which can be implemented by a distance measure or metric like the Euclidean distance. The degree of similarity is expressed, depending on the chosen metric as minimum or maximum response to the input stimulus. The following metrics are commonly employed in pattern recognition tasks, **x** and **w** denote in neural network terminology input vector and matrix of synaptic weights, respectively: Euclidean distance and the related City Block distance:

$$E(x,w) = \|x-w\| = \sum_{i=1}^{n}(x_i - w_i)^2 \quad ; \quad CB(x,w) = \sum_{i=1}^{n}|x_i - w_i| \tag{2}$$

which is in most applications functionally equivalent to the Euclidean distance, and is much easier to implement. Hamming distance:

$$H(x,w) = \sum_{i=1}^{n} x_i \oplus w_i \tag{3}$$

Dot product (counting of correlating active elements for binary values):

$$C = \sum_{i=1}^{n} x_i \cdot w_i \quad ; \quad C_b(x,w) = \sum_{i=1}^{n} x_i \wedge w_i \tag{4}$$

This is the metric available in the basic binary network and its current hardware implementation. Other metrics have to be emulated by resorting to coding artifices. Table 1 illustrates how Hamming distance or City Block distance can be achieved by appropriate encoding.

After distance or similarity evaluation, finding the best match requires a search or competition process. This process can be conducted in two different ways. One way is to look for the minimum (or maximum) value calculated during comparison, irrespective to the absolute amount of similarity. The other way is to determine a fixed level (threshold) of similarity which will classify a stored pattern as a match to a presented pattern. These two approaches can be integrated by defining an initial threshold value equal to the value achieved by comparing two identical patterns. Successive degression of the threshold provides one or as many matching vectors out of the set of stored vectors as required. Thus, a n-nearest neighbor classifier can be realised. The following

Table 1. Feasibility of various metrics by encoding

Target Metric	Implemented by	Bits	Bits	Original	Encoded
Hamming	Dot product	n	2n	000	010101
				001	010110
				...	...
				111	101010
City Block	Dot product	n	$2(2^n - 1)$	000	00000001111111
				001	00000011111110
				...	...
				111	11111110000000
City Block	Hamming	n	$2^n - 1$	000	0000000
				001	0000001
				...	...
				111	1111111

equation mirrors this idea for dot product and binary values, whereby s is a step function and $C_b(x, w_j) \in \{0, 1\}$

$$C_b(x, w_j) = s\left(\left(\sum_{i=1}^{n} x_i \wedge w_{ij}\right) - \theta\right) \quad (5)$$

Crucial to processing speed is the efficient and parallel implementation of this competition mechanism.

CHIP IMPLEMENTATIONS

Based on the work of Steinbuch Palm (1980, 1982) proposed a simple binary associative memory for application in speech recognition using sparsely coded vectors. According to the special requirements of the application and data representation, VLSI-implementation was done by Darmstadt University of Technology (THD) respective to the constraints costs, design time and a network scalability in the order of several thousand neurons. The implementation features bitserial data processing, standard RAM chips for weight storage and standard cell association units. These units are called BACCHUS chips and host 32 neurons each (Huch et al. 1990, Poechmueller and Glesner 1990, Poechmueller 1991) Core of the neuron is an 8-bit counter that integrates similarity evaluation and competition. This is accomplished by preloading the counter with the value of maximum similarity as initial threshold. During evaluation the counter will be decremented for each correlating element of input and weight vector. A perfect match will result in a zero transition of the counter, which is indicated by an active neuron output. Simultaneously decrementing all counters will provide more and more hits, until the desired amount of nearest neighbors is found. Three implementations of this basic concept were carried out, called BACCHUS-I, BACCHUS-II and BACCHUS-III. These implementations differ in pin-count, gate-count, and in hit location identification mechanism. In the first design, a ripple mechanism was employed to get hit identification, in the latest design this was replaced by a more efficient priority encoding mechanism.

Table 2 gives the main data of the three chips as well as the estimate of a new design that will be desribed in a following section. BACCHUS-I and III chips were used in systems described in the following sections.

SYSTEM IMPLEMENTATIONS

Three implementations of associative memory systems utilizing BACCHUS-I or BACCHUS-III

Table 2. Design parameters of associative memory chips

Chip	Clock	Area	Gate equiv.	Transistors	Pins
BACCHUS-I	10 MHz	$77.32 mm^2$	5585	22340	108
BACCHUS-II	10 MHz	$47.58 mm^2$	5767	23068	68
BACCHUS-III	10 MHz	$39.7 mm^2$	6523	26159	68
ARAMYS-II	20 MHZ	/	≈ 15000	/	68

chip were undertaken, one by Palm (1991) and two by König et. al. (1991). The implementation of Palm called PAN-IV is by far the biggest of the three and employs the associative memory for speech recognition with sparsely coded binary vectors, whereas the other two implementations target on pattern recognition applications using arbitrarily coded binary or integer vectors.

The Bacchus-I System

Four BACCHUS-I chips are integrated with their corresponding memory on module boards. These module boards are operated by a controller attached to a standard PC as host computer. Thus, the associative network is available to the host as a neural or associative coprocessor. The controller provides a very low level access to the network, so that the host computer is in charge of all the main functions, which is of course detrimental to processing speed. A system comprising four module boards with a weight vector length of 2kbit was set up and employed for compression of video phone images by vector quantization (König and Glesner 1991, König et al. 1991). This prototype system is illustrated in Figure 1.

Figure 1. Prototype of binary associative memory system

The Aramys-I System

The ARAMYS-I System (Autonomous Real time Associative MemorY System) features a more sophisticated controller utilizing a signalprocessor, and the BACCHUS-III chip. A priority encoding scheme is implemented on chip, board, and controller level of this new system, thus providing higher processing speed and system performance by efficient and selective hit evaluation. The reduced carrier size allows the integration of eight memory chips on a single module board. Weight vector memory is expanded from 2kbit to 32kbit, supporting processing of realistic size data vectors, or of several different vectors stacked on top of each other for time-multiplexed use of neurons. This supports the mapping of several logical networks onto a single physical network, and the building of hierarchies in associative processing. In the preceding BACCHUS-I system implementation the host was required to exert constant supervision of the network during neural data processing. In this advanced approach, the host will be relieved from this cumbersome task and will be free to work concurrently during neural data processing. The controller itself is an autonomous processing unit, featuring an TMS 320C25 signalprocessor. Figure 2 illustrates the block diagram of the controller. Host interfacing and communication is accomplished employing

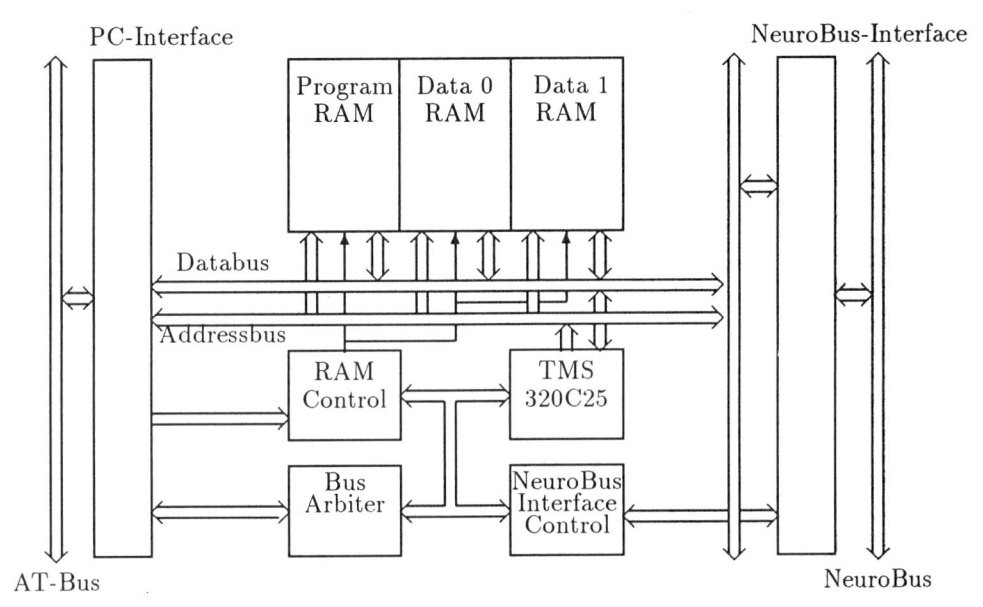

Figure 2. ARAMYS-I Network Controller

RAM memory of the TMS, that is accessed via the arbiter by DMA access. Input vectors are fed to the system by writing them into this RAM area, in addition to some control information contained in communication structures(, e.g., vector size, return data etc ...). The TMS processes this vector, returns the desired results, and indicates completion to the host via the arbiter.

Further additional memory space of the TMS can be used to store arbitrary information in a table associated with the key or index returned by hit evaluation. Thus, hetero-associative processing is feasible. Interpolating associative memory function can be achieved by averaging the associated information of n-nearest neighbors to a given input vector.

As formatting of input data, preprocessing requirement, form and size of hetero-associative information may vary from application to application only the most basic network functions are implemented by standard boot routines, which are downloaded during initialization.
Optionally, more sophisticated routines can be downloaded from the host in TMS RAM dedicated to program code, thus replacing the simple standard routines.

A system implementation was completed with network controller and one module board, comprising eight BACCHUS-III chips, corresponding weight memory, and control circuitry. The ARAMYS-I system was tested and also employed for the vector quantization application, displaying more than twice the processing speed of the preceding system.

The system was further devised and used as a flexible testbed and environment for advanced associative chips. A second board, comprising just four chips and an XILINX FPGA 3090 was set up for this task as development board. Figure 3 shows the ARAMYS-I development board.

Target applications of this development system are in the field of visual quality inspection, turbine control, and image coding.

ASSESSMENT OF IMPLEMENTATIONS

This section is dedicated to a brief assessment of the presented implementations respective to their applicability and efficiency. The current architecture and application is optimized for processing sparsely coded vectors. Input vectors provide addresses to the memory matrix corresponding to the indices of active elements in the input vector. Active output elements are identified and their location is encoded as an address, which is transferred to the controller and thus to the host.

Counter size of 8 bit is also dimensioned for the sparsely coded case. Assuming arbitrary binary vectors, for instance obtained by thresholding imageblocks in OCR, 8 bit vectors will in many cases not be sufficient to compare vectors of realistic size. This argument holds even more if coding schemes given in Table 1 are applied. These will result in vectors with 50% elements active and in vector lengths l growing exponentially obeying $l = 2 \cdot (2^n - 1)$. Counters of 8 bit will certainly not suffice and larger counters in the order of 16 or 20 bit are mandatory.

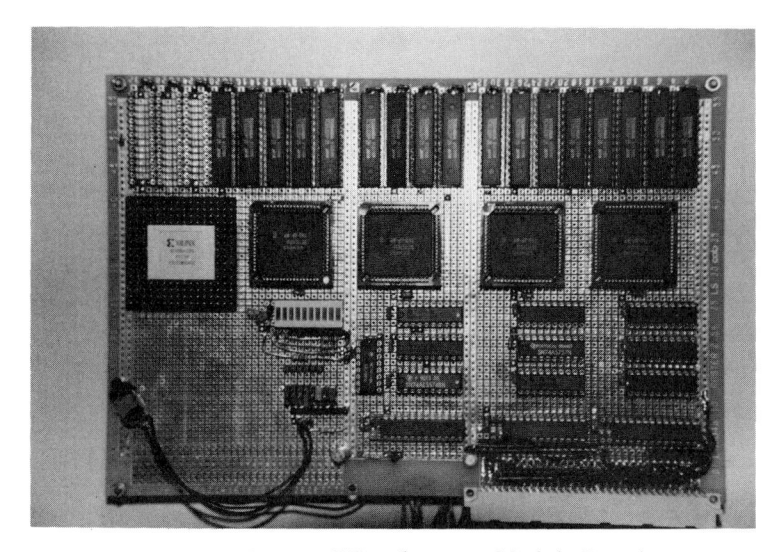

Figure 3. Aramys-I Development Module Board

Processing is parallel for all neurons, but serial for vector elements. So principally system efficiency is independent of the number of reference vectors, but is determined by vector length. Thus, the City Block distance should be implemented in a more efficient way.

. Individual access to counters is not provided in the current implementation. This would be very usefull to accomplish weighted interpolation for heteroassociative recall.

The best match mode requires the successive degression of the threshold until the desired set of best match vector (n-nearest neighbors) are found. In the current implementation all best match vectors become active again after each threshold lowering and thus have to be processed. Introduction of an optional inhibit mechanism will block all neurons after they were active and evaluated once, so that only the most recent ones have to be processed and added to an external list on the controller. Further, saturation technique excluding an overflowing counter from competition will prevent faulty evaluation.

The final issues concerns the overall system architecture. The controller still constitutes the bottleneck in the processing scheme. The bus width of the neuro bus is a multiple of the TMS bus width. Thus, it can be served only in several command cycles of the TMS, which take several external clock cycles. The TMS is clocked with four times the rate of the network, but still the network is idle for approx. 80% of its potential processing time. Therefore the design of a special inherently parallel controller is mandatory to fully exploit the capability of the associative memory.

ARAMYS-II ARCHITECTURE

An augmented architecture, denoted as ARAMYS-II has been developed. It is based on a new algorithm that allows the efficient calculation of the City Block distance while retaining advantage and structure of the simple counter based approach. An associative memory chip is under design that implements Binary Dot Product, Hamming distance, and City Block distance. Also some concepts addressed during assessment are implemented in the current design. It features 16 bit counters, inhibit mechanism, and overflow prevention. The block diagram is given in Figure 4. The interfacing circuitry of the ARAMYS-II chip is compatible with its ancestor BACCHUS-III,

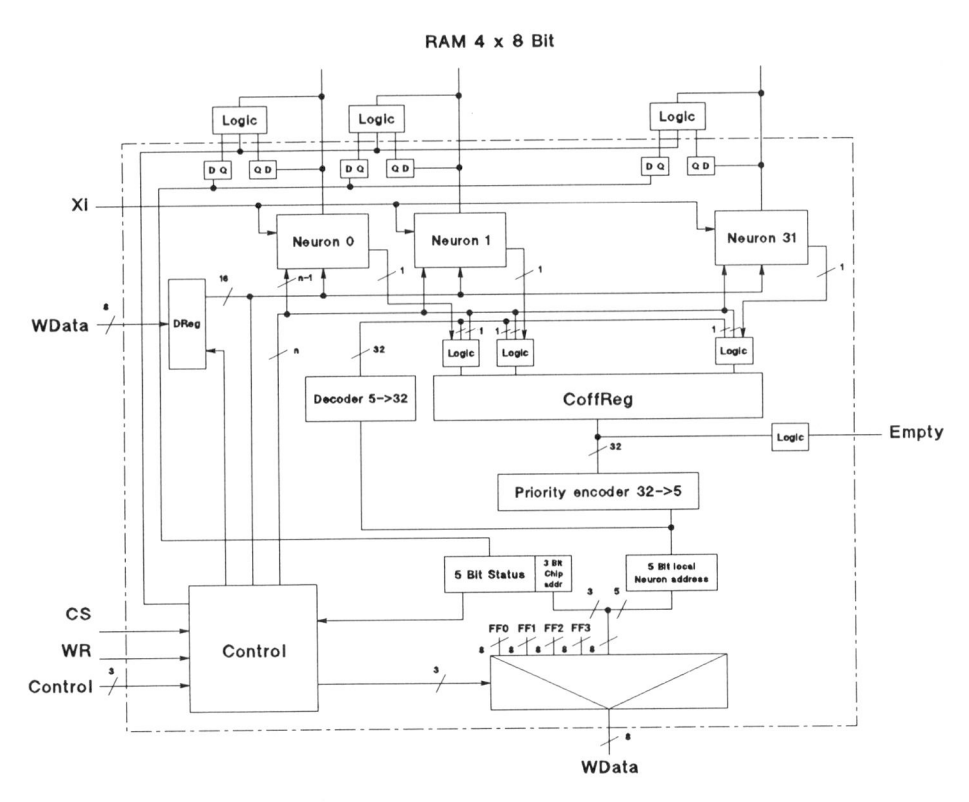

Figure 4. ARAMYS-II Chip Architecture

thus achieving compatibility to the ARAMYS-I set-up. Core of the new design is the dedicated comparison unit shown in Figure 5. This unit essentially consists of a 16 bit up/down counter, that features 8 distinct counting inputs with a weighting of $2^0 - 2^7$. Countersize and data width are not restricted to the chosen values. The counter, two latches, some random logic, and a global shift register allow comparison of two 8 bit values in 9 clock cycles. Using binary dot product the same operation takes 510 clock cycles.
The algorithm integrates subtraction of w_{ij} and x_i, accumulation of absolute differences for all

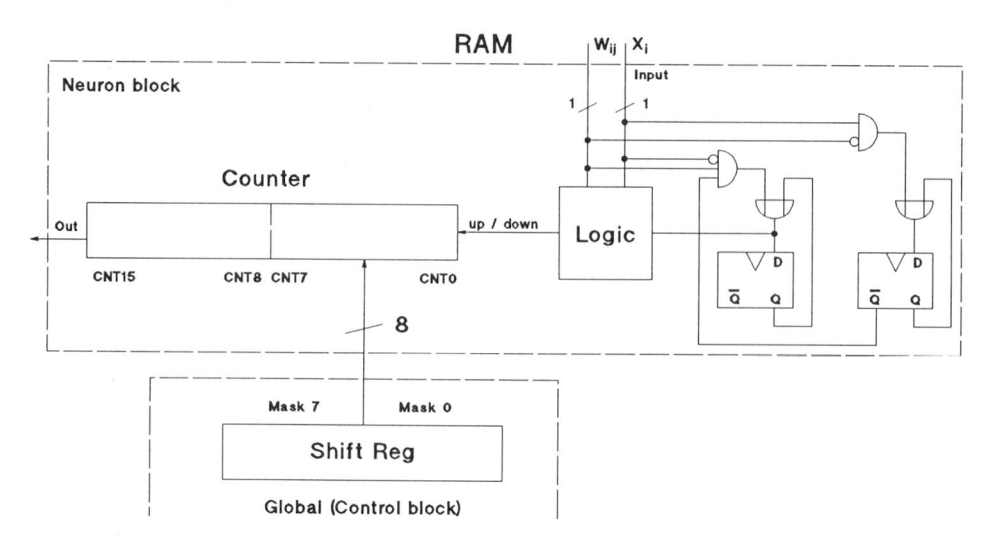

Figure 5. ARAMYS-II Comparison Unit

vector elements, and competition of all neurons. Subtraction is achieved by a stepwise approach, using up counts and down counts. Counting takes place from the MSB to the LSB by means of the shift register which successively activates the counter input corresponding to the binary weight of the current bit position. The main idea is illustrated in Table 3. The absolute value

Table 3. Stepwise subtraction by counting

x_i	w_{ij}	Count action	Reversed action
1	1	No count	No count
1	0	Up count 2^i	Down count 2^i
0	1	Down count 2^i	Up count 2^i
0	0	No count	No count

of the difference can be generated by a little artifice. Monitoring the first pair $x_i \neq w_{ij}$, the counting action is as given in the third column of Table 3 if $x_i > w_{ij}$ holds for the first pair, else counting is reversed as given in the fourth column. Accumulation takes place automatically in the counters. Competition is achieved by decrementing all counters concurrently, until one or several zero transitions are reached. This algorithm can be implemented by a very small circuit ideally suited for VLSI-implementation. Design of the ARAMYS-II chip is close to its completion. Currently, timing simulation and optimization of the individual functional blocks is carried out. The design effort takes place within the scope of Eurochip ES2 manufacturing, using the SOLO 2030 standard-cell design tool. To provide a more flexibel and less costly test platform for new implementations the ARAMYS-I development board was devised. The ARAMYS-II chip was redesigned as a FPGA implementation and verified *in situ*. That way new algorithms can be tested exploiting an existing environment. Objective of this development is the completion of the ARAMYS-II system. This system will retain all the advantages of its predecessors, while additionally providing efficient integer processing.

APPLICATIONS

Some applications make use of the simple operation mode for correlating binary data, e.g., OCR.

But most applications work on integer or pixel data, e.g. image coding and analysis. Three classification tasks gave incentive to the development of the ARAMYS-II architecture.

Image coding by vector quantization using trained Kohonen Maps for the coding process requires the comparison of an image block with the whole codebook in real-time. A single codebook can be stored in the weight matrix (König and Glesner 1991) or several codebooks of various sizes can share the available hardware in a more sophisticated coding approach (König, Reinke and Glesner 1992).

Operating point estimation in turbojet compressor units requires the comparison of the current operating point with a field of operating points to determine proximity to the stall region in real-time (König et al. 1992).

In visual industrial quality control a special histogram based technique (Korn 1988, Korn 1989) can be combined with an associative memory. The gradient of an image is determined and split in two subimages of magnitude and direction. Histograms are calculated for regions of interest. A large database can be created and the associative memory provides by correlation the most similar pattern. This has been verified by simulations. Rotational invariance can be achieved by extending the scheme to correlation of an input vector with all rotated patterns stored in the weight matrix. Thus, the ARAMYS-II system allows the parallel computation of the most similar pattern for an arbitrary rotation of the input vector.

CONCLUSION AND FUTURE WORK

This paper gave a survey on a simple binary associative network and its application and implementation. Interest focused on its application to classification problems, where this network type can be used as n-nearest neighbor classifier. An extension of the basic form was introduced that bears strong relation to LVQ and SOFM as well as to the Nestor RCE approach. Status on a new design effort on chip and system level was given.

Future work will concentrate on the definition of a more powerful architecture. A dedicated parallel controller chip would greatly increase processing bandwidth. Furthermore, a Full-Custom design approach including integration of RAM memory on-chip could be considered in conjunction with MCM (Multi Chip Module) technique. Implementation of an on-chip learning capability based on iterative selection or programming algorithms, or incremental learning algorithms will be considered.

ACKNOWLEDGEMENTS

This work was supported in part by the national research project "Informationsverarbeitung in neuronaler Architektur" (No. ITR 8800) from the Federal Ministry of Research and Technology (BMFT).

REFERENCES

Huch, M., Poechmueller, W., Glesner, M., "BACCHUS: A VLSI Architecture for a Large Binary Associative Memory", in *Proc. of the International Neural Network Conference 90*, Paris, 1990

König, A., Poechmueller, W., Glesner, M., "A Flexible Neural Network Implemented as a Neural Coprocessor to a von Neumann Architecture", in *Proc. 2nd International Conference on Microelectronics for Neural Networks NeuroMicro 91*, pp. 455-460, 1991

König, A., Glesner, M., "An Approach to the Application of Dedicated Neural Network Hardware for Real Time Image Compression", in *Proc. 1rst International Conference on Artificial Neural Networks ICANN'91*, Vol.II, pp. 1345-1348, 1991

König, A., Reinke, M., Glesner, M., "A Fully Neural Approach to Image Segmentation and Image Coding", *Proc. International Joint Conference on Neural Networks IJCNN'92 Beijing*, Vol. I, pp. 631-635, 1992

König, A., Windirsch, P., Glesner, M., Wang, H., Hennecke, D. K., "An Approach to the Application of Neural Networks for Real-Time Operating Point Estimation in Turbojet Compressor Units", *Proc. International Joint Conference on Neural Networks IJCNN'92 Beijing*, Vol. II, pp. 64-69, 1992

Kohonen, T., *Self-Organization and Associative Memory*, Third Edition, Springer 1989

Korn, A.,"Toward a symbolic representation of intensity changes images", *IEEE Transactions on Pattern Analysis and Machine Intelligence*, Vol. PAMI-10, pp. 610-625, 1988

Korn, A., "Zur Beschreibung und Erkennung von Bildstrukturen durch Auswertung von Richtungshistogrammen", Fortschrittsberichte, Band 114, Reihe 10, W. Schwerdtmann (editor), VDI-Verlag Duesseldorf, pp. 44-51, 1989

Palm, G., "On Associative Memory", *Biol. Cybernetics*, pp. 19-31, 1980

Palm, G., *Neural Assemblies, an Alternative Approach to Artificial Intelligence*, Springer, 1982

Palm, G., Palm, M., "Parallel Associative Networks: The PAN-System and the BACCHUS-Chip", in *Proc. 2nd International Conference on Microelectronics for Neural Networks NeuroMicro 91*, pp. 411-416, 1991

Poechmueller, W., Glesner, M. "A Cascadable VLSI Architecture for the Realization of Large Binary Associative Networks",in *Proc. International Workshop on VLSI for Artificial Intelligence and Neural Networks*, pp. 265-274, 1990

Poechmueller, W., Glesner, M., "Supervised Classification with a Binary Associative Memory", *24th Annual Hawai International Conference on System Sciences HICSS'24*, Vol. I, pp. 253-259, Koloa, Hawai, January 8-11th, 1991

Poechmueller, W., Glesner, M., "Informationsverarbeitung in neuronaler Architektur (INA ITR8800M/O"- Entwurf eines Chips für den Aufbau eines schnellen neuronalen Netzwerks in Hardware, Technical Report, 1991

Waldschmidt, K., "Associative processors and memories: Overview and Current Status", Comp-Euro 87, VLSI and Computers, pp. 19-20, 1987

A DATAFLOW APPROACH FOR NEURAL NETWORKS

Thomas F. Ryan, José G. Delgado-Frias, Stamatis Vassiliadis, Gerald G. Pechanek and Douglas M. Green

INTRODUCTION

Artificial neural networks have been introduced as a novel computing paradigm (Kohenen 1988). Processing (or retrieving) in neural networks requires a collective interaction of a number of neurons. Output of neurons are computed based on the inputs from other neurons, weights associated with such inputs, and a non-linear activation function. Specifically, most artificial neurons follow a mathematical model that is expressed as:

$$Y_i(t+1) = F\left(\sum_{j=1}^{N} W_{ij} Y_j(t)\right) \quad (1)$$

where W_{ij} is the weight, $Y_j(t)$ is the neuron input, N is the number of neurons connected to neuron i, and F is a non linear function which is usually a sigmoid (Hopfield 1984) as shown below.

$$F(x) = \frac{1}{1+e^{-x}} \quad (2)$$

Artificial neural networks are parallel interconnected networks of simple elements (Kohenen 1988). The implementation of artificial neural networks requires parallel hardware in order to achieve high performance which is needed for a number of applications. In general, there are two approaches to such hardware implementations: specialized machines (Blayo and Hurat 1989, Duranton, Gobert and Mauduit 1989, Kung and Hwang 1989, Weinfeld 1989, Hammerstrom 1991, Vassiliadis, Pechanek and Delgado-Frias 1991, Pechanek, Vassiliadis and Delgado-Frias 1992, and Delgado-Frias, et.al. 1993) or more general purpose parallel machines (Watson and Gurd 1982, McGraw 1989, Delgado-Frias, Ahmed and Payne 1991). The former approach usually provides high performance. This specialized hardware approach may offer little flexibility to implement different types of neural networks, may not be efficient for sparsely connected models, or may not have support for learning algorithms. On the other hand, general purpose computers offer much greater flexibility which allows one to implement a large variety of neural networks and learning algorithms. This approach may suffer from a degradation in performance. In this paper, we explore dataflow computing as an approach for implementing neurocomputing.

In a multiprocessor system, a computational task is collectively executed by a number of processors. There are different types of parallel systems and each of which has advantages and disadvantages depending on the problem that needs to be solved. Among these are dataflow computers. Dataflow computers differ from other parallel architectures in that there is no instruction pointer in a dataflow machine. The dataflow approach possesses the following characteristics:
- The dataflow graph is the machine form of the program.
- MIMD (Multiple Instruction Multiple Data) architecture executes the graph.
- Task synchronization is achieved through of data (or message) passing.

In dataflow computers, the data is transferred among dataflow nodes by means of tokens. These tokens are usually tagged to support multiple sets of inputs per node. Figure 1 shows a dataflow node with three inputs which have tagged tokens denoted as bullets. Tagged tokens need to be matched in order to have the proper set of inputs for the node. Once a complete set of inputs is matched, the node absorbs the inputs and computes the node's function on these inputs. This matching and computation process is called firing.

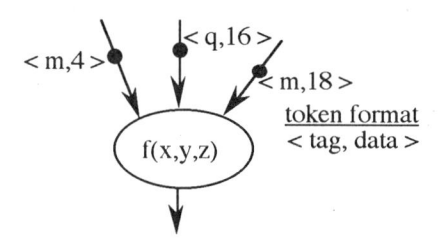

Figure 1. Three input dataflow node

In a dataflow graph, a number of nodes may have a complete set of data on which to operate; these nodes are referred as fireable nodes. These nodes express the maximum parallelism that can be obtained using this computing approach. When a node is fired (or executed) a result is produced in a form of a token. Such tokens are sent to the appropriate node(s) as input.

In this paper, we present a novel approach where dataflow computing is used as the methodology for implementing neurocomputing. In our study, it has been observed that dataflow computing approach offers a number of advantages. These advantages include:
- *Flexibility*. The number of neurons as well as the interconnections among them can be arbitrary. Dataflow computing provides no practical restrictions on the need for flexibility.
- *Learning approaches in same machine*. The same dataflow approach can be used in the retrieval and learning phases of the neural network. This in turn facilitates effective use hardware resources.
- *Maximum parallelism*. All of the computational resources of a dataflow machine could be potentially used when computing neural networks.

This paper has been organized as follows. Two dataflow architectures that have different constraints in the construction of the dataflow graphs are described. Then the mapping of a multi-layered neural network onto a dataflow graph is presented. Lastly, we include simulations that are used to evaluate the potential of the proposed approach.

DATAFLOW ARCHITECTURES

In dataflow computers, the nodes do not perform the matching of tokens themselves. Instead there is a matching unit that stores the tokens and finds matches among them. When a match is found, the matched tokens are moved from storage to the correct node for firing. As there are different dataflow architectures, there are different approaches to the design of the matching unit. Two architectures which are of primary interest to this work are the Manchester dataflow computer (Watson and Gurd 1982) and the SUNY dataflow computer (Delgado-Frias, Ahmed and Payne 1991)

Manchester Dataflow Architecture

A prototype of the Manchester dataflow machine has been implemented to examine the feasibility of a tagged token machine. The matching of tagged tokens is realized in the match store unit. The match store unit implements the firing rules discussed below and uses an associative search and match process to allow for efficient matching and storing of tokens. However due to the limitation of the match store unit a maximum of 2 inputs are allowed for any individual node. As tokens arrive at the match store unit, the unit determines if the token is unary or binary. If the token does not need to be matched (unary), then it is sent to the correct node for firing. If the node needs a match, then two scenarios exist; either the token that matches the incoming token is stored or the token is the first of the pair. The match unit takes the tag and sends it to a hardware hashing function which gives the location of eight different memory locations with the same hash address. These locations are then searched in parallel for a match (Payne and Delgado-Frias 1991). If no match exists, then the token is stored in one of the empty storage locations. If there are no empty storage locations at the hash address, then the token is stored in an overflow location and the address of the token is recorded. If a match is found, then the token is removed from storage and it is sent along with the incoming token to the correct node for firing. The matching unit works in this manner due to the high cost of associative memory at the time this specific machine was built (Watson and Gurd 1982).

At the present time, only one computing ring has been built. The processing unit is a collection of identical processors (Watson and Gurd 1982). One match store unit is responsible for matching tokens from more than one processor.

SUNY Dataflow Architecture

The SUNY dataflow architecture consists of 32 homogeneous processors with a 5 cube interconnection network (Delgado-Frias, Ahmed and Payne 1991). Each of the 32 nodes (Processing Units or PUs) in the hypercube consists of a Processing Element (PE), Match Processing Unit (MPU), and a router. Both the PE and MPU have storage area. The PE, MPU and router operate concurrently thus allowing not only parallelism with the 32 processors but also parallelism within each of the 32 PUs. The MPU is capable of matching n-tuple input dataflow nodes instead of the maximum of 2 found in the Manchester machine. The MPU examines the fields within the token's tag to determine how many tokens are needed for a complete match. Each new entry is stored in a content addressable memory (CAM). An entry is made in the CAM for that token and the data section is then stored in the MPU RAM. The MPU will allocate a block of memory for all of the tokens to be matched. Subsequent tokens will be stored in the RAM's allocated memory. The CAM entry stores the header information and a pointer to the block of RAM. One of the fields in the header is a counter (number of needed tokens). For every token that arrives, the counter is decremented and when the counter reaches 0, all of the tokens are removed from the RAM and sent to the PE's memory for firing.

The PE operates sequentially and fires nodes as the tokens arrive in its token pool. After the node has fired, the result along with new header information is sent to the router. The router then determines where to send the new token. The token can be sent to any of the 32 processors. The router has 6 input and 6 output queues (1 for each output line plus its own PU).

NEURAL NETWORKS MAPPED ONTO A DATAFLOW GRAPH

In order to show the potential of dataflow as a means to implement neural networks, we have selected a multilayered neural network and the back propagation learning algorithm as examples.

In a multilayered neural network, there are three types of layers: input layer, hidden layer and output layer. Figure 2 shows a multilayered neural network. In the forward operation this network receives an input pattern. The neurons in the input layer process the input and produce an output vector which is processed by the hidden layer. The output vector from the hidden layer, is processed and the output from the neural network is produced by the output layer. In this process, a neuron needs all its inputs in order to produce an output and the outputs from the neurons form the inputs into the next layer. This leads to a model of neural network computing as a dataflow graph. Each neuron is modeled as a dataflow graph as shown in Figure 3. This figure shows the model that is applicable to the Manchester machine. Each dataflow node has a maximum of two inputs.

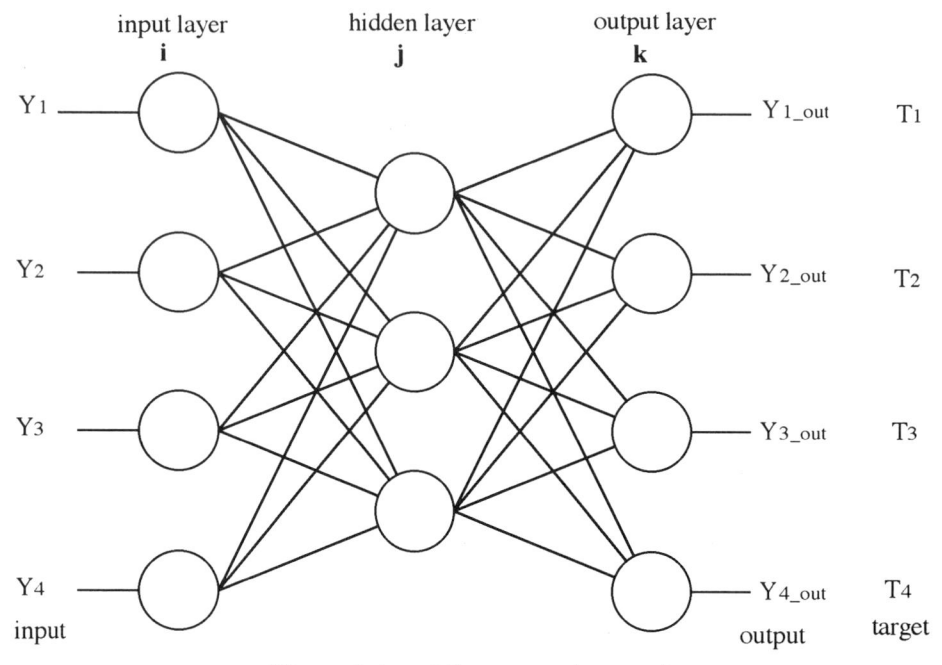

Figure 2 A multilayer neural network

The back propagation learning algorithm requires two vectors, which are the input and the target (or teacher) vectors. The basic back propagation algorithm is expressed mathematically by the following formulas (Wasserman 1989).

$$\delta_{p,k} = out_{p,k}(1 - out_{p,k})(target_p - out_{p,k}) \quad (3)$$

$$\Delta w_{pq,k} = \eta \delta_{q,k} out_{p,j} \quad (4)$$

$$w_{pq,k}(n+1) = \Delta w_{pq,k} + w_{pq,k}(n) \quad (5)$$

where p is a neuron in the hidden layer, q is a neuron in the output layer, k is the output layer, $w_{pq,k}$ is the weight from neuron p to neuron q, and η is the training rate.

Each neuron in the output layer passes its δ value to all neurons that connect to it. The δ value is multiplied by the weight between the two neurons and the sum of all these multiplications is calculated. A new δ value is determined by the following equation.

$$\delta_{p,j} = OUT_{p,j}(1-OUT_{p,j})(\sum_q \delta_{q,k} w_{pq,k}) \quad (6)$$

where $OUT_{p,j}$ is the value of OUT for neuron p in layer j (j≠k). Then using equations 4 and 5 the neuron can update its weights. This process continues until all neurons have updated their weights (Wasserman 1989).

To implement back propagation in dataflow, a graph should be constructed. This graph is very similar to the forward graph previously discussed. In the back propagation algorithm, the graph has the inputs from the target vector. Figure 4 shows a graph for back propagation at the output layer of neuron q. Since the forward and backward operations require similar graphs, we decided to use a large portion of forward graph for the back propagation algorithm. This approach reduces not only the graph complexity, but also the overhead due to the transferring of data.

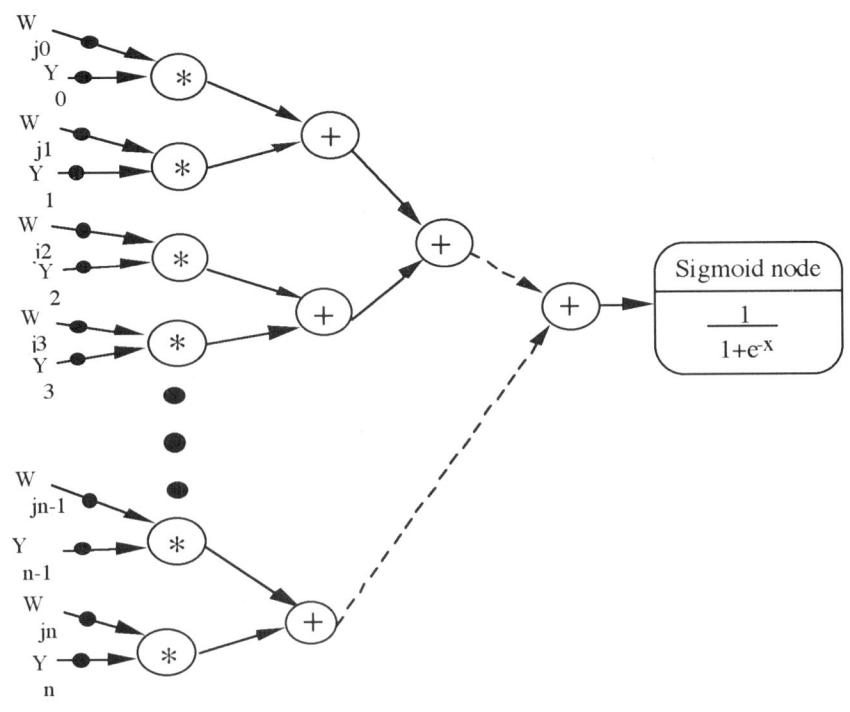

Figure 3 Dataflow graph of a neuron (forward mode)

SIMULATOR

As shown above, neural networks are visualized as graphs and dataflow computers rely almost exclusively on these graphs. The multi-layer network simulator can accept any number of neurons and any number of layers greater than 1. The simulator uses the back propagation algorithm for learning and the simulator is modeled after the Manchester dataflow architecture. Given the number of neurons in each layer, the number of layers and the total number of neurons, the simulator places the neural network onto a dataflow graph.

Dynamic building of simulator

The first function of the simulator is mapping the neural network into a dataflow graph. This is accomplished by building an adder tree for each layer. The adder tree is responsible for calculating $\sum w_{ji} y_i$ where w_{ji} is the weight of the synapse from neuron i to neuron j and y_i is the input to neuron j from neuron i. Since the Manchester-like dataflow architecture is being simulated, there are at most two inputs to any node. Thus the adder tree is a balanced binary tree and it is created with recursion. The leafs of the adder tree are multiplication nodes, the rest of the nodes are all addition nodes. A picture of the dataflow graph for the

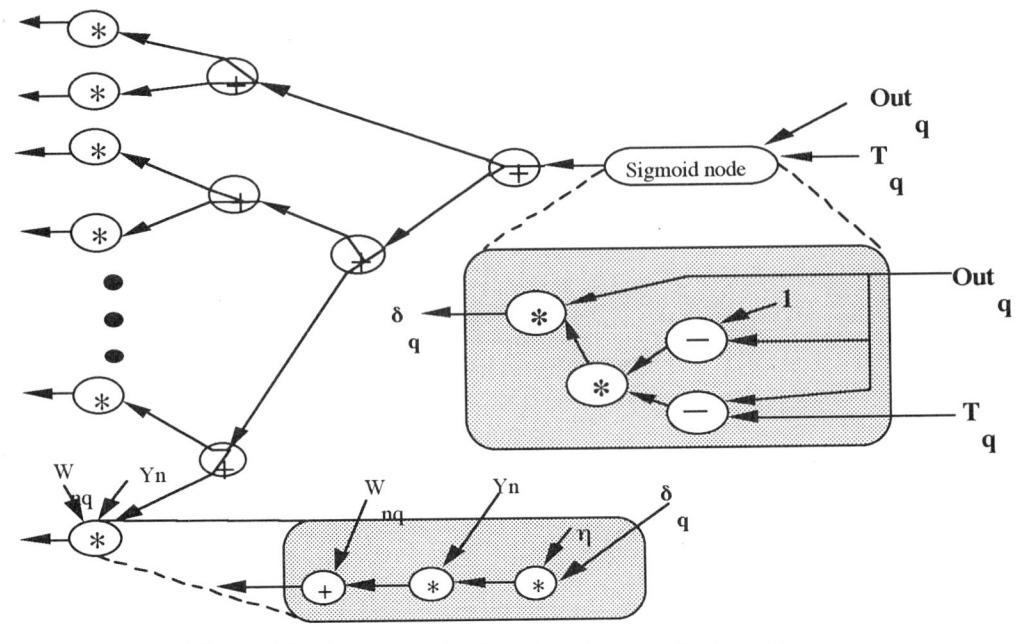

Figure 4 Back propagation learning algorithm in the adder tree

adder tree is shown in Figure 3. The adder trees are calculated for each neuron in the network.

Next, the activation function, which is usually a threshold or a sigmoid function, needs to be calculated. Since the activation function is known beforehand, this dataflow graph is copied from a library of available dataflow functions. The output of the adder tree is the input to the activation function data flow graph. The advantage of having a library of activation function graphs is that there is more than one possible activation function. Thus the only part of the code that needs to be changed is the activation function code. Also since the activation functions have one input and one output, the ports for the functions can be calculated without regard to the activation function itself.

After the activation function dataflow graph is attached to the adder tree, another set of trees is created; they are referred as communicating trees. These trees are responsible for taking the output of the neurons and distributing them to the next layer. The outputs are placed on one of the input ports for each multiplication node. Again, like the adder tree, the d-link tree is a balanced binary tree. In the adder tree, information travels from the leaves to the root. In the d-link tree, data travels from the root to the leaves. The communicating tree is created for every node except for the nodes in the output layer.

The last function that needs to be added to the data flow graph is the learning function. Like the activation function, the learning function is known and therefore is a statically programmed data flow graph. The training vector is only found on the neurons in the output layer. The output of the output layer is the input to the learning function. With all these functions in place, the multi-layer neural network is now realized.

Retrieval

The network runs in two directions -forward and backward. During forward propagation, the network works as described below. For each neuron in each layer, the Σwy and activation function is calculated and then the output of each neuron becomes the inputs for the neurons in the next layer.

After the data has made its way to the output layer, the errors are computed for the output layer in the learning subgraph. The results are then written to the output of the learning subgraph back to the main graph. When there are no more changes to the main graph, a flag is set that changes the flow of the graph. Instead of going from input to output, the data will move from output to input.

Backward Propagation

The responsibilities of the nodes change when the data flows in the backward direction. The learning subgraph is completely bypassed and acts as a data line. Information on its output line is passed directly to its input line. The sigmoid on the output layer works exactly like the learning subgraph. The data on its output line is passed directly to its input line. The plus nodes in the adder tree no longer function as addition but as communicating links. The information from the add trees is passed to multiplication nodes found on the leaves of the add tree. The multiplication nodes have a drastically different function in the backward mode.

In the backward mode, the multiplication node is responsible for modifying the corresponding weight in the weight vector. The multiplication node is replaced by a static graph which calculates the function below.

For all activation functions, except for the one in the output layer, the new function is to collect and sum the data that arrives on its output line. The number of expected inputs to the sigmoid is known; thus when all the values have arrived and been summed together, the results are passed to the add tree. This cycle continues until the values reach the input layer. Here the multiply nodes do not pass their new output on to the previous layer, since there is no previous layer. Again, when there are no more changes, the simulator reverses itself and runs in the forward mode once again. This cycle continues until the network is trained.

Reason for design

The network is designed in this fashion for a number of reasons. First, since it simulates a backward propagation learning network, pipelining and new tags are not utilized, thus the input lines in the forward pass can be used as output lines in the backward pass. Second, the forward graph, with all of its connections, already exists. Thus there is no need to create a second graph for the purpose of modifying the weights. The data can simply be passed back through the existing data flow graph. The only change is that some of the nodes in the graph take on a different function in the backward pass. This also saves memory since a second fully configured graph does not have to be built.

Using the backward propagation learning method has caused some problems. The simulator works correctly in both the forward and backward passes. However, the backward propagation algorithm is a gradient decent algorithm and this function is susceptible to finding local minima. Once a local minima is found, the function gets trapped and the correct results may be unobtainable.

CONCLUDING REMARKS

In this paper a novel approach for artificial neural network computing has been described. The dataflow computing approach provides advantages on flexibility, more efficient implementations of the sparsely connected neuron, and implementable learning algorithms in the same machine. This approach is being investigated on other types of neural networks; such as stochastic neural networks where it exhibits a good performance.

It was shown that the dataflow graphs for multilayer networks and back propagation learning can be combined. This approach could allow efficient use of the dataflow resources; since the two approaches generate similar computing loads.

Future research in this field will include the use of other neural networks in order to fully understand the capabilities and limitations of this approach as compared with other approaches as special purpose computers.

REFERENCES

Blayo, F. and Hurat, P., "A VLSI Systolic Array Dedicated to Hopfield Neural Networks," in *VLSI for Artificial Intelligence*, J. Delgado-Frias and W. Moore (Eds), Kluwer Academic Publishers, 1989.

Delgado-Frias, J. G., Ahmed, A. and Payne, R., "A Dataflow Architecture for AI," in *VLSI for Artificial Intelligence and Neural Networks*, J. G. Delgado-Frias and W. R. Moore (Eds), New York: Plenum, 1991.

Delgado-Frias, J. G., Vassiliadis, S., Pechanek, G.G., Lin, W., Barber, S., and Ding, H., "A VLSI Pipelined Neuroemulator," in *VLSI for Neural Networks and Artificial Intelligence*, J. G. Delgado-Frias and W. R. Moore (Eds), New York: Plenum, 1993.

Duranton, M., Gobert, J. and Mauduit, N., "A Digital VLSI Module for Neural Networks," *Neural Networks from Models to Applications*, L. Personnas and G. Dreyfus (EDS.) Paris: I.D.S.E.T., 1989.

Hammerstrom, D., "A Highly Parallel Digital Architecture for Neural Network Emulation," in *VLSI for Artificial Intelligence and Neural Networks*, J. G. Delgado-Frias and W. R. Moore (Eds), New York: Plenum, 1991.

Hopfield, J., "Neurons with Graded Response Have Collective Computational Properties Like Those of Two-State Neurons," *Proceedings of the National Academy of Sciences*, pp. 3088-3092, May 1984.

Kohonen, T., "An Introduction to Neural Computing," *Neural Networks*, vol. 1, pp. 3-16, 1988.

Kung, S. Y. and Hwang, J. N., "A Unified Systolic Architecture for Artificial Neural Networks," *Journal of Parallel and Distributed Computing*, Vol. 6, pp. 358-387, 1989.

McGraw, J. R., "Dataflow Computing: System Concepts and Design Strategies," Designing and Programming Modern Computer Systems, vol. 3, S.P. Kartashev and S.I. Kartashev (Eds), New York: Prentice Hall, 1989.

Payne, R. and Delgado-Frias, J. G., "MPU: A N-Tuple Matching Processor," *IEEE Int. Conf. on Computer Design: VLSI in Computers and Processors (ICCD'91)*, October 1991.

Pechanek, G. G., Vassiliadis, S. and Delgado-Frias, J. G., "Digital Neural Emulators Using Tree Accumulation and Communication Structures," *IEEE Transactions on Neural Networks*, Vol. 3, no. 6, pp. 934-950, November 1992.

Vassiliadis, S., Pechanek, G. G. and Delgado-Frias, J. G., "SPIN: A Sequential Pipelined Neurocomputer," *IEEE Int. Conference on Tools for Artificial Intelligence*, pp. 74-81, San Jose, Calif., November 1991.

Wasserman, P. D., *Neural Computing: Theory and Practice*, New York: Van Nostrand Reinhold, 1989.

Watson, I. and Gurd, J., "A Practical Data Flow Computer," *IEEE Computer*, vol. 15, no. 2, pp. 51-57, February 1982.

Weinfeld, M., "A Fully Digital Integrated CMOS Hopfiled Network Including the Learning Algorithm," *VLSI for Artificial Intelligence*, J. Delgado-Frias and W. Moore (Eds), pp. 169-178, Kluwer Academic Publishers, 1989.

A CUSTOM ASSOCIATIVE CHIP USED AS A BUILDING BLOCK FOR A SOFTWARE RECONFIGURABLE MULTI-NETWORK SIMULATOR

Jean-Dominique Gascuel, Eric Delaunay, Lionel Montoliu, Bahram Moobed, and Michel Weinfeld

INTRODUCTION

Artificial neural networks open wide areas of investigation leading to efficient information processing systems, which are expected to behave in a way somehow similar to their biological counterparts. Nevertheless, most efforts have been devoted to finding paradigms, such as multi-layer or dynamic networks, to name a few, and to strive for improvements in their internal algorithms and structures. These efforts have produced very interesting results, but still a fundamental property of neural networks remains, namely the unavoidable residual errors. These errors, resulting from several constraints, altogether theoretical and practical, make the use of a unique neural network for decision processes or pattern recognition safe only on the long run, that is to say when many experiments have been performed on real unknown sets of data, results statistically conforming to the training. Similarly, the complexity of what can be performed by any network is obviously limited: making the learning rules and internal structure more complicated to cope with more difficult tasks, lead to networks which are not only impossible to implement, but even to simulate without excessive difficulty.

Biological and psychological evidence shows that the natural information processing is done in a very large amount of functional units, each one highly specialized in some particular simple task, these units working mostly in parallel, but also with mutual time delays or synchronization. Moreover, from the sensory inputs to the innermost representations in the brain, the information undergoes continual processing through these concurrent units and subsystems (Freeman 1981). Another important property of this processing is that it is error prone, but it includes enough various correcting mechanisms so that it can assure a global acceptable behavior. On a short time scale, these mechanisms are amenable to ordinary feedback, whereas on a long time scale they possibly imply structural modifications. Both can be grouped under the important notion of auto-adaptivity.

Hence, it is only natural to think of artificial neural processing architectures in terms of many low-level networks, each having a rather high connectivity, more loosely interconnected to form some kind of oriented trellis (Weinfeld 1990). These trellis, being adaptive, can modify themselves by local learning in the networks that are their basic constituents, and also in their local or global topology by modification of the paths between these networks. The networks themselves need not be all the same kind, need not be all adaptive, and some constituents can even be only simple non-connectionnist structures, such as for instance majority voters, ordinary memories or logical operators.

As it is well known that the simulation of the simplest networks is already a non trivial task (especially when taking learning into account), it seems difficult to expect simulating the above mentioned structures without a very high amount of efforts and computational resources. A better and promising solution consists in hardwiring the basic networks, letting to software only the care of interconnecting them. Later, whenever complex interesting architectures have been found, it will prove more profitable to hardwire them to obtain the maximum processing power and more compactness.

To put these ideas in practice, we have built a software reconfigurable simulator, based on a custom associative memory chip that we have previously designed. We shall describe successively the chip, then the simulator.

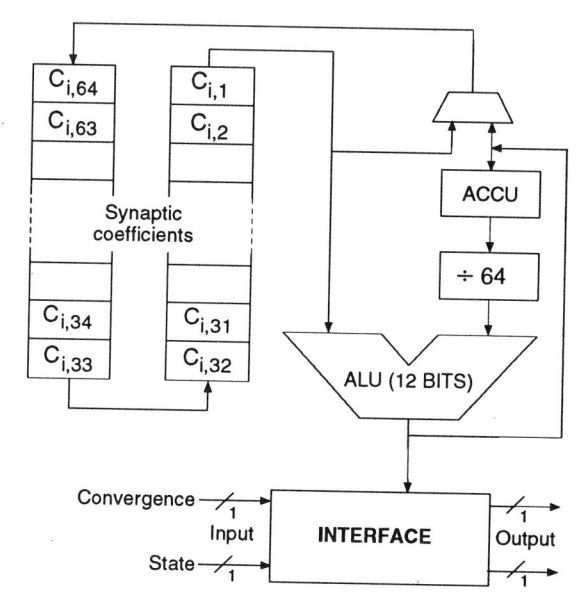

Figure 1 SIMPLIFIED SCHEMATICS OF A NEURON

THE AUTO-ADAPTIVE ASSOCIATIVE CHIP

General properties

The chip represents a fully digital 64 binary neurons Hopfield type network. Several unique and original features imbedded in this circuit make it a good candidate for higher level auto-adaptive architectures (Weinfeld 1988):
 - internal learning, implementing the Widrow-Hoff scalar local rule, which iteratively converges towards the non-local projection rule, allowing storage of moderately correlated patterns;
 - internal detection of convergence on spurious states, allowing automatic routing of the information processed by the network to other levels of a multi-network architecture;
 - internal random perturbation of the network (roughly equivalent to some kind of annealing), which can be triggered by the spurious states detection, allowing a significant overall improvement of the attractivity of the network stored memories.

The architecture is a linear systolic loop [Weinfeld 1988, Kung and Hwang 1988), in which each neuron has its own synaptic coefficients (one column of the synaptic matrix) stored locally in a circular shift register (**Figure 1**). Input/output operations are performed in parallel, the serialization of the data taking place inside the chip ; it is possible, for use with external 16 bits buses, to multiplex the 64 bits in four 16 bits blocks. The circuit totals 420,000 transistors for an overall surface of 1 cm^2, and has been manufactured using a 1.2 μm CMOS double metal technology. It is encapsulated in a PGA176 package (**Figure 2**).

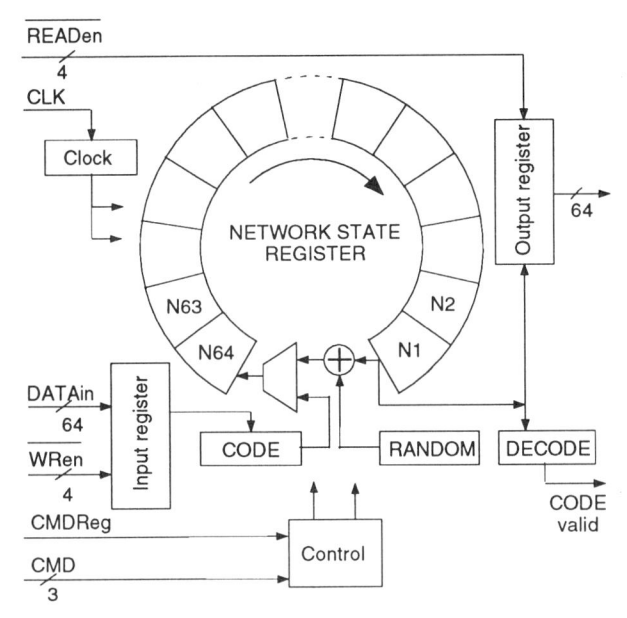

Figure 2 BLOCK DIAGRAM OF THE NETWORK

Automatic recognition of spurious attractors

The network his able to signal by itself whether it has or not succeeded in recognizing a pattern (since a network gives always an output for every input stimulus), producing a binary signal indicating whether the final state is a stored pattern or a spurious state. This property is provided by using a small part of the vectors as a label, deduced from the rest of the vector by a coding through a cyclic redundancy code. This coherence, only found on attractors corresponding to learnt prototypes, provides a 98% confidence identification with a 6 bits code field. (Alla *et al* 1990, Gascuel and Weinfeld 1991, Johannet *et al* 1992).

Speed of operation

The overall duration of one updating cycle, with a basic internal clock cycle of 100 ns (20 MHz external clock), is 6.4 μs. In the retrieval phase, if we suppose that typically four cycles are needed (this number depends somehow on the initial Hamming distance of the stimulus to the corresponding stored pattern), counting the extra cycle during which convergence is assessed, one pattern recognition is performed in less than 30 μs. In the learning phase, the speed is lower for two reasons. The first is that the calculation of the coefficients increments needs two cycles: one for the calculation of each neuron activation, and the second for the calculation of the increments themselves. The second reason is that one has to reach the convergence of the coefficients to stable values corresponding to the exact learning rule: the numbers of iterations needed is roughly proportional to the number of stored patterns. Thus, in the same conditions as above, learning 15 moderately correlated patterns may take approximately 3 ms, which is close to an absolute maximum, corresponding to a rather high learning ratio ($\approx 25\%$). This speed is compatible with most real-time stand alone applications, and with the timing of the simulator which will be described now.

THE SIMULATOR

General features

Since the higher level architectures to be implemented must be very diverse, we need a great deal of flexibility: the number of basic networks to be interconnected is variable, the interconnection must be completely configurable, which means that for instance each component (bit) of the input or the output vector of a network must be separately accessed and connected to any other component. So we have chosen a flexible environment: four chips are wired onto a standard intelligent interface board, this board inserted in a microcomputer (**Figure 3**). Several chips can 'talk' to each other through the microcomputer bus, and if some microcomputers are linked through a local area network, chips on different machines can as well be 'connected' together and also communicate with memory buffers and/or disk files. Using standard interfaces, software and operating systems, relieves us from all low-level implementation chores, allows minimal development efforts, and provides good opportunities for future upgrades (chip as well as host machine).

The simulator allows to implement sequentiality in the information processing, that is for instance making some chip wait relevant inputs before beginning a relaxation, and so on. It is also possible to operate several independent instances of the simulator, from different machines on the same network, provided there are enough chips available: each instance 'books' the necessary chips among those seen as being free on the local area network.

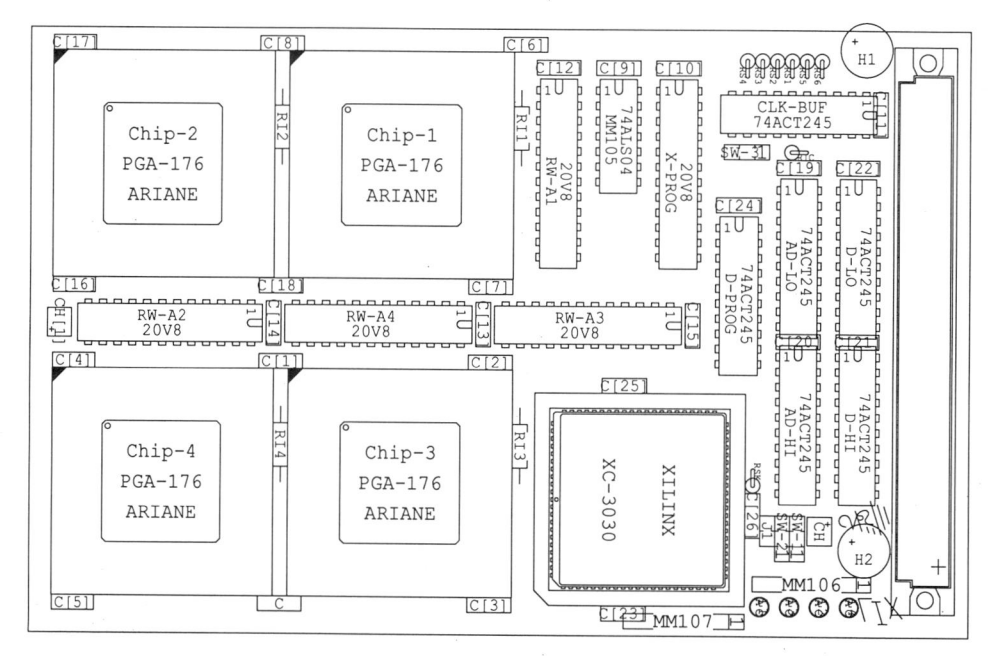

Figure 3 LAYOUT OF THE FOUR NETWORKS PRINTED BOARD

The speed of the simulator depends on many factors: number and location of the chips, complexity of the virtual communication links between them. In no case is the intrinsic chip speed a limitation, but rather the message-passing overhead. Nevertheless, the expected goal is reached: the software load is shifted onto a higher level of complexity than the basic neural network, which has become an almost ordinary component.

The basic interfacing

It consists in a MCP™ board (Macintosh Coprocessor Platform) supplied by Apple Computer Corp. This board includes a M68000 microprocessor, memory, and specific interface ASICs. Custom circuitry connected to the microprocessor bus on the board can be accessed by the host microcomputer through a board-resident real time system called A/ROSE™ (Apple Real Time Operating System Environment), exchanging information using a message-passing scheme. There is enough space on the board to accommodate four independent associative chips, occupying different addresses in the memory space of the M68000. Using a Macintosh II family computer allows to interface five such boards in one machine, that is 20 chips. Several machines linked by the LocalTalk™ proprietary network can thus allow the use of many more chips.

The software

The simulator software is organized into several distinct layers, each of them having a dependency only with its predecessor.

Board-level. The board-resident software has the following main features:
   - time sharing management of concurrent tasks residing on the board's processor ;
   - passing of messages between different tasks, whether these tasks are local to the board or run in another board in the same machine.

The reentrant main program interfaces each of the four chips with a message queue, which includes data and commands:
   - test and initialization of the chip ;
   - reception of commands to be executed together with the relevant data ;
   - micro-code generation ;
   - output of results.

Since the switching time of a task is approximately 135 µs, it would be time-consuming to switch constantly from one task to another, each corresponding to different chips. Thus, the chip managing task works as a blocked task, in the learning as well as in the relaxation phase. Information is grouped into packets (580 bytes long, as determined by the scheduling machine resident task, see later), so the processor is allowed to switch between tasks only during the interval between two packets. This strategy allows some kind of parallelism between the control of a specific chip, and the packets transmission between chips on the same board, or between boards.

Since the routing of the data must be done at the bit level, the 64 bit vector present at the output of a chip, together with two control bits (spurious state detection bit and identity bit) are cut into nine bytes, each treated as independent, and encapsulated into nine different packets. These packets are broadcast to all other chips that are needing even only one bit of a particular byte. At the input of a circuit, the task has to reconstruct a 64 bits vector from information gathered into many different packets ; at this moment, it is possible to perform some logical operations between the received bits, if necessary. This overall strategy ensures a good compromise between the granularity of the transmissions (the coarser, the faster) and the granularity of the data, at bit level.

Machine level: the forwarder and the transport layer. The A/ROSE environment supplies a background task in the host machine. This particular task interfaces the messages issued by the boards to the LocalTalk local area network, allowing the extension of the simulator between different machines. Each machine hosts such a task, all the remote ones being clients of the local one.

The man-machine interface. We use the basic system of the Macintosh family to implement a graphic interface, providing tools to configure the required structure, without too much concern for the physical location of the needed chips (except when overall speed is to be taken into account, see below). When the connectivity of the simulator has been set (**Figure 4**), the data needed for learning, or the stimuli to be processed can also be entered through this interface. To avoid too much overhead, it is also possible to declare the structure and the data using a simple language, and store or retrieve data and results in ordinary disk text files. This last feature can be used as a mean of communication between the whole simulator and any other Unix machine, through Ethernet networking.

Performances

We have tested a network made of four chips, each of them being in a different machine, with a connectivity represented by eight links of one byte wide. One packet takes approximately 500 ms to travel, this time being imposed by the LocalTalk network (as said earlier, the intrinsic chip speed is not at all a limiting factor). So clearly the user must use some care when configuring his architecture, to confine the heavily linked chips in the same machine. Nevertheless, the simplicity of the simulator makes it a much more comfortable and flexible solution than any full software simulation of similar multi-networks. For the future, it is probable that fully wired boards using integrated chips (before the advent of fully integrated systems), will be the solution for optimal speed and compactness, but still the use of an efficient simulator, such as the one presented here, should be necessary as a first step.

SOME EXAMPLES OF HIGH LEVEL ARCHITECTURES

A rather simple example of multi-network association is a so-called 'multi-agent' classifier. In most connectionnist classification schemes, a single multi-layer network is used, with the well-known backpropagation learning. Its use implies a complete sequence of presentation of many patterns belonging to each class to be later separated. The structure of the network results from a rather tedious procedure, where the size and number of layers must be found somehow by trial and error to satisfy the learning and generalization criteria. Later, the inclusion of new knowledge may require a complete re-learning sequence, and even perhaps a modification of the structure of the network.

We suggest using several associative networks, each one of them learning various examples of the same class. In the recognition phase, an unknown input is fed in parallel to all of them, the answer comes from the network that recognizes a known pattern, whereas all or a majority of the other networks relax on a spurious attractor. In no case the particular pattern on which one network has relaxed, is of interest. The answer of the whole system is a bit pattern composed of the known pattern/spurious attractor bit of each network, this bit pattern containing in general only one bit raised to 1, all the others being at 0. The ambiguity, where more than one bit is on, is a significant case that must be normally taken into account by the rest of the system. Among the advantages of this kind of classifier, we have just seen the simplicity of ambiguity detection. Another important advantage is that a new class can be learnt by just adding a new network and having it learn this particular class, without disturbing any of the other networks. This procedure can even be made partly or totally automatic.

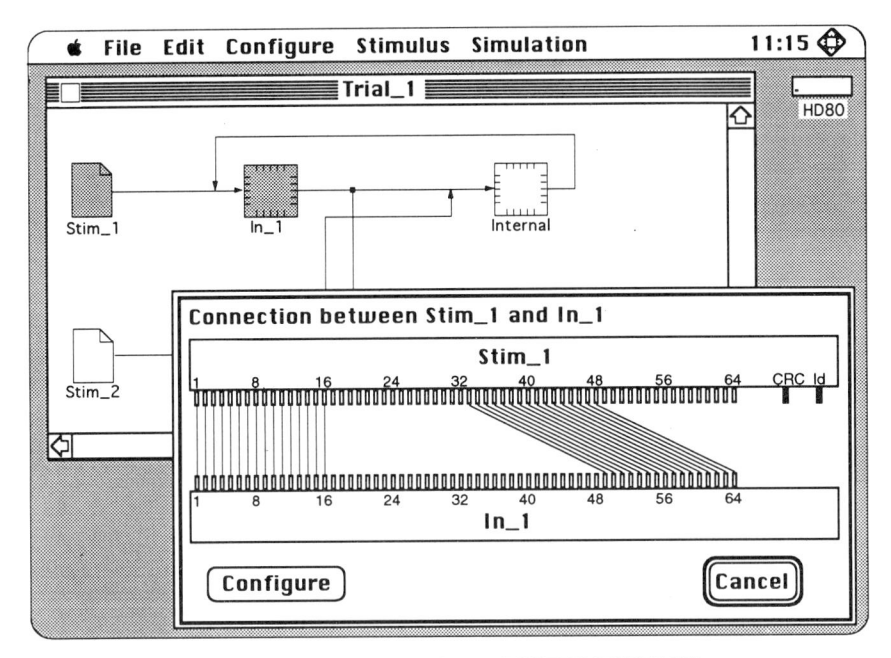

Figure 4 AN EXAMPLE OF THE SOFT INTERFACE

Another example of elementary network association we are currently working on (intended for image recognition), consists in a quad-tree structure, each node being an identical classifier made of several associative network, as described above. Each level in the tree corresponds to a different resolution of the picture, the recognition beginning with the root network, at the lowest resolution. All clusters store identical elementary features, as for instance left-right or top-down contrasts, vertical, horizontal and oblique lines, etc., and receive all together the fraction of the picture corresponding to their level in the tree. The response of the whole system is a vector, the components of which (the identification bits of each network) characterize the presence of an elementary feature at a certain level of resolution. This hierarchical description of a picture is of course much more efficient and compact than any "flat" representation of equivalent or even longer number of bits. In addition, it is obtained from the independent and simultaneous relaxation of many associative networks, so it is fast.

CONCLUSION

We have shown briefly that it is possible to interconnect elementary networks to obtain more complex architectures, able to deal with information processing problems that isolated networks have difficulties to cope with. Provided that these elementary networks have the necessary properties for implementing auto-adaptive systems, and that they do not induce computational overhead by themselves, a software reconfigurable simulator is capable to address a higher level of complexity, probably closer to the complexity of some cognitive systems (even if many higher levels are still to explore). To assure this goal, integrated custom neural chips are obviously one good solution, since the advances of technology make them easier to produce and to use.

REFERENCES

Alla, P.-Y., Dreyfus, G., Gascuel, J.-D., Johannet, A., Personnaz, L., Roman, J., Weinfeld, M., "Silicon integration of learning algorithm and other auto-adaptive properties in a digital feedback neural network", *Proc. 1st Workshop on Microelectronics for Neural Networks*, Dortmund 1990. U.Ramacher and U.Rückert (eds.), Kluwer Academic 1991.

Freeman, W.J., "A physiological hypothesis of perception", *Perspectives in biology and medicine*, 24, pp.561-592, 1981.

Gascuel, J.-D. and Weinfeld, M., "A 64 neurons digital CMOS fully connected neural network with in-circuit learning capability and automatic identification of spurious attractors", *International Joint Conference on Neural Networks*, Seattle, 1991.

Johannet, A., Personnaz, L., Dreyfus, G., Gascuel, J.-D., Weinfeld, M., "Specification and implementation of a digital Hopfield-type associative memory with on-chip training", *IEEE Trans. on Neural Networks,* July 1992.

Kung, S.Y. and Hwang, J.N., "Parallel architectures for artificial neural nets", *Proc. IEEE International Conf. on Neural Networks*, vol. II, pp. 165-172, 1988.

Weinfeld, M., "A fully digital CMOS integrated Hopfield network including the learning algorithm", *International Workshop on VLSI for Artificial Intelligence*, Oxford, 1988, in *VLSI for artificial intelligence*, pp. 169-178, J.G.Delgado-Frias and W.Moore (eds.), Kluwer Academic, 1989.

Weinfeld, M., "Integrated artificial neural networks: components for higher level architectures with new properties", *Proc. NATO Advanced Workshop on Neurocomputing*, F.Fogelman and J.Hérault (eds.), Springer Verlag, 1990.

PARALLEL IMPLEMENTATION OF NEURAL ASSOCIATIVE MEMORIES ON RISC PROCESSORS

U. Rückert, S. Rüping and E. Naroska

INTRODUCTION

The implementation of neural networks by means of multiprocessor architectures is a promising compromise between flexible modelling - the system is still program controlled - and a complete implementation by means of special-purpose VLSI chips. The use of general purpose microprocessors has further advantages, e.g. they are relative cheap compared to ASICs of low or medium quantity, they are currently available, and they utilize the highest integration level of state of the art VLSI technologies.

Neural network simulation mostly uses vector-matrix or matrix-matrix operations (e.g. see Garth 1990 and Ramacher 1991). The input/output variables and the weights are usually of type 'real'. Therefore the most compute-intensive parts of simulation programs are floating point multiplications, and processors with a high floating point performance are required for fast neural net simulations. DSPs (Digital Signal Processors) and RISCs (Reduced Instruction Set Computers) are often used for this application (Treleaven 1989).

A special type of neural networks are neural associative memories (NAMs) based on distributed storage of information with binary inputs and outputs. Many different models of NAMs have been discussed and analysed in literature (Willshaw 1960, Palm 1980, Kohonen 1984) and it turns out that the majority of algorithms do not require the weight precision offered by floating point calculation. Integer arithmetic and a dynamic range for the weights (synapses) of 1 to 16 bit seems to be sufficient for NAM applications without noticeable reduction of functionality. Hence, NAM implementations impose different requirements on computational hardware as for neural network implementation in general.

In this paper we report on a case study addressing the problem of implementing NAMs on general purpose microprocessors. After a short introduction to the architecture of NAMs we will give an efficient algorithm for their parallel implementation on single microprocessors. This algorithm was mapped to well known standard microprocessors (MC88xxx, ARM2, Sparc, RS/6000, MC68xxx and Intel 286/486) which will be compared in respect to the number of connections per second (CPS) in the recall phase of the NAM. The paper concludes with a discussion of the main results of this case study.

ASSOCIATIVE MEMORY WITH NEURAL ARCHITECTURE

The structure of a neural associative memory is shown in Figure 1. The basic operation of NAMs (Kohonen 1984) are pattern mapping (heteroassociative recall) and pattern completion (autoassociative recall). In addition NAMs have the capability of fault tolerance as described in Palm (1982), Rückert and Surmann (1991), and Kohonen (1984).

There are two phases in working with NAMs. The first is called learning phase in which the weights ($w_{1,1}$; $w_{1,2}$; ... $w_{m,n}$) are updated according to a learning rule and the specific set of z input-output pairs (X^h, Y^h) which have to be stored (h = 1, 2, ...,z). Each of the weights is responsible for several input-output associations (distributed storage of information).

The second phase is the recall phase, where the input X^h is presented and the memory creates the output Y^h. For each output component y_j an activity a_j is calculated by the form:

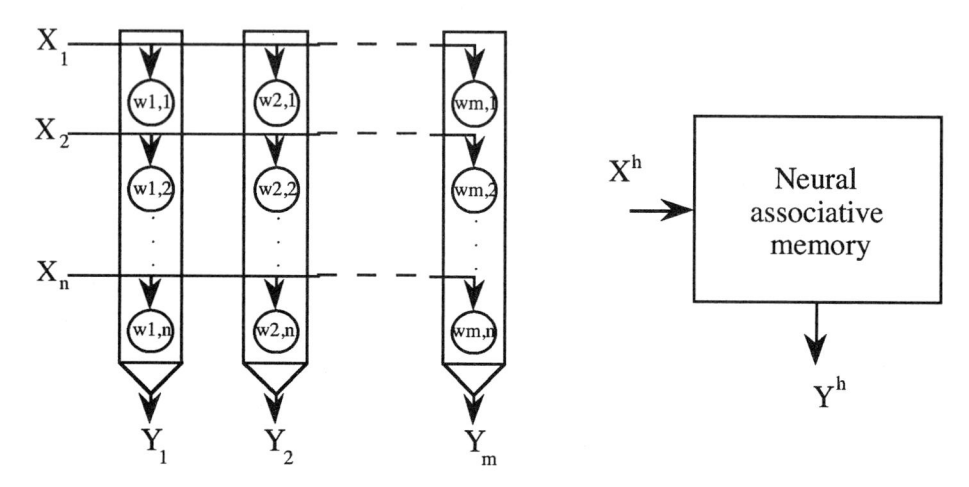

Figure 1 Structure of a neural associative memory

$$a_j = \sum_i w_{i,j} \cdot x_i^h$$

The associated binary output vector is obtained by a threshold operation. In general the components of the input vector and the output vector as well as the weights can be of arbitrary type. Because of the computational demands and due to storage efficiency (Palm 1990) we restrict our considerations on NAMs with binary input and output vectors and reduced weight precision of 1 to 16 bit.

Another important fact with regard to NAMs is the coding of the input- and output vectors. The theory shows, that the best storage efficiency can be reached by using sparse coding (Palm 1980, Palm 1982, Meunier et al 1991). For sparsely coded vectors the probability p for a component that takes the value '1' (representing the active state) is rather small, e.g. only l = log(n) (k = log(m)) components of the input (output) vectors are active at any time.

In summary, the dominant characteristics of NAMs in regard to implementation are a regular as well as modular architecture, a reduced weight precision and sparsely coded I/O patterns. The latter effects mainly the communication and dataflow management as will be discussed below.

MULTIPROCESSOR ARCHITECTURE

For applications like information-retrieval or associative knowledge processing (Palm et al 1991), NAMs with several thousand neurons are needed. This requires a high performance and a large size of memory of the used computers. Especially for real time applications, the solution can be a parallel architecture. For NAMs a highly scalable and simple architecture is shown in Figure 2.

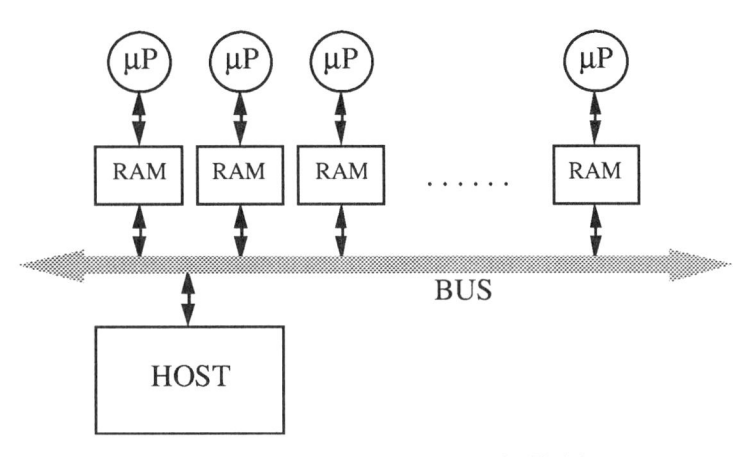

Figure 2 Multiprocessor architecture for NAMs

Several microprocessors with local memory are connected to a common bus. The controlling and management of the system is done by a host computer. For the application of NAMs there is in general no need for communication between the microprocessors. The host schedules the tasks for the processors and broadcasts the I/O-vectors and weights to the different memory parts. During the recall and learning phases all processors can work in parallel.

Each processor has to create the output for a specific number of neurons. When the calculations are finished, the host can fetch the results. Because of the sparsely coded I/O patterns the speed up in the recall phase for this multiprocessor architecture is almost linear the more mircoprocessors are added.

IMPLEMENTATION

Distribution of the weight-matrix

The distribution of the weight-matrix depends on the size of the matrix, the precision of the weights and the number of parallel processor units. In this paper we investigate processor types mainly with a databus size of 32 bit. For a weight precision of w_b bit one word of a memory segment contains $32/w_b$ weights. It is very useful to split the matrix into vertical slices, so that each microprocessor (one per slice) can calculate its output locally. The new weights during the learning phase and also the activity during the recall phase can be calculated for each slice without having information about the other slices.

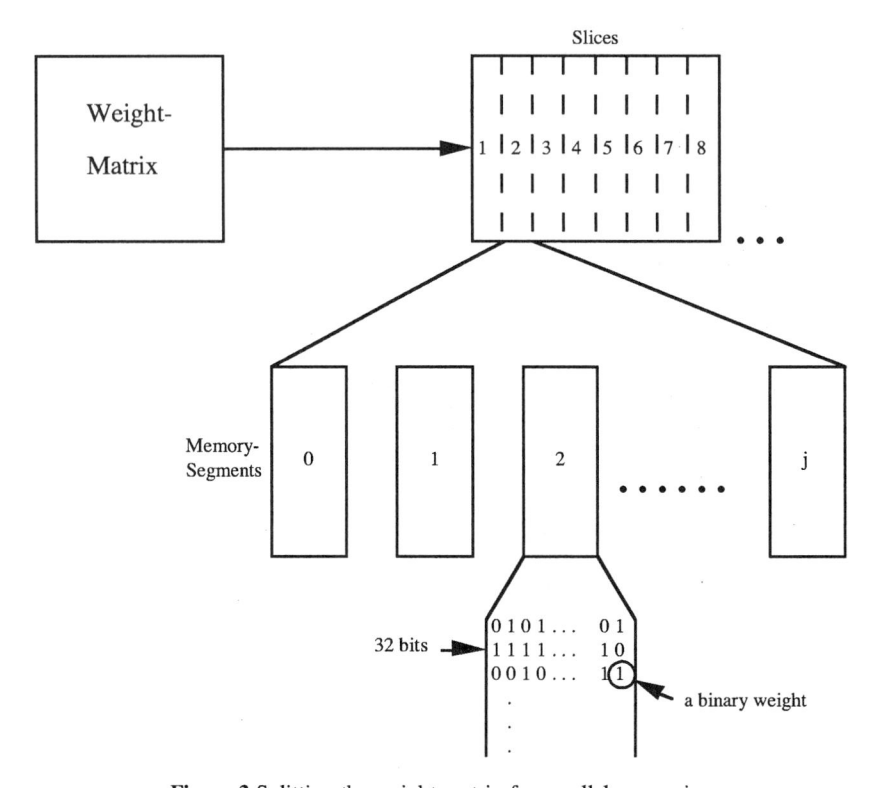

Figure 3 Splitting the weight-matrix for parallel processing

The different processors work on several memory segments as shown in Figure 3. If the system has processors with similar performance, the number of segments per slice should be about the same.

As mentioned before, sparse coding of input and output vectors will result in a relatively simple communication management. Instead of transfering a n-bit vector only l addresses of size log(n) have to be transfered (Palm and Palm 1991, Rückert et al 1992).

The overflow algorithm

In order to utilize the full performance of a given microprocessor we developed the new so called "overflow algorithm". The main idea of this algorithm is to handle a subset of $32/w_b$ weights contained in each memory word in parallel. Figure 4 shows the structure of the memory words.

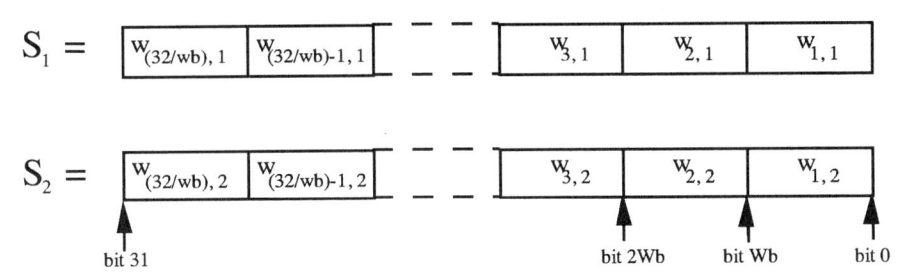

$$S_1 = \boxed{W_{(32/wb), 1} \mid W_{(32/wb)-1, 1} \mid - - \mid W_{3,1} \mid W_{2,1} \mid W_{1,1}}$$

$$S_2 = \boxed{W_{(32/wb), 2} \mid W_{(32/wb)-1, 2} \mid - - \mid W_{3,2} \mid W_{2,2} \mid W_{1,2}}$$

↑ bit 31   ↑ bit 2Wb   ↑ bit Wb   ↑ bit 0

Figure 4 Internal structure of two memory words (S_1, S_2)

The problem is that adding the two words S_1 and S_2 can create an overflow between two adjacent weights. Therefore it is necessary to split S_1 into $S_{1,1}$ and $S_{1,2}$ as well as to split S_2 into $S_{2,1}$ and $S_{2,2}$. This is shown in Figure 5. $S_{1,1}$ contains the weights with an odd index i. The space between the weights is filled with zeros. This extra space can now be used for the overflow bits. $S_{1,2}$ will do the same for the weights with an even index i. Therefore the word must be shifted w_b bits to the right to bring the weights into the correct position. S_2 is handled in the same way. The splitting can be done by using two standard microprocessor commands, "AND" and "SHIFT".

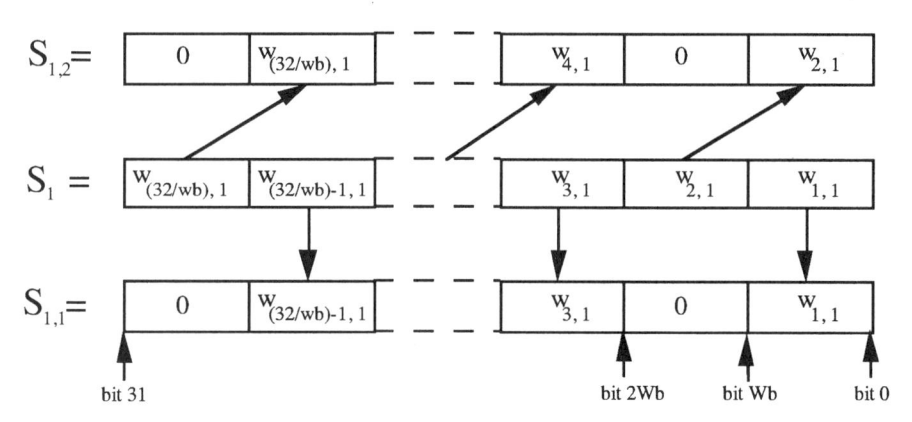

$$S_{1,2} = \boxed{0 \mid W_{(32/wb), 1} \mid - - \mid W_{4,1} \mid 0 \mid W_{2,1}}$$

$$S_1 = \boxed{W_{(32/wb), 1} \mid W_{(32/wb)-1, 1} \mid - - \mid W_{3,1} \mid W_{2,1} \mid W_{1,1}}$$

$$S_{1,1} = \boxed{0 \mid W_{(32/wb)-1, 1} \mid - - \mid W_{3,1} \mid 0 \mid W_{1,1}}$$

↑ bit 31   ↑ bit 2Wb   ↑ bit Wb   ↑ bit 0

Figure 5 Splitting S_1 into $S_{1,1}$ and $S_{1,2}$

After splitting $S_{1,1}$ and $S_{2,1}$ can be added. A possible overflow will be stored in the space between the weights. A single "ADD" instruction is now able to handle $32/(2 \cdot w_b)$ weights in parallel. The result of $S_{1,1} + S_{2,1}$ is shown in Figure 6.

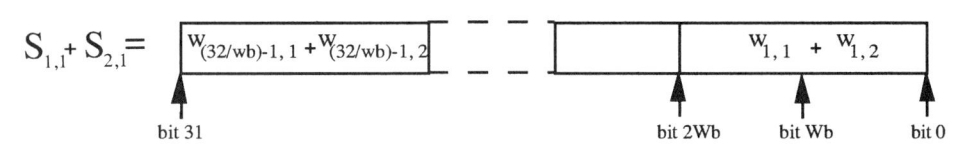

$$S_{1,1} + S_{2,1} = \boxed{W_{(32/wb)-1, 1} + W_{(32/wb)-1, 2} \mid - - \mid \mid W_{1,1} + W_{1,2}}$$

↑ bit 31   ↑ bit 2Wb   ↑ bit Wb   ↑ bit 0

Figure 6 Adding $S_{1,1}$ and $S_{2,1}$

The number of additions which are possible without getting overflow problems is given by:

$$\# \text{add} = \frac{2^{2 \cdot w_b} - 1}{2^{w_b} - 1}$$

When this number is reached, splitting can be done again and adding the weights can proceed.

Assuming $w_b = 4$ bit and a word length of 32 bit the microprocessor has to isolate and add 8 weights serially. Using the overflow algorithm, the microprocessor operates on 4 weights in parallel and the cycle "isolate and add" has to be done only twice. In general a single microprocessor is able to handle about

$$c\# = \frac{w_{proc}}{2 \cdot w_b}$$

columns (neurons) of a NAM in parallel. The parameter w_{proc} is the width of the data bus of the processor (at present mainly 32 bit). In other words standard microprocessors are capable of emulating about $c\#$ processors with a smaller data path w_b in parallel. Obviously, for binary weights the highest parallelism is achieved.

COMPARISON OF DIFFERENT GENERAL PURPOSE PROCESSORS

Description of the measurements

For testing the performance of the different processors a complete association of a 1024x1024 bit matrix is done. The input vector has 20 active components ($l = 20$) and the output vector contains $k = 1024/(w_b \times 32)$ active bits. That means, if w_b is equal to 1 the output vector has a dimension of 1024 and contains 32 active components. If w_b is equal to 4 the dimension is 256 and the number of active components is 8. This condition guarantees that the storage requirement for NAMs with different weight precision ($w_b = 1, 2, 4, 8, 16$) is the same.

All vectors are stored as a list of addresses which describe the positions of the active bits. It is important to mention that the output vectors are also stored in this way, because the transfomation from the bit form to the address form takes some time and is included in our time measurements.

Table 1 shows a list of all measured (and estimated) times for the different processors expressed as MCPS (Million Connections Per Second). There are three types of values summarized in this table. They are marked by the letters A, C and E. The letter A means that the program is written in Assembler, C means that the program is written in C-language and E means that the MCPS value was estimated by the clock rate, the necessary commands and the number of clock cycles per command.

Obviously, the processors investigated in this case study are not a complete set of all general purpose microprocessors. The listed processors were available at our institute.

Comparing the different results it is important to remind the way how the values were measured. A C-program is usually not as optimized as an assembler program. To investigate this, Table 2 shows a comparison between the resulting MCPS for a C-program and for an assembler program.

Table 1 Results for different weight precisions

	$w_b = 1$	$w_b = 2$	$w_b = 4$	$w_b = 8$	$w_b = 16$	
ARM2	7,9 (A)	5,2 (A)	2,9 (A)	1,6 (A)	0,9 (A)	MCPS
MC68000	2,0 (A)	1,7 (E)	1,0 (E)	0,5 (E)	0,4 (E)	MCPS
80286	2,1 (A)	2,2 (E)	1,2 (E)	0,6 (E)	-	MCPS
MC68030	20,6 (A)	17,8 (A)	8,4 (A)	5,0 (A)	2,7 (A)	MCPS
MC88100	27,8 (A)	18,8 (E)	10,0 (E)	5,7 (E)	2,9 (E)	MCPS
i486-50	34,1 (C)	29,3 (C)	13,5 (C)	7,5 (C)	4,1 (C)	MCPS
RS/6000	18,0 (C)	15,8 (C)	8,4 (C)	4,9 (C)	2,8 (C)	MCPS
Sparc2	14,5 (C)	13,8 (C)	6,6 (C)	3,8 (C)	2,0 (C)	MCPS

A : Assembler   C : C programming language   E : estimated

Table 2 Comparison of C and Assembler programs

	Assembler	C	
ARM2	7,9	2,1	MCPS
MC68030	20,6	9,3	MCPS
MC88000	27,8	10,2	MCPS

It seems to be a fact that a program written in assembler is about 2 to 4 times faster than a C-program for this special application of NAMs.

On the other hand it is important which compiler is used. Table 3 shows a comparison of two different C-compilers on a i486-50MHz computer. In this example the program produced by the second compiler is nearly two times faster than the other.

Table 3 Comparison of two C-compilers on a i486-50MHz computer

	N = 1	N = 2	N = 4	N = 8	N = 16	
Zortec	20,9	16,8	8,83	4,92	2,67	MCPS
Gnu	34,1	29,3	13,5	7,53	4,13	MCPS

DISCUSSION

Comparing the different processors can not be done simply by comparing the MCPS values. As shown in Table 2 and Table 3 it is very important to remind the type of computer language and also the special product that is used. For this application of NAMs some compilers do a much better job than others, which does not mean that they are much better products because they surely have their disadvantages for other applications.

The first 5 processors listed in Table 1 are measured using an assembler program. Therefore the simulation times are almost minimized. The last 3 processors are measured by running a C-program, so the listed performances are expected to increase for running assembler programs.

It is an interesting result that the i486-50MHz processor is about 2 times faster than the Sparc processor (40 MHz). A possible explanation is the usage of different C-compilers. With the Zortec compiler the i486-50MHz has a performance of 20,9 MCPS which is about the same as the RS/6000 and a little faster than the Sparc. It might be possible that another compiler on the SUN workstation could increase the performance. Other explanations might be the different sizes of cache memory or the influence of the operating system (Unix, MS-DOS, OS/2, ...) on the performance which is difficult to estimate. Nevertheless the i486-50MHz seems to be a powerful processor for this special type of application. The instructions that are used for the overflow algorithm are very simple (ADD, AND, SHIFT) and are executed by the i486-50MHz in a few clock cycles. This could explain the fact that this special CISC processor is as fast as the RISC processors.

Comparing the Motorola microprocessor families MC68k (CISC) and MC88k (RISC), the RISC architecture is the better choice for this special application, as might be expected. With a clockrate of 16 MHz the MC88100 is about 1.3 times faster than the MC68030 with a clockrate of 50 MHz ($w_b = 1$, Tab. 1). However CISC processors as the 68k and i486 are mass products and hence cheaper than RISC processors, in general.

A comparison of different generations of the same microprocessor family (68xxx, 80x86) shows a considerable speed up of 10 or more. This should be considered while developing application specific integrated circuits (ASICs), especially in respect to the expected cost/performance ratio of the complete system.

Another interesting result is that the gain in performance from $w_b = 2$ to $w_b = 1$ is not as high as might be expected (a factor of 2). The reason for that is the time needed for the transformation of the output vectors from bit to address form. The largest gain results for the MC88100. This is mainly because the processor has appropriate instructions for this transformation.

Nevertheless, in this paper it is not our intension to find out the best microprocessor for simulation of NAMs. This seems to be impossible at the moment because a general benchmark is missing for that work. Such a benchmark should provide the same presuppositions for all processors. There are a lot of dependencies which influence the performance and which could not be simply estimated (compiler, cache memory, operating system, ...). At present we are working on a specification of an appropriate benchmark for NAMs based on distributed storage and sparsely coded I/O patterns.

Overall the speeds available from general purpose microprocessors are sufficient for a range of NAM applications. For example, giving a 4096x4096 NAM with binary weights ($w_b = 1$) in which about 320000 patterns ($l = 12$, $k = 3$) can be stored with low error probability (Rückert 1991), a single i486-50MHz microprocessor (34 MCPS, Tab. 2) performs about 700 associations per second. The associations per second are given by:

$$\frac{\text{Assoc}}{\text{sec}} = \frac{\text{MCPS}}{l \cdot n}$$

It should be possible to increase the performance by working on assembler level. If even higher speed is required a multiprocessor architecture (Fig. 2) could be used. Last but not least we have to be aware of the fact that the third RISC processor generation offers a 64 bit data path, an internal clock rate above 100 MHz and a superpipelined and superscalar architecture (Weiss 1992).

CONCLUSION

The reported case study on simulation of NAMs on the basis of general purpose microprocessors is interesting in three aspects at least.

Firstly, adapting the implementation of NAMs to the characteristics of the hardware gave us an efficient algorithm with which it is possible to handle about $c\# = w_{proc}/2w_b$ neurons in parallel by a single processor. For a low weight precision ($w_b = 1, 2$) a speed up factor of about 10 is available.

Secondly, in order to achieve peak performance low level work on assembler program level is rewarded. Our assembler programs are about 2 to 4 times faster than C-programs for this special application of NAMs.

Thirdly, the speeds available from general purpose microprocessors are sufficient for a rage of NAM applications. Hence, it seems to be advisable to prove for each application of NAMs, whether a high cost ASIC solution is necessary or a low cost solution with general purpose microprocessors would meet the requirements.

ACKNOWLEDGEMENT

This work has been partly supported by the german ministry of research and technology BMFT, contract number 01-IN 103 B/O.

REFERENCES

Garth, S., "Simulators for neural networks", in Advanced Neural Computers, R. Eckmiller (ed), Elsevier Science Publishers, 1990, pp 177-183.

Kohonen, T., "Selforganisation and Associative Memory", Springer Verlag, Berlin, 1984.

Meunier, C., Yanai, H.F., Amari, S.I., "Sparsely coded associative memories: capacity and dynamical properties", Network, November, 1991, pp 469-487.

Palm, G., "On Associative Memory", Biol. Cybern. 36, 1980, pp 19-31.

Palm, G., "Neural Assemblies", Springer Verlag, Berlin, 1982.

Palm, G., Local Learning Rules and Sparse Coding in Neural Networks, in Advanced Neural Computers, R. Eckmiller (ed.), North-Holland, 1990.

Palm, G., Rückert, U., Ultsch, A., "Wissenverarbeitung in Neuronaler Architektur", Tagungsband zum 4. Int. GI-Kongress: Wissensbasierte Systeme - Verteilte kuenstliche Intelligenz, Muenchen, 1991.

Palm, G., Palm, M., Parallel Associative Networks: The PAN-System and the Bacchus-Chip, in Proc. of the 2nd Int. Conf. Microelectronics for Neural Networks, U. Ramacher, U. Rückert, J.A. Nossek (eds.), Kyrill&Method Verlag, München 1991.

Ramacher, U., "Guide lines to VLSI design of neural nets", in VLSI Design of Neural Networks, U. Ramacher, U. Rückert, Kluwer Academic, 1991, pp 1-17.

Rückert, U., "VLSI Design of an associative memory based on distributed storage of information", in VLSI Design of Neural Networks, U. Ramacher, U. Rückert, Kluwer Academic, 1991, pp 1-17.

Rückert, U., Surmann, H., "Toleranz of binary associative memory towards stuck-at-faults", in Proceedings of the International Conference on Artficial Neuronal Networks (ICANN-91), Helsinki, 1991, pp 1195.

Rückert, U., Funke, A., Pintaske, C., "Acceleratorboard for Neural Associative Memories", to be published in Microelectronics for Neural Networks, Neurocomputing, Vol. 4, No. 6, 1992.

Treleaven, P.C., "Neurocomputers", International Journal of Neurocomputing, Issue 1, 1989, pp 4-31.
Weiss, R., "Third-generation RISC processors", EDN March 30, 1992, pp 97-108.
Willshaw, D.J., Bunemann, O.P., Longuett-Higgins, H.C., Non-holografic associative memory, Nature 222, 1969, pp 960.

RECONFIGURABLE LOGIC IMPLEMENTATION OF MEMORY-BASED NEURAL NETWORKS: A CASE STUDY OF THE CMAC NETWORK

Aleksander R. Kolcz and Nigel M. Allinson

INTRODUCTION

The CMAC (Cerebellar Model Articulation Controller) is a neural network architecture based on the functional structure of the mammalian cerebellum and proposed by J. Albus (1975). The cerebellum plays the major role in coordination and control of fine motor movements in organisms, but despite the large number of neurones it contains (for some species, more than half the total number in the brain), only a small percentage of them are activated during any control action. This property has inspired a mathematical model of a neural controller in which the output is computed as a sum of a small and fixed number of weights, selected from a large available memory. In effect, save for the address generation (associative mapping), the result can be computed with very few weight accumulation cycles, which is of great importance in any real-time neural network applications. The output of a CMAC node is not thresholded and, indeed, this architecture is intended for applications where multivalued scalar or vector outputs are required.

A complex system can be realised with only a few CMAC nodes of this kind, and the appropriate amount of weight memory. It is an example of a neural network where a large share of computation is implicitly enclosed in RAM memory with only a small amount of logic performing the necessary access operations. As such, the whole structure of the CMAC node can be mapped onto several programmable logic devices, which have recently been developed. The number of equivalent logic gates in these devices reach tens of thousand making them applicable even for large designs. The reconfigurability of field programmable gates arrays offers the additional possibility to change the internal logic architecture of a chip during normal operation. This feature seems especially well suited for neural network implementations, where different models are best suited for different tasks.

An attempt has been made to implement the CMAC architecture using programmable logic devices and embed it into a larger system. As a result the entire associative mapping and accumulation logic has been fitted on a Xilinx FPGA device with 64Kbytes off-chip weight memory. Based on this design, an AT-board has been built, incorporating the CMAC and interface logic, with the MC68000 as the training controller. The board is intended to be used as a hardware accelerator in evaluating the applicability of the CMAC in various neural network real-time applications.

THE CMAC ARCHITECTURE

The function performed by the CMAC is a mapping from a N-dimensional discrete vector space into a multivalued scalar or vector space, namely,

$$c: S[s_0, s_1, \ldots, s_{N-1}] \to y = \sum_{j=0}^{K} w_j.$$

Each of the input vector variables may come from a distinct range of integers, but it will be assumed that their domains are uniform, thus

$$s_0, s_1, \ldots, s_{N-1} = 0, 1, \ldots, (L-1).$$

The final result is computed as a sum of K weights, that are selected by an associative mapping. The constant K is the CMAC association parameter which remains constant for a given configuration. The weight selection mapping determines the network generalisation properties, i.e., input points being close in the generalised Hamming metric will have several weights in common. In particular, points differing in one variable by one position would have exactly $(K-1)$ weights from the same set. On the other hand, points in which one or more variables on the respective positions differ by more than K levels will not have any weights in common. Thus K determines the receptive field of the CMAC in each input dimension.

One of the most important features of this architecture is that a fixed number of weights is selected for *any* input vector value, irrespective of the input vector dimensionality or the range of variables. This property allows very fast computation which is important in real-time applications. The parameter that is dependent on the input space, however, is the weight memory size.

To achieve a mapping transformation with the outlined properties each variable of the input vector is first assigned a K-dimensional code-word,

$$s_i \to M_i[m_{i0}, m_{i1}, \ldots, m_{iK-1}]; \qquad i = 0, \ldots, (N-1),$$

where

$$m_{ij} = K \cdot \left[\frac{s_i + K - j - 1}{K} \right] + j; \qquad m_{i0}, m_{i1}, \ldots, m_{iK-1} = 0, \ldots, (L+K-2).$$

Subsequently, the code-words, M_i, corresponding to all input vector variables are combined to create a K-dimensional virtual address vector, $V[v_0, v_1, \ldots, v_{K-1}]$. During this process the respective positions of the vector, V, are created by concatenation of their corresponding code-word components in M_i,

$$v_i = m_{i0}m_{i1}\ldots m_{iK-1} \qquad\qquad i = 0,\ldots,(N-1).$$

Consequently, the mapping from the input vector to the virtual address vector is given by,

$$S[s_0,s_1,\ldots,s_{N-1}] \to V[v_0,v_1,\ldots,v_{K-1}]$$

The virtual addresses posses all the neighbourhood generalisation properties, but they are very long in terms of the number of bits needed to represent them, i.e., $N \cdot \log_2(L+K-1)$ bits for each address, v_j. As the amount of physical memory, P, available is limited, the address words have to be transformed to conform to the system capabilities,

$$V[v_0,v_1,\ldots,v_{K-1}] \to A[a_0,a_1,\ldots,a_{K-1}].$$

This is usually done by random hash coding with satisfactory results. The errors introduced by such an approach come from cases where distinct addresses map to the same memory location, and are known as "hash collisions".

Training

The supervised training of the CMAC is implemented using the Widrow LMS adaptive algorithm (Widrow, 1985). Given a training set, consisting of input vectors and desired outputs, the weights in the weight memory are modified after each training step, l, according to,

$$w_j^{l+1} = w_j^l + \frac{\alpha \cdot error}{K};$$

where

$$error = correct\_value - \sum_{j=0}^{K-1} w_j \text{ (selected weights)},$$

and α is the adaptive algorithm learning factor. Training stops when the error falls below a set threshold over the whole training set. No particular weight initialisation is necessary for this algorithm, so they are usually cleared to zero at the start of training.

Fig. 1 shows an example of the CMAC learning a two-dimensional function (original in the upper-left corner), with $K=16$, and 200 training samples. The CMAC approximation is given in the bottom-right corner, and the remaining image shows the resulting error surface.

HASH COLLISIONS AND THE BLOCKED MEMORY SCHEME

Ignoring the hash collisions, unique mapping of the input vector into a set of K weights is possible provided that

$$L^N < \left(\frac{P}{K} - 1\right)^K.$$

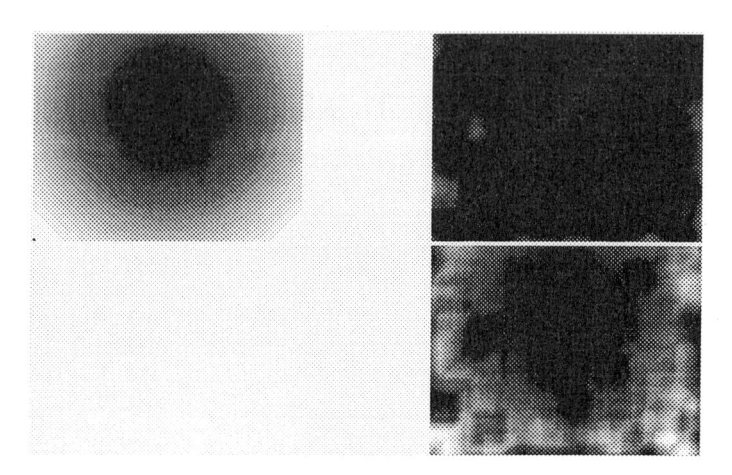

Figure 1 Trained after 20 iterations

Collisions may be thought of as a source of noise in the input data, and manifest themselves in two forms:

1. Intravector collisions, occurring when addresses, v_j and v_k, from the same vector, V, map onto the same location.

2. Intervector collisions, taking place when the same location is assigned to addresses corresponding to different input vectors that do not lie in their mutual neighbourhood. These collisions are governed by the binomial distribution with the single trial success probability $p = \frac{K}{P}$. Thus, the probability of l collisions in the mapping of two distinct addresses on the memory space is given by,

$$P(l) = \binom{K}{l} \cdot p^l \cdot (1-p)^{K-l}.$$

The intravector collisions can be avoided when a blocked rather than contiguous weight memory space is adopted. In such case, virtual addresses corresponding to different positions in the same vector, V, are mapped into separate memory blocks, where block size is,

$$B = \frac{P}{K}.$$

At the same time, the probability of intervector collisions remains unaffected. An important consequence of the blocked memory scheme is that no longer have the code-word components, m_{ij}, to come from separate range of values for different code positions j. Therefore, the number of bits necessary for their representation can be reduced to $\log_2\left(\left\lceil\frac{L}{K}\right\rceil+2\right)$, where $\lceil . \rceil$ is the rounding-up operator. This reduction makes the manipulation of virtual address words easier, especially for large values of N. The modified algorithm for computing the code-word components is given by,

$$m_{ij} = \left[\frac{s_i + K - j - 1}{K}\right] + j\,; \qquad m_{i0}m_{i1}\ldots m_{iK-1} = 0,\ldots,\left\lceil\frac{L}{K}\right\rceil + 1.$$

HARDWARE IMPLEMENTATION

General Architecture

The central part of a CMAC processing unit consists of the weight address generation block. Once the addresses are generated, the result is produced in a straightforward manner by summing the weights. The amount of weight memory necessary for unique mapping of a N-dimensional input vector consisting of L-valued variables into a set of K weights requires at least

$$K\cdot\left\lceil 1 + L^{\frac{N}{K}}\right\rceil$$

words in the weight memory. For $L=256$, $N=16$ and maximum $K=8$ this corresponds to 512 Kwords. This figure rather discourages integration of the weight memory and address generation logic on a single chip. It is exacerbated when high precision weights are represented as long words. However, if standard semiconductor memory devices are used as the weight store, the remaining logic can be easily integrated on a single device.

Address generation

Given a N-variable input vector the CMAC must generate K weight addresses. According to the two-stage mapping scheme, the virtual address is generated first. Its value for the jth address position is given by,

$$v_j = m_{0j}m_{1j}\ldots m_{N-1j}; \qquad j = 0,\ldots,(K-1).$$

For larger values of N it would be difficult to concatenate the constituting elements, m_{ij}, into a whole word, v_j, before further transforming it into a_j, because of the large size of v_j. For that reason, the process of assembling the address, a_j, is done sequentially by combining the generation of code-word components, m_{ij}, and their hash coding in each step. The exclusive-or random function has been chosen to facilitate the process of the random address mapping.

The code-word components are computed according to,

$$m_{ij} = \left[\frac{s_i + K - j - 1}{K}\right] + j; \qquad \text{with values in range } 0,\ldots,\left[\frac{L}{K}\right]+1.$$

Direct XORing of these components would result in limiting the 'in-block' or offset address length to $\log_2\left(\left[\frac{L}{K}\right]+2\right)$ bits. Therefore, after generation, each of the m_{ij} components is rotated in a barrel-shifter so that non-zero bit values are present in the higher order positions. The number of shift positions is determined by the input variable index, i. Subsequently, the rotated values are XORed to produce the offset part of the final address,

$$a_j^{off} = m_{0j}' \oplus m_{1j}' \oplus \ldots \oplus m_{N-1j}' \qquad j = 0,\ldots,(K-1),$$

where m_{ij}' designates a rotated m_{ij} and $\oplus$ is the XOR operator. The weight memory is partitioned into K blocks, with the current block number given by the value of the code position index, j. Thus, the final weight address is created by concatenating these two parts,

$$a_j = j | a_j^{off}.$$

To summarise, the hardware algorithm to compute the address, a_j, can be expressed in the following form:

1. The address register is cleared.
2. The input vector variables s_i ($i=0, \ldots, N-1$) are read sequentially and the jth code-word components are computed according to,

$$m_{ij} = \left[\frac{s_i + K - j - 1}{K}\right] + j,$$

and are rotated by i-dependent number of positions.

3. The resulting values are XORed with the contents of the address register.
4. The offset address together with the base address (containing the current value of j) is used to access the weight memory.

Ordinarily, the division by K and rotation would require several clock cycles for either operation, assuming sequential logic implementation. Considering that the same process must be repeated $K \cdot N$ times for every single result produced by the CMAC, it is desirable to realise these operations using combinational logic. The complexity of combinational shifters and rotators is quite high, and therefore some trade-offs are necessary. Additionally, implementing the division as shifting puts a constraint on the values of K to powers of two. In the present design the number of possible rotate positions in the barrel-shifter has been limited to seven. It should be noted that implementing barrel shifters in MOS VLSI circuits usually leads to very regular structures.

Weight accumulation

Since no multiplication is necessary, the weight accumulation is reduced to a simple summation. It has been assumed that the weights are in 8-bit format and the result is presented as a 16-bit value.

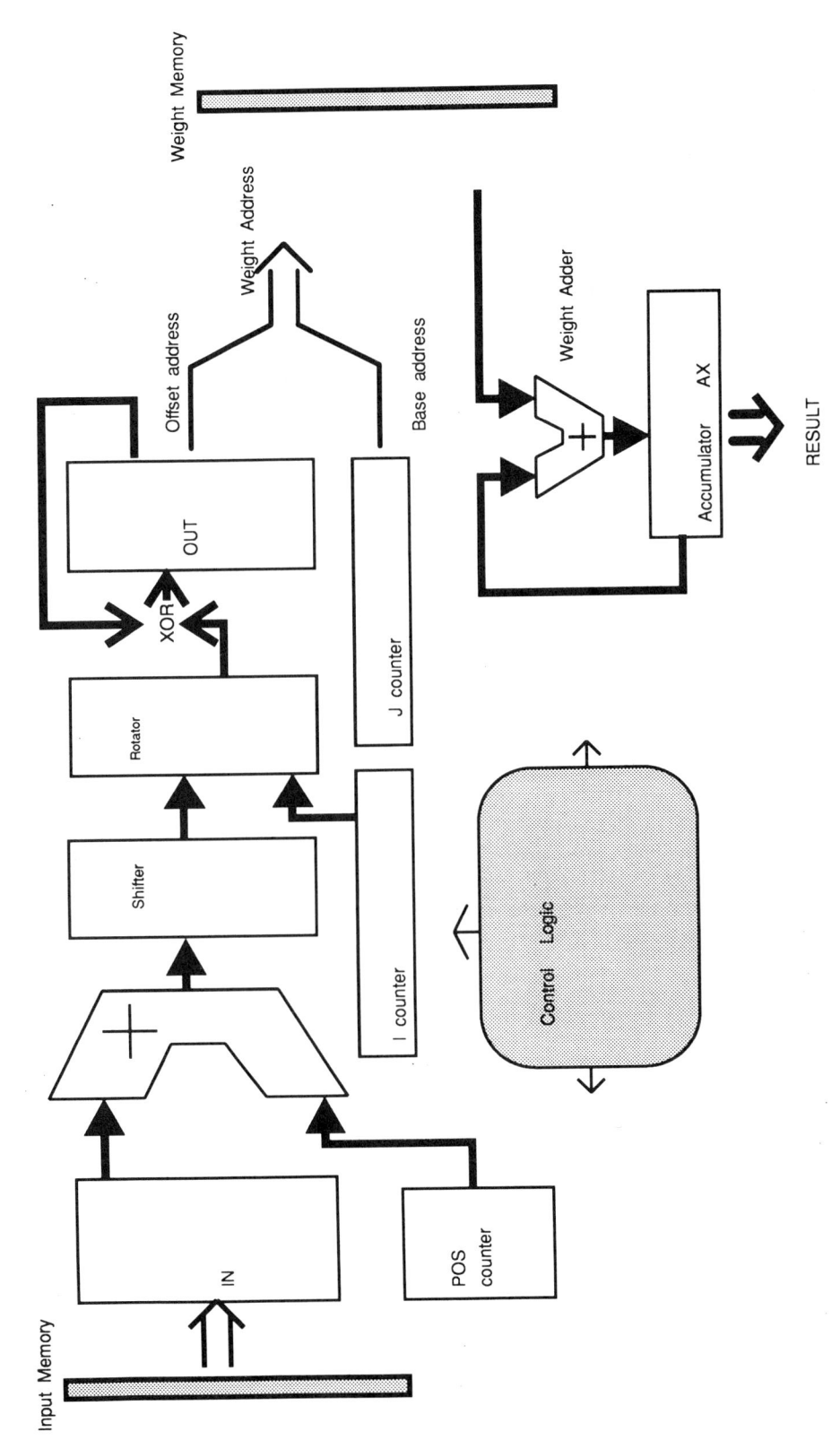

Figure 2. The CMAC block diagram.

Training

The calculations of error and weight modification are performed off-chip. During the adaptation phase the external controller monitors the weight address lines, and when the CMAC generates a valid address, the controller uses it to access and modify the weight. The end-of-memory-access signal for the CMAC is then delayed until the controller completes its operation (i.e., updates the weight).

Configuration and I/O interface

The operation of the chip depends mainly on input dimensionality, N, and the association constant, K. These parameters are configurable (within their respective ranges) and can be changed during normal operation. The configuration data resides in three registers,

    N register : contains the input vector dimension;

    K register : contains the CMAC association constant;

    logK register : contains the $\log_2 K$ used as the divisor.

These, and other functional parts of the chip are presented in Fig. 2. Table 1 explains the function performed by the these blocks and Fig. 3 shows their approximate layout in the FPGA implementation.

Table 1 The CMAC internal blocks description

Block name	Function performed	Bits
Input Adder	$s_i + j + K - 1$	8
Shifter	division by K	9 to 8
Rotator	rotates the shifter output by 0..7 positions depending on the 3 LSB of the index i	8
Offset Register	XORs its contents with the rotator output. Cleared at the beginning of each address computation cycle. Its size enforces the weight memory block size (256 locations)	8
Weight Adder	AX + mem[Base I Ofst]	16
AX	cleared at the beginning; accumulates the weights and contains the result at the end of computation	6
POSition counter	contains the (j + K - 1)value which is added to all input vector variables S[i] for a given address computation.	8
J counter	assigned (K-1) at the beginning; decremented after each address position computation; signals the end of computation when 0. Provides the base weight address.	8
I counter	assigned (N-1) at the beginning of each address position computation cycle; decremented after each input variable processing cycle ; signals the end of address position computation when 0.	8

Performance Evaluation

Perhaps the best measure of performance for this architecture is the response time or latency. Since the CMAC requires K weights to produce its result, the address computation and weight accessing are repeated K times. If T_i denotes the time necessary to access one input vector variable and calculate its contribution to the current weight address, and T_w designates the time necessary to access and accumulate a single weight then the latency must follow the inequality,

$$Latency \leq K \cdot (N \cdot T_i + T_w).$$

Overlapping of the address generation and weight accumulation phases is possible, resulting in shorter response time especially for large values of K. The two phases are performed under control of separate but synchronised state machines. After computing the first address the two machines run concurrently, which results in the following expression for the response time,

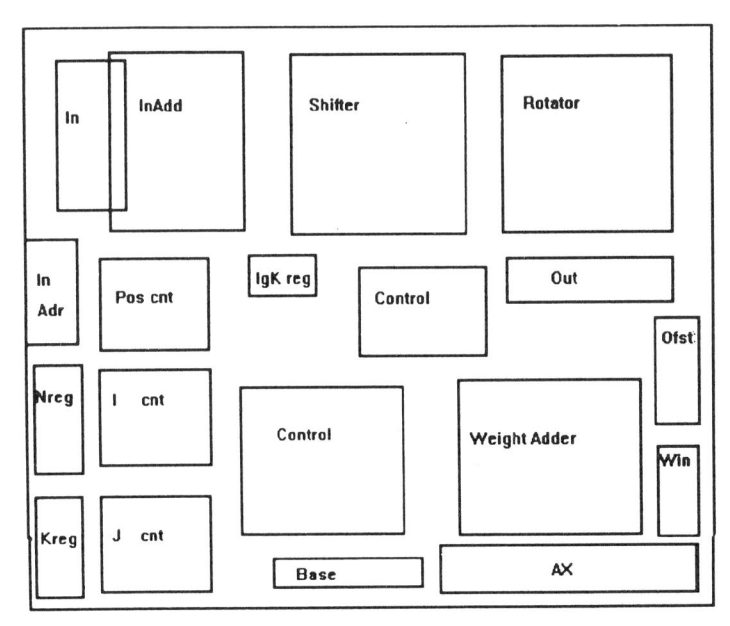

Figure 3 The approximate layout of logic blocks on the FPGA

$$N \cdot T_i + T_w + (K-1) \cdot \max(T_w, N \cdot T_i).$$

It can be expected that $N \cdot T_i > T_w$, hence

$$Latency > K \cdot N \cdot T_i + T_w.$$

The speedup achieved by phase overlapping is thus,

$$1 + \frac{K-1}{1 + K \cdot N \cdot \frac{T_i}{T_w}}.$$

SYSTEM INTEGRATION

The CMAC node, in the form described above, has been mapped onto a Xilinx XC3064 FPGA. This device has been selected because it allowed all the logic to be placed on a single chip and added benefit of in-use reconfigurability. This last option is useful if several different CMAC organisations with the same output functionality are to be considered. During the design process the major functional blocks were separately created as macros and laid out as in Figure 3. About 70% Configurable Logic Blocks (CLBs) of the device have been utilised, leaving some room for further modifications. Moving to larger FPGA versions could enable integration of two CMAC nodes in the same device.

The CMAC chip has been incorporated into an AT-card. The board contains a MC68000 microprocessor as the weight modification controller, 64Kbytes of weight memory store, 512 bytes of input and host interface buffer memories, and two Altera EPLDs (EPM5032 and EPM5128) as the timing and interface logic. The processor and FPGA clock operates at 10MHz, and 20MHz is used for the timing logic.

CONCLUSIONS

A method to map the CMAC node structure into hardware has been presented. In the current implementation only one address generation/weight accumulation block have been placed on a chip. Consequently, the output is a scalar value. However, if a vector result is required, additional weight accumulator sections could be incorporated with the same address generation block. In this case, the accessing of weights for the respective dimensions could be done in parallel. The possible limitation would be the number of additional I/O pins required for the address and data lines. At some point line multiplexing would be probably unavoidable. If VLSI implementation of this architecture were to be attempted, consideration should be given to including the weight updating controller logic on the chip and making it common to several CMAC nodes.

REFERENCES

Albus, J.S., "A new approach to manipulator control: The cerebellar model articulation controller (CMAC)" in *J. Dynam. Sys. Measurement Control,* 97(3), pp. 220-227, 1975.

Ellison, D.,"On the Convergence of the Multidimensional Albus Perceptron" in *Int. J. Robotics. Res,* 10(4), pp. 338-357, 1991.

Kohonen, T., *Content Addressable Memories,* Berlin: Springer-Verlag, 1990.

Widrow, B. and S.D. Stearnes, *Adaptive Signal Processing,* Englewod Cliffs, *N.J.*: Prentice-Hall, 1985.

A CASCADABLE VLSI DESIGN FOR GENET

Chang J. Wang and Edward P. K. Tsang

INTRODUCTION

This paper presents a VLSI design for a competitive neural network model, known as GENET (Wang and Tsang 1991), for solving Constraint Satisfaction Problems (CSP). The CSP is a mathematical abstraction of the problems in many AI application domains. In essence, a CSP can be defined as a triple (Z, D, C), where Z is a finite set of variables, D is a mapping from every variable to a domain, which is a finite set of arbitrary objects, and C is a set of constraints. Each constraint in C restricts the values that can be simultaneously assigned to a number of variables in Z. If the constraints in C involve up to but no more than n variables it is called an n-ary CSP. The task is to assign one value per variable satisfying all the constraints in C (Mackworth 1977). In addition, associated with the variable assignments might be costs and utilities. This turns CSPs into optimization problems, demanding CSP solvers to find a set of variable assignments that would produce a maximum total utility at a minimal cost. Furthermore, some CSPs might be over-constrained, i.e. not all the constraints in C can be satisfied simultaneously. In this case, the set of assignments to a maximum number of variables without violating any constraints in C, or the set of assignments to all the variables which violates a minimal number of constraints might be sought for.

Problems that may be formalized as the above CSPs are legion in AI applications. For instance, line labelling in vision (Waltz 1975), temporal reasoning (Tsang 1987, Dechter et al 1991), and resource allocation in planning and scheduling (Prosser 1990, Dincbas et al 1988), to name a few, are all well known CSPs.

The conventional approaches for solving CSPs are based on problem reduction and heuristic search (Mackworth 1977, Haralick and Elliott 1980). A few techniques in such categories have been used in commercial constraint programming languages, such as CHIP, Charme and Pecos. Prolog III and a functionally similar language CLP($\mathcal{R}$) are based on linear programming techniques for handling continuous numerical domains. For symbolic variables, CHIP uses the Forward Checking with Fail First Principle (FC-FFP) heuristic (Haralick and Elliott 1980). CHIP's strategy has worked reasonably well in certain real life problems (Dincbas et al 1988). However, since CSPs are NP-hard in nature, the size of the search space for solving a CSP is inherently $O(d^N)$, where d is the domain size (assuming, for simplicity, that all domains have the same size) and N is the number of variables in the problem. Thus, search techniques, even with good heuristics, inevitably suffer from exponential explosion in computational complexity. Hence, these techniques will not be effective for solving large problems, especially, when timely responses from the CSP solvers are crucial (e.g. in interactive and real time systems). Although speed could be gained in problem reduction and search-based CSP solvers by using parallel processing architecture, as Kasif (1990) points out, problem reduction is inherently sequential, and it is unlikely that a CSP can be solved in logarithmic time by using only a polynomial number of processors.

Swain and Cooper (1988) proposed a *connectionist* approach that can fully parallelize the problem reduction in binary CSPs, implementing an arc-consistency maintenance algorithm which attempts to reduce the size of the variable domains by deleting values which can never be part of any solution (cf. Mackworth 1977). In this approach, a CSP is represented as a network. The nodes are organized in a two-dimensional structure, with each column corresponding to one variable and containing all the nodes that correspond to some value that can be assigned to the variable. An arc in the network represents the compatibility between the connected two nodes. It employs one JK flip-flop for each node, thus $N \times d$ JK flip-flop are required for N variables each with a domain size d; and one JK flip-flop for each arc, thus $N^2 \times d^2$ JK flip-flop are required. Accordingly, enough combinational logic would be required to control setting/resetting the flip-flops.

Initially, all the node flip-flops are set *on* and arc flip-flops are set according to the constraints in C - *on* for compatible values and *off* otherwise. Then, the network iterates to delete the nodes, by resetting the node flip-flops, that are constrained by every possible assignment of its neighbour variable. This is known as the relaxation procedure. However, solutions to the CSP are not directly generated by this procedure. The time complexity for the relaxation procedure is $O(Nd)$ in terms of cycles.

Guesgen (Guesgen and Hertzberg 1992) extends Swain and Cooper's approach to facilitate the generation of solutions from the converged network. This approach distributedly encodes the possible solutions in bit strings stored in the remaining nodes. Additional $O(N^3 \times d^3)$ JK flip-flops are required for this purpose. Thereafter, solutions may be found by succession of bit-wise AND operations over N binary strings, selected one from each set of the nodes that represent the values assignable to a single variable. If the result of the AND operations is non-zero, a solution is thus found as the selected nodes. Unfortunately, the number of possible selections of the bit-strings is an exponential function of the *compatible* variable domain size, assuming a uniform domain size for each variable after the network converges. It should be noted that arc-consistency maintenance may not always reduce the variable domain sizes significantly, and the space complexity of $O(N^3 \times d^3)$ will quickly reach the capacity limit of the current VLSI technology.

To summarize, the constraint programming languages and systems implemented on conventional workstations are too slow for solving problems of realistic size, e.g. those that may have hundreds of variables with a domain size of a few tens. The *connectionist* approach based on VLSI technology is not scalable due to its massive internal connections for setting/resetting the J-K flip/flops and, hence, will be severely limited by the chip size currently available. In addition, none of these approaches can handle over-constrained CSPs which are at the heart of many real-life problems. To address these problems, a generic neural network approach, known as GENET, that effectively realizes a stochastic heuristic search, has been proposed to speed up the performance of CSP solvers (Tsang and Wang 1991). In the next section, we shall briefly describe the GENET model. This will be followed by a presentation of a cascadable VLSI design for implementing GENET.

GENET

GENET was inspired by the early attempt to apply neural network technique to solve binary CSPs (Adorf and Johnston 1990), which has led to the discovery of the Heuristic Repair

Method (Minton and Johnston 1990). The Heuristic Repair Method is based on a heuristic called the *Min-conflict heuristic*. Initially, variables are picked at random to be assigned values which violates the least number of constraints (the Min-conflict heuristic). Then the program iterates to check the compatibility of the assignments, dividing the variables into two sets: $\mathcal{A}$, containing the variables whose assignment violates no constraints, and $\mathcal{B}$, containing the variables whose assignment violates some constraints. A variable in $\mathcal{B}$ is then selected to be *repaired*. It will be assigned a value that violates the least number of constraints (Min-conflict). After the repair, the variables are re-examined and transferred from $\mathcal{A}$ to $\mathcal{B}$ or vice versa, according to their status. This procedure iterates until set $\mathcal{B}$ becomes empty (which is not guaranteed to happen). Then set $\mathcal{A}$ contains a solution to the CSP.

The Heuristic Repair Method has been demonstrated to be capable of solving the million queens problem in minutes when simulated on a single processor. However, a vital problem with the Heuristic Repair Method is that it can easily be trapped in local minima. Hence, it may only be useful for solving loosely constrained CSPs where many solutions exist and are relatively evenly distributed in the search space.

The GENET approach is based on the principle of Min-conflict, but incorporates a learning algorithm to escape local minima. GENET is a competitive neural network model that has binary neurons (*on/off* state) and symmetric connections. The connectivity is sparse because only constrained neurons are connected with a negative weight, initially set to be *-1*. The neurons in the network are clustered to represent variable domains. For instance, if the CSP to be solved has N variables and each variable may be assigned with one of d values at one time, then the network will consist of N clusters, each of d neurons, and the connections will be set up according to the constraints in the original CSP. **Figure 1** shows a cluster of neurons, where the input signals are the output states of the connected neurons that represent incompatible (constrained) value assignments.

The neuron states are updated in convergence cycles. In each convergence cycle, every neuron adds up the weights connected to active neurons. The neurons in the same cluster compete with each other and the one that receives a maximum input, meaning that less constraints would be violated if it is turned on rather than others, will be selected to turn on and the others are turned off. In case of tie situations, preference is given to the *on* neuron, if there is one in the tie, else a random choice will be made. The convergence cycles continues until all the *on* neurons have a nil input, which means that no constraints are violated if the variables are assigned the values indicated by the current network state.

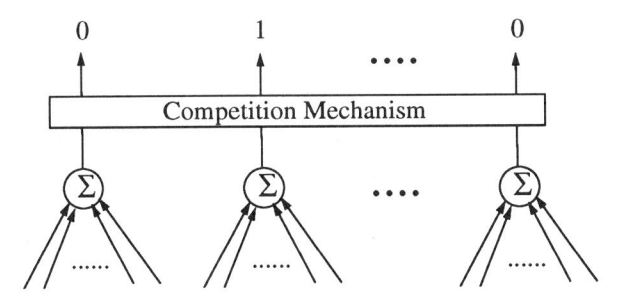

Figure 1 An example of a cluster of neurons in GENET, assuming the second neuron receives the maximum input.

This convergence procedure resembles the Min-Conflict heuristic but is deterministic in identifying local minima. The network is in a local minimum, if a convergence cycle fails to update any neuron's state whilst some of the *on* neurons still receive negative inputs. This means that some constraints are violated by the variable assignments represented by the current network state. In this case, a learning procedure is applied to penalize the connection weights that link two active neurons. The time required by GENET to solve a CSP is measured in terms of number of convergence cycles it takes for the network to settle down with a solution.

Extensive experiments have been carried out over several thousands of randomly generated CSPs with various tightness of constraints and various problem sizes. This approach has also been tried on the car-sequencing problem (Wang and Tsang 1991). Although the completeness is not guaranteed in GENET, our experiments have shown that GENET always finds solutions for solvable problems. More interestingly, when GENET is given insoluble problems, optimal solutions are always found, where optimality is measured by the number of constraints being violated. This is verified by a branch-and-bound program. The result of our experiments can be summarized as follows.

First of all, given a fixed tightness among the constraints, the number of convergence cycles required to solve significantly large problems is limited. For example, with the tightness used in our tests, problems with 200 variables which have uniform domain size of 6, require no more than 200 cycles to be either solved or concluded over-constrained. It should be pointed out that problems of this size have a potential search space of $O(6^{200})$, which is the largest problem we could simulate within a tolerable time on a Solbourne 902/5E with two SPARC processors. If a full parallelism of $N \times d$ is supported by VLSI implementation of GENET and if a convergence cycle time were to take a few hundreds of nanoseconds, such a problem can be solved in terms of tens of microseconds. For reference, a program which employs the FC-FFP heuristic (a strategy integrated in CHIP) would take over 40 minutes CPU time when run on the Solbourne 902/5E. This implies that the VLSI implementation of GENET would provide a potential speed gain in an order of 10^6 to 10^8 over existing CSP languages run on commercial workstations. The potential speed-up of such a high order means that a very high overhead is affordable should full parallelism be not possible. In fact, a parallelism of d, which is highly conceivable with current VLSI technology, will lead to a speed-up of $O(10^3)$ over the sequential heuristic search approach. This will be explained in detail later.

Secondly, in our experiments, the weight values never fall below *-50* and the total input to a neuron never fall below *-100*. If this proves to be the general case, then 8-bit weights and 8-bit accumulation registers should provide sufficient precision in digital implementation. This also gives guidelines for implementations using analog VLSI technology. Below, we discuss VLSI implementation of GENET based on these guidelines.

A CASCADABLE VLSI ARCHITECTURE FOR GENET

The computation in GENET consists of two parts. First, every neuron will have to sum up weighted input signals, and then all the neurons in the same cluster will compete with each other to select the winner. Since the GENET is a binary neural network, the former has become a simple summation of the weights connected to *active* (*on*) neurons in the network. This would significantly simplify the design. To implement GENET, the weight storage, summation mechanism in each neuron, and the competition mechanism in each cluster will have to be considered.

In the current literature on implementing neural networks, various VLSI technologies, such as analog (e.g. Graf and Jackel 1989), digital (e.g. Yasunaga et al 1990), hybrid (e.g.

Morishita et al 1990) and pulse-modulated (e.g. Murray 1989 and Tomberg and Kaski 1990), have been reported for implementing synapse memories, neuronal activation functions, and the multiplication and summation mechanism in a parallelized fashion. Any of these techniques may be borrowed to implement GENET as and when they are commercially mature. For competitive neural networks, Lazzaro et al have designed a sub-threshold competition mechanism (WTA) that has been used in VLSI sensory systems.

A requirement essential to GENET applications is that the competition mechanism should allow the size of each cluster to vary dynamically at run time. This is because the size of a cluster depends on the domain size of the variable it represents, which varies from application to application. Therefore, the size of a cluster as well as the connections between clusters has to be programmable. It would be difficult for sub-threshold WTA to meet these requirements. For these reasons, we investigated the design of a multiple signal comparator (N-Comparator) as the competition mechanism. The function of this comparator is to take n input signals and identify which of them is the maximum, or a maximum one if there is a tie. The input signals may be analog or digital depending on the technology adopted for implementing other parts of the neurons. However, the output signals are logic 1 or 0, one for each input. This comparator will replace the upper box of **Figure 1**. The critical requirement of the design is that such comparators should be cascadable via switches so as to vary the number of signals to be compared simultaneously.

Analog N-Comparator

The design of the analog comparator is based on comparing voltage signals. **Figure 2** shows a single N-comparator designed in CMOS technology. The input V_1 through to V_n are analog voltage signals to be compared and the output O_1 through to O_n are analog voltage signals that amplify the difference between the maximum input signal and the rest of the input signals. Cascading a few stages of such comparators can amplify this difference so much that the final output signals will become logical signals, indicating clearly which input signal wins the competition.

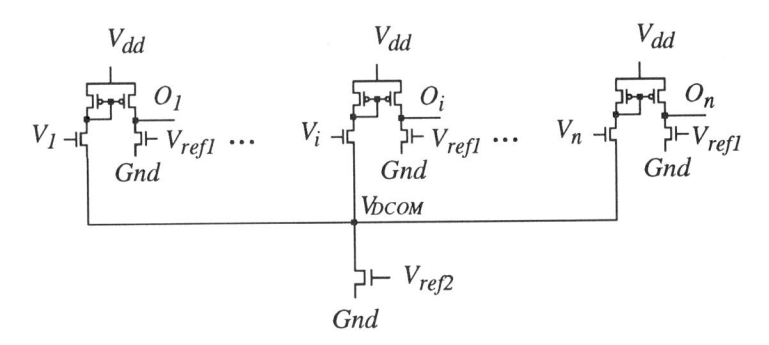

Figure 2 Analog multiple signal comparator

The sensitivity of this N-comparator, i.e. the minimum voltage difference among the input signals that may drive their corresponding outputs into logical *on* or *off*, correlates with the number of signals, i.e the value of n, due to the common current sink. SPICE simulation shows that for a single stage 4-comparator, using standard 2μ CMOS technology and $V_{dd}=5V$, the sensitivity is about $0.7V$, but non-linear across the operating range ($0.86V$ to $4.25V$ in our simulation). In terms of quantized analog signals, this means that four different levels of voltage signals, each with a $\pm 0.22V$ safety margin, can be reliably compared. This corresponds to a two-bit comparator, with a D/A at each input line.

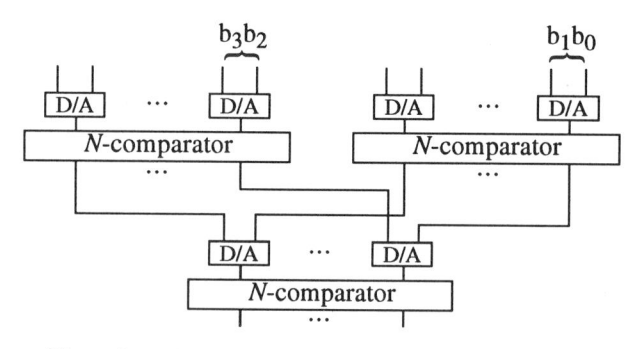

Figure 3 Cascading 2-bit comparators for 4 bit signals

Whilst the sensitivity can be greatly improved by adding more stages to amplify the voltage difference or using bipolar technology, for true analog input signals, it is sufficient to show that, for digital signals, an analog comparator with the above sensitivity can always be cascaded to realize higher precision, as shown in **Figure 3**. This also means that higher precision digital values may be compared in a kind of serial fashion; and the comparison speed only depends on the number of bits in the comparands, rather than the number of the signals being compared simultaneously.

Digital N-Comparator

The digital N-comparator compares n digital values, simultaneously, bit by bit from the *msb* to *lsb*, and the maximum value will be identified by the success of a series of bit-wise comparisons. This can be realized by using a bus structure with *p*-MOSFETs to drive the bus high and a common current sink to shunt the bus low.

Starting from the *msb*, each signal puts a bit onto the corresponding bus line and, after some time delay for the bus to settle down, compares its bit value with that on the bus line. If they are the same, it continues to do the same with the next bit. Otherwise, it will stop putting its lower bits onto the bus and signal a failure in the competition. If a signal succeeds in every bit position, this signal contains a maximum value. **Figure 4** show the logic for a

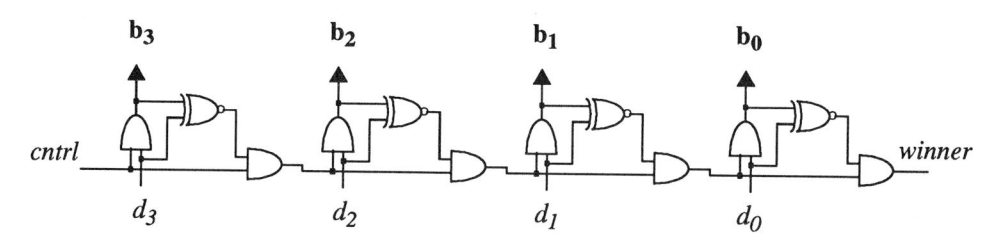

Figure 4 Control logic for one signal in a digital multiple signal comparator

signal of 4-bit value, where $b_3b_2b_1b_0$ are connected to the bus lines; $d_3d_2d_1d_0$ represent the 4-bit digital input signal; the signal *cntrl* may prohibit the input signal from participating the competition; and the signal *winner*, if true, indicates this input has won the competition.

The advantage of this design is two-fold. Firstly, the value n can be large due to the digital bus structure. Secondly, cascading such comparators by connecting their buses together via switches would allow dynamic clustering at run time. The latter would facilitate programmable network configuration. The speed of comparison only depends on the number of bits in the values being compared, three gates' delay plus a bus charge/discharge per bit, as may be seen in **Figure 4**. SILOS simulation under Cadence™ showed that an 8-bit *63*-Comparator can reliably complete the comparison in well under *40ns*. An added benefit is that the signal present on the bus is in fact the maximum value of the digital numbers being compared. This simplifies the detection of the winning value, which is required to signal the successfully converged network state.

Note. In the above description of the *N*-comparator, both analog and digital, the necessary control logic and status registers are omitted for the sake of simplicity. The omitted control logic includes: a) the circuit that detects the *nil* value of the winning input, indicating that no constraints were violated; b) the circuit that determines which of the winning input signals, in the case of a tie, is going to turn *on* its corresponding output state; and c) the circuit that detects a *no-change* in a convergence cycle, i.e. the selected neuron is already *on* in the previous cycle.

It should also be pointed out that, although the N-comparators as described above are designed to select the maximum value, similar techniques may be applied to design them for selecting the minimum value. It is also possible to compare positive values with negative ones, with a treatment of the sign. Since all the weights in GENET are negative, the N-comparator will actually be comparing the magnitude of negative values. This can be realized either by designing the comparator to select the minimum value, equally applicable to analog as well as digital design; or selecting the maximum 1's complement of the values, in the case of a digital design; or selecting the maximum voltage on reversely charged capacitors, in the case of an analog design. Therefore, we will ignore this problem in our discussions.

GENET Module

Having discussed the competition mechanism, we can now describe the building block of GENET architecture - the GENET module. The GENET module is based on the structure of a cluster of neurons. The architecture of the GENET module is shown inside the box in **Figure 5**. It consists of *n* neurons connected to an *N*-comparator. Each neuron has its own weight memory and a summation mechanism, the ellipses labelled Σ. The output signals of the *N*-comparator are latched into state registers, boxes labelled *S*, and fed back to the comparator for its control logic to determine the signal *no_change*. The module outputs the code of the active neuron in the cluster, and takes as input the codes of the active neurons in other clusters. The signal *no_violation* indicates the summation of the input to the winning neuron is nil, meaning no constraints have been violated. The signal *no_change* indicates that, in the current convergence cycle, the neuron selected to turn on was already on in the previous cycle. These two signals indicate the status of the cluster after a convergence cycle, e.g. whether this cluster may conform a part of a solution or a potential local minimum. If it is the latter, a learning cycle may be triggered, if necessary. The GENET module can be cascaded, by connecting the *link_control* signals, to extend the number of neurons per cluster without reducing the performance significantly.

A complete GENET network can be realized by organizing the GENET modules to suit the structure of the CSP to be solved. **Figure 6** shows an example of the overall architecture of GENET in application, where each row represents a cluster and a number of GENET modules may be cascaded for a large cluster. A chain of such GENET modules may be implemented on the same chip and dynamic configuration of the cluster size at run time is simply to program the switches. This is perfectly feasible with the digital comparator, but there might be some technical difficulty with the analog one.

For a realistic measurement of the GENET performance, let's assume that the weights

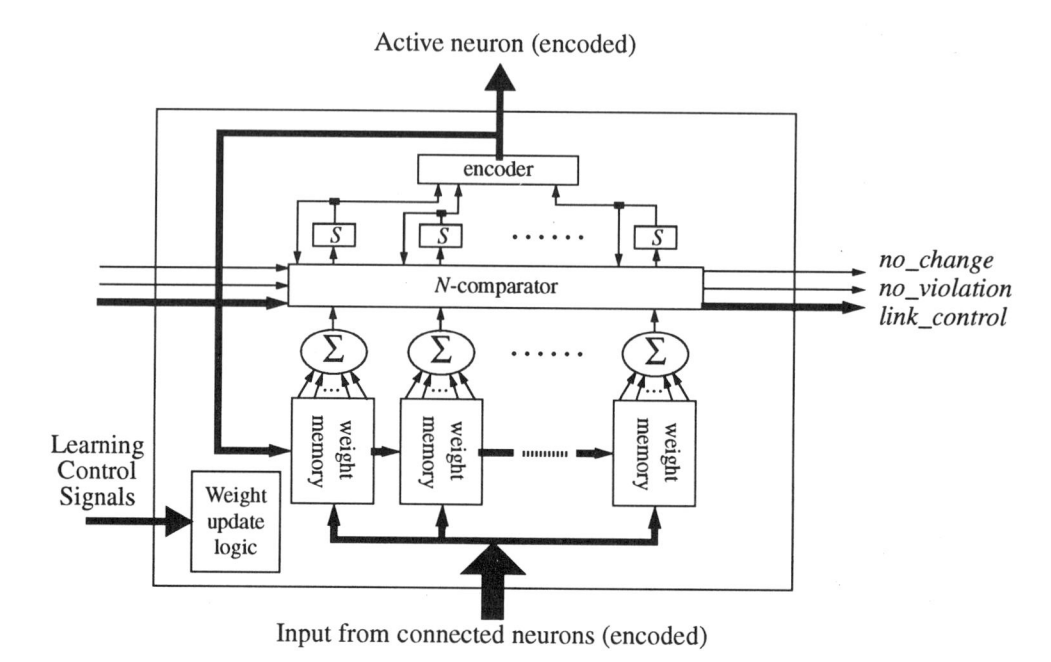

Figure 5   Architecture of a GENET module

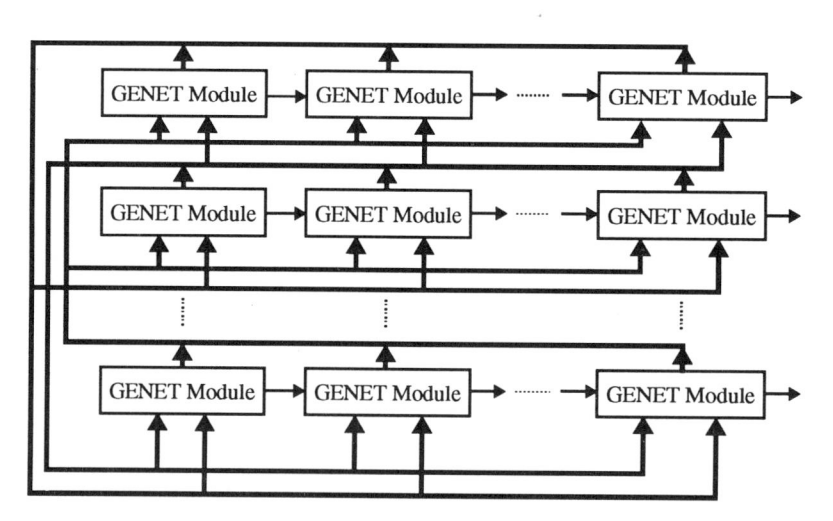

Figure 6   Overall architecture of GENET

are 8 bits and stored in off-chip fast SRAM, e.g. the 16K×4 bit INTEL C51C98-25 with 25ns access time. It means that the real memory access time can be made about 40ns, when address decoding overhead is included. Further, assume the digital comparator is integrated with a bit serial adder for the purpose of summing up the input signal to each neuron. Hence, only one input line is needed for one signal being compared. Take a CSP that has 200 variables for example, in a convergence cycle, each neuron will sum up 200 weights, because there is only one neuron in each cluster turned on. To sum up 200 weights, with each of 8 bits sequenced in, will need 200×8×40ns=64µs. To be more pragmatic, let's assume the 200 input signals are multiplexed via 20 lines. This means one convergence cycle would take 640µs. The competition time is ignorable in this case.

According to our simulation, the above problem can be solved in no more than 200 cycles, i.e. less then 200×640µs=128ms, while the FC-FFP heuristic program will take over 40 minutes of CPU time to solve the same problem. This means that the GENET approach may give a potential speed-up in the order of 10^5 against the heuristic search program. This estimation assumes a *full* parallelism of $N\times d$ neurons. In fact, even a parallelism of d would give a potential speed-up of the order of 10^3. This estimation overlooks other overheads that may occur in practice, such as updating weights in a learning phase, which may be twice the time of a convergence cycle. However, we may still conclude that the GENET, when implemented in VLSI technology with a moderate parallelism, e.g. tens of GENET modules each of tens of neurons which is highly feasible with the current VLSI technology, will out-perform the existing CSP solvers significantly.

CONCLUSIONS

Real life Constraint Satisfaction Problems are typically large and complex, and can not be satisfactorily solved by the current commercially available constraint programming languages and systems within a tolerable period of time. The competitive neural network model GENET has been developed to tackle these problems. The GENET approach provides an effective parallel algorithm for solving CSPs. When implemented in VLSI technology with a realistic, moderate parallelism, a significant speed-up over the existing constraint programming languages and systems is possible.

Acknowledgment

The authors would like to acknowledge the contribution of Kate W. C. Sin who investigated GENET behaviour with over-constrained CSPs, Nordin B. Salleh who simulated the analog design using SPICE simulator, and Hai Cheong Wong who simulated the digital design using SILOS simulator under Cadence™ EDGE®.

References

Adorf, H.M. & Johnston, M.D., "A discrete stochastic neural network algorithm for constraint satisfaction problems", Proceedings, International Joint Conference on Neural Networks, 1990.

Dechter, R., Meiri, I. & Pearl, J., "Temporal constraint networks", Artificial Intelligence, 49, pp. 61-95, 1991.

Dincbas, M., Simonis, H. & Van Hentenryck, P., "Solving car sequencing problem in constraint logic programming", Proceedings, European Conference on AI, pp. 290-295, 1988.

Dincbas, M., Van Hentenryck, P., Simonis, H., Aggoun, A. & Graf, T., "Applications of CHIP to industrial and engineering problems", First International Conference on

Industrial and Engineering Applications of AI and Expert Systems, June 1988.

Graf, H. P. and Jackel, L.D., "Analog Electronic Neural Network Circuits", IEEE Circuits and Devices Magazine, pp. 44-55, July 1989.

Guesgen, H.W. & Hertzberg J., "A Perspective of Constraint-based Reasoning", Lecture Notes in Artificial Intelligence, Springer-Verlag, 1992.

Haralick, R.M. and Elliott, G.L., "Increasing tree search efficiency for constraint satisfaction problems", Artificial Intelligence 14, pp. 263-313, 1980.

Kasif, S., "On the parallel complexity of discrete relaxation in constraint satisfaction networks", Artificial Intelligence (45), pp. 275-286, 1990.

Lazzaro, J., Ryckebusch, S., Mahowald, M. A., and Mead, C. A., "Winner-Take-All Networks of O(N) Complexity", in *Advances in Neural Information Processing Systems I*, Touretzky, ed., San Mateo, CA: Morgan Kaufmann, 1989, 703-711.

Mackworth, A.K., "Consistency in networks or relations", Artificial Intelligence 8(1), pp. 99-118, 1977.

Minton, S., Johnston, M.D., Philips, A. B. & Laird, P., "Solving large-scale constraint-satisfaction and scheduling problems using a heuristic repair method", American Association for Artificial Intelligence (AAAI), pp.17-24, 1990.

Morishita, T., Tamura, Y. and Otsuki, T., "A BiCMOS Analog Neural Network with Dynamically Updated Weights", IEEE Int. Solid-State Circuits Conf. Dig. Tech. Papers, pp. 142-143, Feb. 1990.

Murray, A. F., "Pulse Arithmetic in VLSI Neural Networks", IEEE Micro Mag., pp. 64-74, Dec. 1989.

Prosser, P., "Distributed asynchronous scheduling", PhD Thesis, Department of Computer Science, University of Strathclyde, November 1990.

Swain, M.J. & Cooper, P.R., "Parallel hardware for constraint satisfaction", Proc. AAAI, pp. 682-686, 1988.

Tomberg, J. E. and Kaski, K. K. K., "Pulse-Density Modulation Technique in VLSI Implementations of Neural Network Algorithms", IEEE J. of Solid-State Circuits, vol. 25, no. 5, pp. 1277-1286, Oct. 1990.

Tsang, E.P.K., "The consistent labelling problem in temporal reasoning", Proc. AAAI Conference, Seattle, pp. 251-255, July 1987.

Tsang, E. P. K., & Wang, C. J., "A generic neural network approach for constraint satisfaction problems", Proc. NCM'91 Applications of Neural Networks, to be published in Series in Neural Networks by Springer Verlag, 1992.

Waltz, D.L., "Understanding line drawings of scenes with shadows", in WINSTON, P.H. (ed.) The Psychology of Computer Vision, McGraw-Hill, New York, pp. 19-91, 1975.

Wang, C. J., & Tsang, E. T. K., "Solving constraint satisfaction problems using neural networks", Proceedings, IEE Second International Conference on Artificial Neural Networks, pp. 295-299, 1991.

Yasunaga, M., Masuda, N., Yagyu, M., Asai, M., Yamada, M., and Masaki, A., "Design, Fabrication and Evaluation of a 5-Inch Wafer Scale Neural Network LSI Composed of 576 Digital Neurons", Proc. Int. Joint Conf. on Neural Networks, Vol. II, pp. 527-535, June 1990.

PARAMETRISED NEURAL NETWORK DESIGN AND COMPILATION INTO HARDWARE

Wayne Luk, Adrian Lawrence, Vincent Lok, Ian Page and Richard Stamper

INTRODUCTION

Most artificial neural networks consist of one or more arrays of components, each of which is obtained by replicating a few simple processing elements connected together in a uniform manner. This paper illustrates the use of Ruby, a language of relations and functions, for describing such networks and for implementing them in hardware. Our objective is to enable designs to be rapidly realised and evaluated.

Ruby has a number of generic relations – such as replication and transposition – that can be used to generate interconnection patterns commonly found in neural systems. It also has a small set of constructors for building composite circuits from simpler ones. These features enable many neural architectures, for instance multi-layer perceptrons and Hopfield networks, to be captured very concisely in Ruby.

We shall also discuss how Ruby can be used to derive, from a simple expression, a complex parametrised representation for a family of architectures. For instance, a parallel design such as the perceptron network shown in Figure 5a can be systematically transformed into a serial architecture like that in Figure 7. This approach permits developing from a high-level description a range of designs with different performance trade-offs, and the features of such designs can be summarised quantitatively – see Table 1 for an example. These tables can be used to find an appropriate implementation for a particular application, given the performance required and the availability of hardware resources.

In the next section we shall provide an overview of our approach, further details of which can be found in Jones and Sheeran (1990) and Luk (1992).

DESIGN REPRESENTATION

A design will be represented by a binary relation of the form $x\ R\ y$ where x and y represent the interface signals and belong respectively to the domain and range of R. For instance, a squaring operation can be described by $x\ sqr\ y \Leftrightarrow x^2 = y$ or, more succinctly, by $x\ sqr\ x^2$.

Transformed or composite circuits are usually described by functions which map one or more relations to a relation. As an example, the converse of R is defined by $x\ R^{-1}\ y \Leftrightarrow y\ R\ x$. It can be considered as a reflected version of R.

Two components Q and R can be connected together if they share a compatible interface s which is hidden in the composite circuit (Figure 1a): $Q; R$ is given by $x\ (Q;R)\ y \Leftrightarrow \exists s.\ (x\ Q\ s)\ \&\ (s\ R\ y)$. For instance, $x\ (sqr;sqr)\ x^4$. This is, of course, just the common definition of relational composition. It is simple to show that relational composition is associative, and that $(Q;R)^{-1} = R^{-1}; Q^{-1}$. A collection of such theorems constitutes a calculus for reasoning about designs, which can usually be used without the need to refer to the meaning of symbols such as Q and R.

As shown later, many useful theorems can be expressed in the form $R = P^{-1}; Q; P$. The pattern $P^{-1}; Q; P$ – in words 'Q conjugated by P' – will be abbreviated as $Q \backslash P$.

a. $Q\,;R$ b. $[Q,R]$ c. $Q \leftrightarrow R$ d. $Q \updownarrow R$

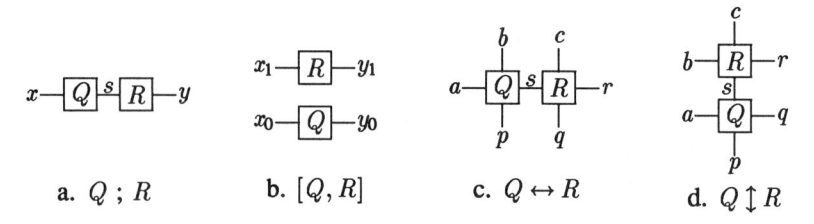

Figure 1 Binary compositions.

Parallel composition of two components Q and R, given by $[Q, R]$ (Figure 1b), represents the combination with no connection between Q and R. Given that a tuple (an ordered collection) of signals are enclosed by angle brackets, parallel composition can be defined by $\langle x_0, x_1 \rangle [Q, R] \langle y_0, y_1 \rangle \Leftrightarrow (x_0\, Q\, y_0)\, \&\, (x_1\, R\, y_1)$; so $\langle x, y \rangle [sqr, (sqr; sqr)] \langle x^2, y^4 \rangle$. One can easily check that $[P, Q]; [R, S] = [P; R, Q; S]$, and that $[P, Q]^{-1} = [P^{-1}, Q^{-1}]$.

There are several operations involving pairs of signals that we will require. First of all, given that ι is the identity relation, we have the abbreviations $\mathrm{fst}\, R = [R, \iota]$, and $\mathrm{snd}\, R = [\iota, R]$. Next, the relation *fork* can be used to duplicate a signal, since x *fork* $\langle x, x \rangle$. The projection relations π_1 and π_2 extract an element from a pair: $\langle x, y \rangle\, \pi_1\, x$ and $\langle x, y \rangle\, \pi_2\, y$. Finally, we need to be able to swap the elements of a pair: $\langle x, y \rangle$ *swap* $\langle y, x \rangle$. Examples of theorems involving these operations include $\mathrm{fst}\, Q\,;\,\mathrm{snd}\, R = \mathrm{snd}\, R\,;\,\mathrm{fst}\, Q = [Q, R]$ and $[Q, R] \backslash swap = swap\,;\,[Q, R]\,;\,swap = [R, Q]$. It should also be clear that $fork; [\pi_1, \pi_2] = [\iota, \iota]$, and that $\pi_1^{-1}; \pi_1 = \iota \neq \pi_1; \pi_1^{-1}$ in general and similarly for π_2.

A rectangular component with connections on every side is modelled by a relation that relates 2-tuples, with the two components in the domain corresponding to signals for the west and north side and those in the range corresponding to signals for the south and east side. Such components can be assembled together by the beside ($\leftrightarrow$) and below ($\updownarrow$) operators (Figure 1c and Figure 1d): $\langle a, \langle b, c \rangle \rangle\, (Q \leftrightarrow R)\, \langle \langle p, q \rangle, r \rangle \Leftrightarrow \exists s.\, \langle a, b \rangle\, Q\, \langle p, s \rangle\, \&\, \langle s, c \rangle\, R\, \langle q, r \rangle$ and $Q \updownarrow R = (Q^{-1} \leftrightarrow R^{-1})^{-1}$. Theorems that have been proved for beside can readily be adapted for below, and vice versa.

It is also useful to have a conjugate operator for pairs: $Q \backslash\backslash [R, S] = [S^{-1}, R^{-1}]; Q; [R, S]$. Given that the conjugate operators have a lower precedence than all other operators except relational composition, one can show that $Q \backslash\backslash R = R^{-1} \backslash swap\,;\,Q\,;\,R$, and that $\mathrm{snd}\, Q^{-1}; R; \mathrm{fst}\, Q = R \backslash\backslash (\mathrm{fst}\, Q)$. We shall also use the abbreviations $\mathrm{fsth}\, R = R \leftrightarrow swap$, $\mathrm{fstv}\, R = R \updownarrow swap$, and $\mathrm{fstvh}\, R = \mathrm{fstv}(\mathrm{fsth}\, R)$.

Repeated compositions

Let us now look at the ways that we describe one- and two-dimensional arrays of components. Repeated relational composition of a given relation R cascades together copies of R (Figure 2a); it is defined inductively by the equations $R^1 = R$ and $R^{n+1} = R^n\,;\,R$.

Repeated parallel composition, map R (Figure 2b), relates two equal-length tuples such that the corresponding elements of the tuples are related by R (note that $\#x$ denotes the number of elements in tuple x):

$$\text{if } \#x = \#y = N \text{ then } x\,(\mathrm{map}\, R)\, y \Leftrightarrow \forall i : 0 \leq i < N.\, x_i\, R\, y_i.$$

For clarity, on some occasions we shall make explicit the number of R's in a map and write it as $\mathrm{map}_N\, R$. This expression can be considered to be an abbreviation of $\mathrm{map}\, R \setminus N$ where N is the identity relation on N-tuples.

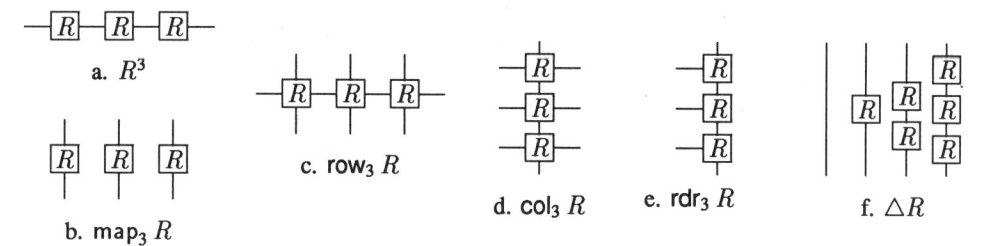

Figure 2 Repeated compositions.

A row of components (Figure 2c) is built from repeated composition of beside, and can be described by

if $\#x = \#y = N$ and $ax = \langle a, x \rangle$ and $yb = \langle y, b \rangle$ then
$$ax \, (\text{row } R) \, yb \iff \exists s. \, (s_0 = a) \, \& \, (s_N = b) \, \& \, \forall i : 0 \le i < N. \, \langle s_i, x_i \rangle \, R \, \langle y_i, s_{i+1} \rangle.$$

A column of components (Figure 2d) can be obtained from $\text{col } R = (\text{row } R^{-1})^{-1}$. A degenerate form of col, called a right-reduction (rdr, Figure 2e), is also frequently used; it describes the result of applying a binary operation on a tuple in a right-associative manner, like $\langle\langle a, b, c \rangle, z \rangle \, (\text{rdr } add) \, x \iff a + (b + (c + z)) = x$. Right reduction can be defined by $\text{rdr } R = \text{col}\,(R \, ; \, \pi_1^{-1}) \, ; \, \pi_1$. The corresponding degenerate version of row, known as left-reduction, is given by $\text{rdl } R = \text{row}\,(R \, ; \, \pi_2^{-1}) \, ; \, \pi_2$.

We shall also need the relation $\triangle R$ (Figure 2f) which relates two equal-length tuples such that their i-th elements relate to each other according to R^i. The $\triangle$ operator is useful for formulating distributive theorems for col: on the assumption that $[A, B]; R; \text{snd } C = R; \text{fst } B$, one can show that

$$\text{col}_n \, (\text{snd}\,B; R) \;=\; [\triangle A, \, B^n] \, ; \, \text{col}_n \, R \, ; \, \text{snd} \, \triangle C. \tag{1}$$

The use of this equation in pipelining designs will be explained later.

Sometimes we shall need to interleave an array of components from two equal-length tuples. This can be achieved by zip, given by $\langle x, y \rangle \, zip \, z \iff \forall i : 0 \le i < N. \, \langle x_i, y_i \rangle = z_i$, on the assumption that $\#x = \#y = \#z = N$. For instance, $\langle\langle 1, 2, 3 \rangle, \langle 4, 5, 6 \rangle\rangle \, zip \, \langle\langle 1, 4 \rangle, \langle 2, 5 \rangle, \langle 3, 6 \rangle\rangle$.

Sequential circuits and serialisation

So far we have been using relations to model a static situation – the steady state behaviour of a circuit at a particular instant of time. To deal with sequential circuits, an expression is interpreted as a relation that relates a *stream* in its domain to a stream in its range. For our purpose, a stream can be considered to be a doubly-infinite tuple containing data at successive clock 'ticks'. Notice that the clock is an abstract means for specifying data synchronisation, and it may be realised either by a global synchronous clock or by some hand-shaking mechanism.

We shall use x_t to denote the t-th element from some reference point – such as the time when the circuit is initialised – in the stream x; given that x_t is a tuple, $x_{t,i}$ is its i-th element. An adder can be described in the stream model as $x \, add \, y \iff \forall t. \, x_{t,0} + x_{t,1} = y_t$.

There are two primitives that do not possess a static interpretation. The first is *delay*, $\mathcal{D}$, defined by $x \, \mathcal{D} \, y \iff \forall t. \, x_{t-1} = y_t$. An *anti-delay* $\mathcal{D}^{-1}$ is such that $\mathcal{D}; \mathcal{D}^{-1} = \mathcal{D}^{-1}; \mathcal{D} = \iota$. A latch is modelled by a delay with data flowing from domain to range, or by an anti-delay with data flowing from range to domain.

For a circuit R which contains no primitives that possess a measure of absolute time, it is the case that $\mathcal{D}; R = R; \mathcal{D}$. With $A = B = \mathcal{D}$ and $C = \mathcal{D}^{-1}$, the pre-condition for Equation 1 becomes valid so that the transformation can be applied to distribute latches among the R's to reduce the longest combinational path. This process is usually called *retiming*, and examples of deriving pipelined circuits based on an algebraic treatment of retiming can be found elsewhere (Jones and Sheeran 1990, Luk 1992).

A serial design R with an internal feedback path can be modelled by the loop construct in Figure 3. One can show that $\nu R = (\text{fstv } R) \backslash\backslash \text{snd } sndfb^{-1}$ where $x \ sndfb \ \langle\langle x, s\rangle, s\rangle$.

$$\langle x, u\rangle \ (\nu R) \ \langle y, v\rangle \Leftrightarrow \exists s. \ \langle\langle x, s\rangle, u\rangle \ R \ \langle y, \langle v, s\rangle\rangle$$

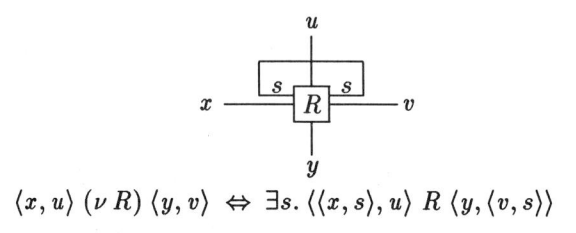

Figure 3 A function that describes designs with feedback.

The intuitive idea behind our serialisation equations, the details of which are included in Luk (1992), is to circulate data through a processor n times to emulate the effect of n cascaded processors. A multiplexer $\underline{cmx_n}$ controls when to accept external data x and feedback data y: $\langle \ldots, \langle x_0, y_0\rangle, \langle x_1, y_1\rangle, \langle x_2, y_2\rangle, \ldots\rangle \ \underline{cmx_3} \ \langle \ldots, x_0, y_1, y_2, x_3, y_4, y_5, \ldots\rangle$, and the relation $\underline{bundle_n}$ describes converting between serial and parallel data: $\langle \ldots, x_0, x_1, x_2, x_3, x_4, x_5, \ldots\rangle$ $\underline{bundle_3} \ \langle \ldots, \langle x_0, x_1, x_2\rangle, \langle x_3, x_4, x_5\rangle, \ldots\rangle$. The relations $\underline{ev_n^{-1}}$ and $\underline{ev_n}$ are used to inject and to reject dummy data when the processor is in feedback mode: $\langle \ldots, x_0, x_1, x_2, \ldots\rangle \ \underline{ev_3}$ $\langle \ldots, x_0, x_3, x_6, \ldots\rangle$. The number of latches in a serialised processor, $slow_n R$, has to be n times of that of the unserialised version R, since it contains up to n interleaved computations with each corresponding to a copy of R. As an example, the following equation can be used to serialise a row of components:

$$\text{row}_n R = \nu \left(\text{fst } \underline{cmx_n} \ ; \ \text{slow}_n R \ ; \ \text{snd}(\mathcal{D}; fork)\right) \backslash\backslash \ [\underline{bundle_n}, \underline{ev_n}] \ ; \ \text{snd } \mathcal{D}^{-1}. \quad (2)$$

Again the corresponding theorem for a column of components can be obtained by substituting row R by $(\text{col } R^{-1})^{-1}$.

DEVELOPING PERCEPTRONS

First, recall that if x is a tuple, then $\langle x, 0\rangle \ (\text{rdr } add) \ \sum_i x_i$. Let $x \ !c \ y \Leftrightarrow x = y = c$, and $sndzero = \pi_1^{-1}; \text{snd }!0$. Then $x \ (sndzero; \text{rdr } add) \ \sum_i x_i$.

Given input x_j and weights $w_{i,j}$ where $0 \leq i < m$ and $0 \leq j < n$, a node in a perceptron computes the output $y_i = th \left(\sum_j w_{i,j} \times x_j\right)$ where th is a threshold function such as the sigmoid function; that is, $\langle x, w_i\rangle \ zmadds \ y_i$ where $zmadds = zip; sndzero; \text{rdr}_n \ madd; th$, and $madd = \text{fst } mult; add$. To pass the value of x to a neighbouring node, we use the wiring cell $wire1 = fork; \text{snd}\pi_1$ to implement a broadcast circuit, so that $\langle x, w_i\rangle \ node1 \ \langle y_i, x\rangle$ where $node1 = wire1; \text{fst } zmadds$.

A layer in a perceptron consists of a row of m nodes, $layer1 = \text{row}_m \ node1 \ ; \ \pi_1$ (Figure 4a), and our first description of a multi-layer perceptron, $mlp1$ (Figure 4b), is assembled by arranging the layers according to left reduction:

$$mlp1 = \text{rdl } layer1 = \text{row}\left(layer1 \ ; \ \pi_2^{-1}\right) ; \pi_2.$$

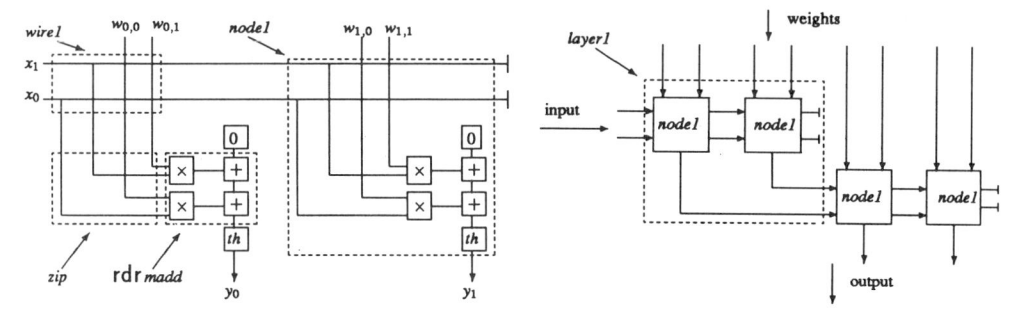

Figure 4a Design $layer1$ ($m = n = 2$).

Figure 4b Design $mlp1$.

Our next task is to distribute the multiply-adders $madd$ among the buses in the broadcast cell $wire1$; this transformation does not substantially improve performance by itself, but it enables further transformations such as pipelining and serialisation to be applied. Using equations such as $wire1 = \text{fst} fork; \text{fstv} wire1; \text{snd}\pi_2$, $zip = (\text{map } wire1)\backslash zip^{-1}$ and $fork; zip = \text{map } fork$, we obtain $layer2$ (Figure 5a) which has a more uniform layout:

$$\begin{aligned} layer2 &= \text{snd} (\text{map } sndzero) \,;\, \text{row}_m \, node2 \,;\, \pi_1, \\ node2 &= wire2 \leftrightarrow (\text{col}_n \, madd2; \text{fst} th); \text{fst}\pi_2, \\ wire2 &= \text{map} (wire1; \pi_2^{-1})\backslash zip^{-1}, \\ madd2 &= \text{fstv} (madd; \pi_1^{-1}); \text{snd}\pi_2. \end{aligned}$$

It can be shown that snd $sndzero$; $node2 = node1$, and the size and performance of $layer1$ and $layer2$ are identical if area and delay of wires are ignored.

Pipelining and serialisation

Since $mlp2 = \text{row} (layer2 \,;\, \pi_2^{-1}) \,;\, \pi_2$, we can use the row version of Equation 1 to pipeline it and Equation 2 to serialise it; one possibility is shown in Figure 5b. There are further opportunities in transforming the architecture of $layer2$, and we shall consider some of these next.

If all the coefficients w_i's are hardwired in $node2$, we can eliminate the $wire2$ block to give $node3$, which behaves like snd (fst $[w_i \,|\, 0 \leq i < n]$); $node2$ while having a simpler structure. Given that icol $\langle P, Q, R \rangle$ describes a column of heterogeneous components with P below Q below R, then

$$\begin{aligned} node3 &= \text{icol} \,\langle madd3_i \,|\, 0 \leq i < n \rangle, \\ madd3_i &= \text{fst} (fork; \text{fst} (\pi_1^{-1}; \text{snd} \,! \, w_i)) \,;\, madd2. \end{aligned}$$

To produce a faster circuit, a theorem similar to Equation 1 can be used to pipeline $node3$; the resulting design, $node4 = \text{icol} \, \langle madd3_i; \text{fst} \mathcal{D} \,|\, 0 \leq i < n \rangle$, is shown in Figure 6.

On the other hand, if we want to reduce the number of multiply-adders in $layer2$, theorems such as Equation 2 can be used to serialise it. Given that $m = ap$ such that $1 < a \leq m$ and $x \, sndfb \, \langle \langle x, y \rangle, y \rangle$, we can reduce the number of columns in $layer2$ by a factor of a by

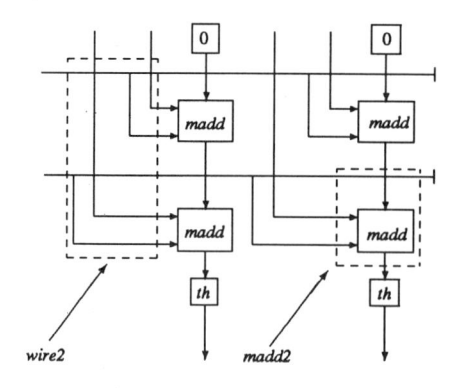

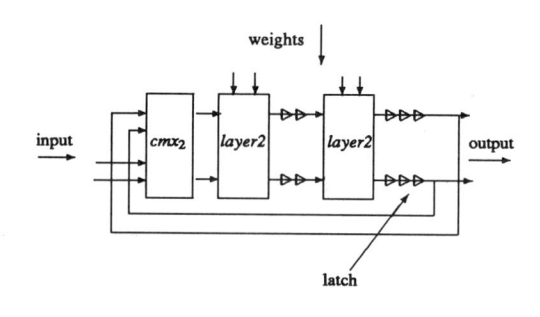

Figure 5a Design $layer2$ ($m = n = 2$).

Figure 5b Pipelined and serialised $mlp2$.

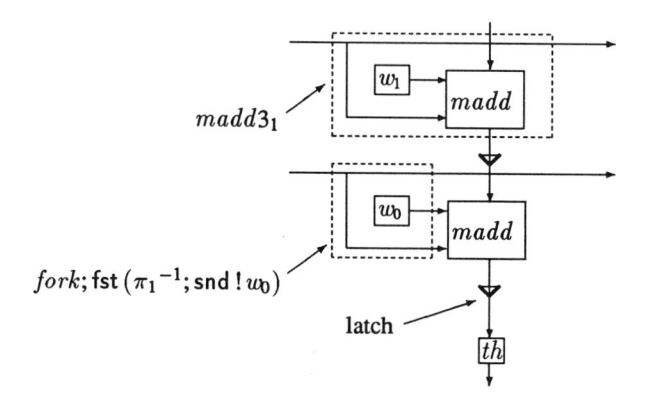

Figure 6 Design $node4$ ($n = 2$).

serialising it horizontally to obtain

$$\begin{aligned}
layer5 &= pre'layer5 \; ; \; \text{row}_p \; node5 \; ; \; \text{snd}\,(\text{map}\,(\text{fst}\mathcal{D};fork^{-1})) \; ; \; \pi_1, \\
node5 &= wire5 \leftrightarrow (\text{col}_n\,(\text{fstv}\,madd2;\text{fst}th));\text{fst}\pi_2, \\
pre'layer5 &= [\text{map}\,(sndfb;\text{fst}\,\underline{cmx_a}),\,\text{map}\,sndzero], \\
wire5 &= \text{map}\,(\text{fstv}\,(wire1;\pi_2^{-1}))\backslash zip^{-1}.
\end{aligned}$$

An example of $layer5$ will look like the one in Figure 7 without the vertical feedback wires and the associated latches. Instantiating $layer5$ with $a = m$ and $p = 1$ gives a design which is similar to that described by Baji and Inouchi (1992).

Another possibility is to reduce the number of rows of cells in $layer2$ by a factor of b (where $n = bq$ and $1 < b \leq n$) by serialising it vertically; this gives

$$\begin{aligned}
layer6 &= \text{snd}\,(\text{map}\,sndzero) \; ; \; \text{row}_m \; node6 \; ; \; \pi_1, \\
node6 &= pre'node6 \; ; \; wire2 \leftrightarrow (\text{col}_q\,(\text{fsth}\,madd2)) \; ; \; post'node6, \\
pre'node6 &= \text{snd}\,(\text{snd}\,(sndfb;\text{fst}\,\underline{cmx_b})), \\
post'node6 &= \text{fst}\,(\pi_2;\text{fst}\mathcal{D};fork^{-1};th).
\end{aligned}$$

Notice that the critical path is also reduced by a factor of b. Instantiating $layer6$ with $b = n$ and $q = 1$ gives a design which is similar to that described by Skubiszewski (1992).

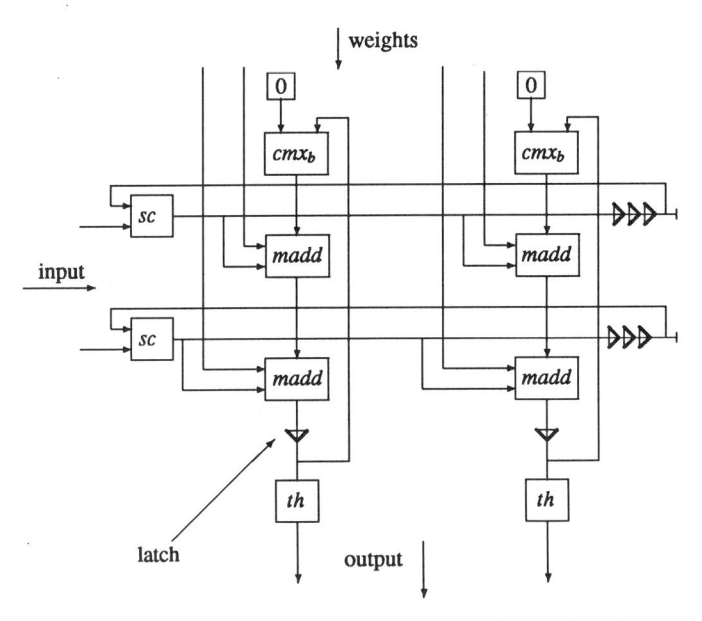

Figure 7 Design $layer7$ ($sc = \underline{scm}_{b,a}$, $m = n = 6$, $a = b = 3$, $p = q = 2$).

Finally, we describe the design $layer7$ (Figure 7), obtained by serialising $layer2$ horizontally by a factor of a and then vertically by a factor of b:

$$layer7 = pre'layer7 \ ; \ \text{row}_p \ node7 \ ; \ \text{snd} \left(\text{map} \left(\text{fst} \mathcal{D}^b; fork^{-1} \right) \right) \ ; \ \pi_1,$$
$$node7 = pre'node6 \ ; \ wire5 \leftrightarrow (\text{col}_q \ (\text{fstvh} \ madd2)) \ ; \ post'node6,$$
$$pre'layer7 = [\text{map} \ (sndfb; \text{fst} \ \underline{scm}_{b,a}), \text{map} \ sndzero],$$

where $scm_{b,a} = \text{slow}_b(\underline{cmx}_a)$ is a component that repeatedly extracts for b cycles the first element of a pair and for the next $(a-1)b$ cycles the second element; for instance

$$\langle \ldots, \langle x_0, y_0 \rangle, \langle x_1, y_1 \rangle, \langle x_2, y_2 \rangle, \ldots \rangle \ \underline{scm}_{2,3} \ \langle \ldots, x_0, x_1, y_2, y_3, y_4, y_5, x_6, x_7, y_8, \ldots \rangle.$$

A design similar to $layer7$ can be obtained by first serialising $layer2$ vertically and then horizontally; it will look like $layer7$ but with more latches on the vertical wires than on the horizontal wires. Note that the multiplier can itself be serialised: one such strategy can be found in Murray *et al* (1987).

The features of our designs are summarised in Table 1; note that T_{ma} and T_{th} correspond to the combinational delay of cell $madd$ and th, and wire delays are assumed to be insignificant. Such tables, when they are reasonably complete, can be used in checking whether designs can be appropriately parametrised to meet requirements for a specific application. Promising designs can then be implemented on Field-Programmable Gate Arrays using the prototype compilers for various dialects of Ruby (Luk and Page 1991).

EXAMPLE

In this section we report some experimental results from software simulations which can be used to guide the construction of hardware accelerators for neural systems.

The benchmark problem we examined was learning the parity of some set number of binary inputs (Tesauro and Janssens 1988); we concentrated on the 4-parity problem. We

Table 1 Comparison of perceptron designs for computing $th\left(\sum_j w_{i,j} \times x_j\right)$, where th is a threshold function and $0 \leq i < m$ and $0 \leq j < n$.

Design	Serialisation factor	Minimum cycle time	Number of inputs and outputs	Number of *madd* in array	Number of *th* in array	Number of latches in array
layer2	1	$nT_{ma} + T_{th}$	$m + n + mn$	mn	m	0
layer3	1	$nT_{ma} + T_{th}$	$m + n$	mn	m	0
layer4	1	$\max(T_{ma}, T_{th})$	$m + n$	mn	m	mn
layer5	a	$nT_{ma} + T_{th}$	$(m + na + mn)/a$	mn/a	m/a	n
layer6	b	$\max(nT_{ma}/b, T_{th})$	$(n + mb + mn)/b$	mn/b	m	n/b
layer7	ab	$\max(nT_{ma}/b, T_{th})$	$(na + mb + mn)/ab$	mn/ab	m/a	$(m/a) + n$

studied three-layer feed-forward perceptrons with 12 units in the hidden layer, using the standard sigmoid threshold function. For simplicity in simulation, no momentum term was used in back-propagation training. For the same reason we used incremental learning, back-propagating error and updating connection weights after each presentation of a training case.

A network simulator was written in C, using a fixed-point representation for integers. The results of all arithmetic operations were clipped to within a range determined by the number of bits being employed. The sigmoid function was evaluated using floating point exponentiation and division, but the result was appropriately quantised. Two questions regarding this fixed-point approximation are: What is the minimum acceptable *range*? What is the minimum acceptable *precision*?

Investigations into the minimum range revealed that no clipping occurred when five bits (including the sign bit) were used for representing integers, and the amount of clipping that occurred when four bits were employed had no significant effect on training.

To investigate precision, we trained nets with different fixed-precisions, but all with 4 bits for sign and integer. Training was considered to have succeeded when all outputs were within 0.4 of their desired values. If success had not been achieved within 2000 epochs, the net was regarded as having failed to train. Table 2 gives the (hypergeometric) mean number of epochs for training, and the number of times that training failed in a series of 100 trials. The number of bits given is the *total* number in the fixed-point representation. Performance is satisfactory with 13 bits or more; with fewer, training fails too often.

Table 2 Effect of precision on training time.

Bits	17	16	15	14	13	12	11	10
Average epochs	192	189	185	183	176	191	280	557
Failures	0	1	2	5	10	27	51	76

Although at least 13 bits are required for training, fewer are needed when applying a net. This was investigated by training a network with 15-bit fixed-point numbers, then truncating those weights successively down to 6 bits (Table 3). The second line of the table gives the percentage of trials in which there was no deterioration in performance. Truncation has a steady cumulative effect down to 8 bits, after which performance collapses. A more sympathetic approach is to train the network to a greater degree; outputs must be within 0.2 of their desired values for success during training, but only within the 0.4 threshold when testing (Table 3, third line). We now see no deterioration in performance down to 9 bits, and the success rate at 8 bits is acceptable. The effect of reducing the range for execution was also explored: a reduction to 3 integer bits always significantly reduced the success rates.

Table 3 Effect of truncation on execution.

Bits	15	14	13	12	11	10	9	8	7	6
% (threshold 0.4)	100	91	89	88	77	60	68	62	11	0
% (threshold 0.2)	100	100	100	100	100	100	100	97	46	0

Thus, for training, we found that a minimum of 13 bits are required for the numerical representation, with 4 integer bits. For execution of a pre-trained network, however, as few as 8 bits may be sufficient. These results agree with those of other studies (Holt and Hwang 1992).

We then used our compiler and timing analysis tools to estimate the speed of multipliers implemented in Field-Programmable Gate Arrays manufactured by Algotronix Limited (Algotronix 1990). Using a simple shift-and-add architecture, the maximum clock frequency was found to range from 1.1 MHz for multiplying two 13-bit numbers, to 1.8 MHz for multiplying two 8-bit numbers, to 7.5 MHz for multiplying two 2-bit numbers. A fully-pipelined multiplier, operating in a bit-serial or in a bit-parallel fashion, can run at 16 MHz; this would be attractive for applications such as video processing which demands high-throughput while tolerating large latency. A more detailed evaluation of various ways of implementing neural structures on a number of hardware platforms is currently being undertaken.

Note that the threshold function th can be implemented as a look-up table. An example of how this was achieved is given in Cox and Blanz (1992).

CONCLUDING REMARKS

We have described a method of developing parametrised descriptions of neural networks with different trade-offs in size and performance. Our framework provides a basis for theories and computer-based tools to systematise and formalise design expertise, so that a variety of architectures can be generated and evaluated rapidly. Future work will include conducting further case studies, enhancing our libraries of components and transformations, and extending them to handle optimisations such as weight sharing (Boser *et al* 1992).

ACKNOWLEDGEMENTS

The support of Rank Xerox (UK) Limited, the U.K. Science and Engineering Research Council (GR/F47077), Scottish Enterprise and Algotronix Limited is gratefully acknowledged.

REFERENCES

Algotronix Limited, *CAL 1024 Datasheet*, 1990.

Baji, T. and Inouchi, H., "Systolic Processor Elements for a Neural Network", US Patent 5,091,864, 25 February 1992.

Boser, B.E., Sackinger, E., Bromley, J., IeCun, Y. and Jackel, L.D., "Hardware Requirements for Neural Network Pattern Classifiers", *IEEE Micro*, February Issue, pp. 32–40, 1992.

Cox, C.E. and Blanz, W.E., "Ganglion – A Fast Field-Programmable Gate Array Implementation of a Connectionist Classifier", *IEEE J. Solid-State Circuits*, vol. 27, pp. 288–299, 1992.

Holt, J.L. and Hwang, J.N., "Finite Precision Error Analysis of Neural Network Hardware", to appear in *IEEE Trans. Neural Networks*, 1992.

Jones, G. and Sheeran, M., "Circuit Design in Ruby", in *Formal Methods for VLSI Design*, J. Staunstrup (ed), North-Holland, pp. 13–70, 1990.

Luk, W., "Systematic Serialisation of Array-Based Architectures", to appear in *Integration, Special Issue on Algorithms and VLSI Architectures*, 1992.

Luk, W. and Page, I., "Parametrising Designs for FPGAs", in *FPGAs*, W. Moore and W. Luk (ed), Abingdon EE&CS Books, pp. 284–295, 1991.

Murray, A.F., Smith, A.V.W. and Butler, Z.F., "Bit-Serial Neural Networks", *Proc. BIPS Conf.*, pp. 573–583, 1987.

Skubiszewski, M., "A Hardware Emulator for Binary Neural Networks", *Proc. FPL 92*, Vienna, 1992.

Tesauro, G. and Janssens, B., "Scaling Relationships in Back-Propagation Learning", *Complex Systems*, vol. 2, pp. 39-84, 1988.

KNOWLEDGE PROCESSING IN NEURAL ARCHITECTURE

G. Palm, A. Ultsch, K. Goser and U. Rückert

INTRODUCTION

In recent years there has been an increasing interest in the use of artificial neural networks (ANNs) for technical applications (e.g. Rogers 1990). Particularly attractive is the application of ANNs in those domains where at present humans outperform any currently available high performance computers, e.g. in areas like auditory perception, vision, or sensory-motor control. Neural information processing is expected to have a wide applicability in areas that require a high degree of flexibility and the ability to operate in uncertain environments where information usually is partial, fuzzy, or even contradictory. The computing power of biological neural networks stems to a large extend from a highly parallel, fine-grained and distributed processing and storage of information as well as from the capability of learning.

At present, ANNs are relatively successful in technical applications dealing with *sub-symbolic raw data*, in particular, if the data are noisy or inconsistent. Such subsymbolic-level processing seems to be appropriate for perception tasks and perhaps even for tasks that call for combined perception and cognition. For example, ANNs are able to learn structures of an input set without using a priori information. Unfortunately they cannot easily explain their behaviour because a distributed representation of the kowledge is used. They only can tell about their knowledge by showing responses to given inputs.

Applications of expert systems have been successful in areas like diagnosis, construction and planning. This shows the usefulness of a *symbolic* knowledge processing and representation on which traditional artificial intelligence (AI) relies. An important property of knowledge stored in symbolic form is that it can be interpreted and communicated by experts or computers. The limits of such an approach, however, become quite evident when sensor data or measurement data, for example from physical processes, are handled. Inconsistent data can force symbolic systems into an undefined state. A further problem in expert system design is the acquisition of knowledge. It is almost impossible for an expert to describe his domain specific knowledge entirely in form of rules or other knowledge representation schemes. In addition, it is very difficult to describe knowledge acquired by experience. One can automatize the knowledge acquisition process by using machine learning strategies (Michalski *et al* 1984); but all these symbolic algorithms need very much domain specific knowledge a priori because they cannot learn the necessary internal representation. Furthermore, most of the known automatic as well as non-automatic strategies are in general very time-consuming and expensive.

Both approaches of modelling brain like information processing capabilities are complementary in the sense that traditional AI is a top-down approach starting from high-level cognitive brain functions whereas ANNs are a bottom-up approach on a biophysical basis of neurons and synapses. It is a matter of fact that the symbolic as well as the subsymbolic aspects of information processing are essential to systems dealing with real-world tasks. When we try to recognize a certain pattern, it helps if we know what we are looking for (Pao 1989). Linking symbolic and subsymbolic information processing is certainly a challenging research task. It is the research goal of our project to find bridges of understanding between symbolic and neural information processing.

The project presented in this paper, titled *Knowledge Processing in Neural Architecture* (WINA), is part of the INA (Information Processing in Neural Architecture) program funded by the german ministry of research and technology (BMFT) in which currently 11 research projects and a total of 43 national research teams are active (Reuse 1992). The partners of the WINA project are the University of Ulm, Dept. of Neural Information Processing (Neuroinformatik), headed by Prof. Palm, and the University of Dortmund, Dept. of Electrical Engineering (Prof. Goser and Dr. Rückert) and the Dept. of Computer Science (Dr. Ultsch). The project coordinator is Prof. Palm covering the project period 1991-1993. The detailed scientific objectives and first results of our project will be presented in the following.

A HYBRID KNOWLEDGE PROCESSING SYSTEM

The overall topic of the research project is the implementation of a hybrid knowledge processing system in which ANNs and conventional rule based components are merged. The architecture of the hybrid system is shown in Fig. 1. A realization of a prototype system has a couple of interesting consequences. For example, interfaces between the different components have to be developed in order to get a working prototype. Theoretical results concerning the application of different ANNs for certain subtasks within such a system could be verified directly. Last but not least, test applications of the prototype system will give an idea of the computational requirements of the system which may motivate parallel hardware implementation of neural subcomponents. In the following paragraphs the components of the hybrid knowledge processing system as shown in Fig. 1 will be discussed in more detail.

Neural Classifier

The main function of this module is the analysis and the preprocessing of large (subsymbolic) data sets. The task is to explore properties and find structures in a given data set. Methods for data analysis are well known from multivariate statistics (e.g. cluster analysis, Jain and Dubes 1989) and pattern recognition (e.g. Tou and Gonzales 1974, Pao 1989). Furthermore, ANNs have been proposed by several authors for this task as well (e.g. Kohonen 1984, Oja 1989, Palm 1982, Pao 1989, Ultsch and Siemon 1990). Connectionist systems are claimed to have advantages over conventional methods in handling noisy and incomplete data. In addition, by utilizing the parallelism inherent in ANNs an efficient implementation for real-time applications is feasible. An important piece of theoretical work required in this research area is development of appropriate evaluation criteria for a comparison of the different methods. This is a theoretical subgoal of our project.

At present, the WINA project is engaged mainly in the investigation of cluster analysis methods. In addition to classical methods known from multivariate statistics, selforganizing feature maps (SOFMs) as proposed by Kohonen (1984) are investigated for the use of exploratory data analysis (Ultsch and Siemon 1990). A SOFM uses an unsupervised learning algorithm to adapt itself suitably to the structure of a given (high dimensional) data space. The algorithm can be thought of as a mapping from R^n to a flattened two-dimensional surface (layer of neurons) such that interesting topological relations and the point density of the vectors are conserved (Kohonen 1984).

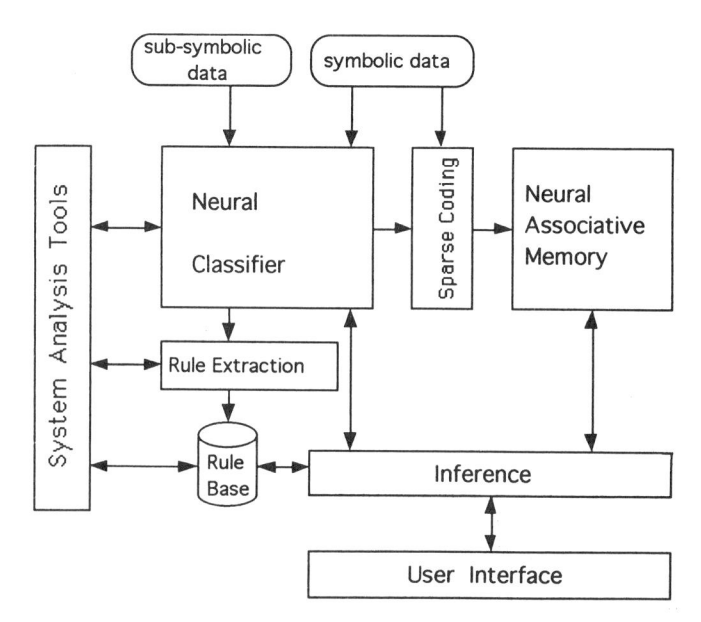

Figure 1 Architecture of the hybrid knowledge processing system

The main idea of our approach is illustrated in Fig. 2. The data set under consideration, for example the three-dimensional test data set (about 100 vectors) as shown in Fig. 2a, has to be trained to a two-dimensional SOFM. In general, a transformation of the vectors in the data set is required so that the vector components have equal scopes, for example by the use of a z-transformation (Ultsch *et al* 1991). Because the SOFM performs a topology conserving mapping we expect input vectors which are close in the input space to be mapped onto neurons which are close in the two-dimensional SOFM. In total, for all input vectors we expect a similar clustering on the SOFM as given in the input space.

In Fig. 2b a trained SOFM (20x20 neurons) for the test data set (Fig. 2a) is shown. As expected, the vectors of different clusters in the input space are mapped to different regions of the SOFM and the topology relationship of the clusters is preseved, too. Furthermore, we can see that the point density distribution of the input vectors is represented on the SOFM as well. Clusters with a high vector density take a larger area on the SOFM. No input vectors are mapped to the black area on the SOFM.

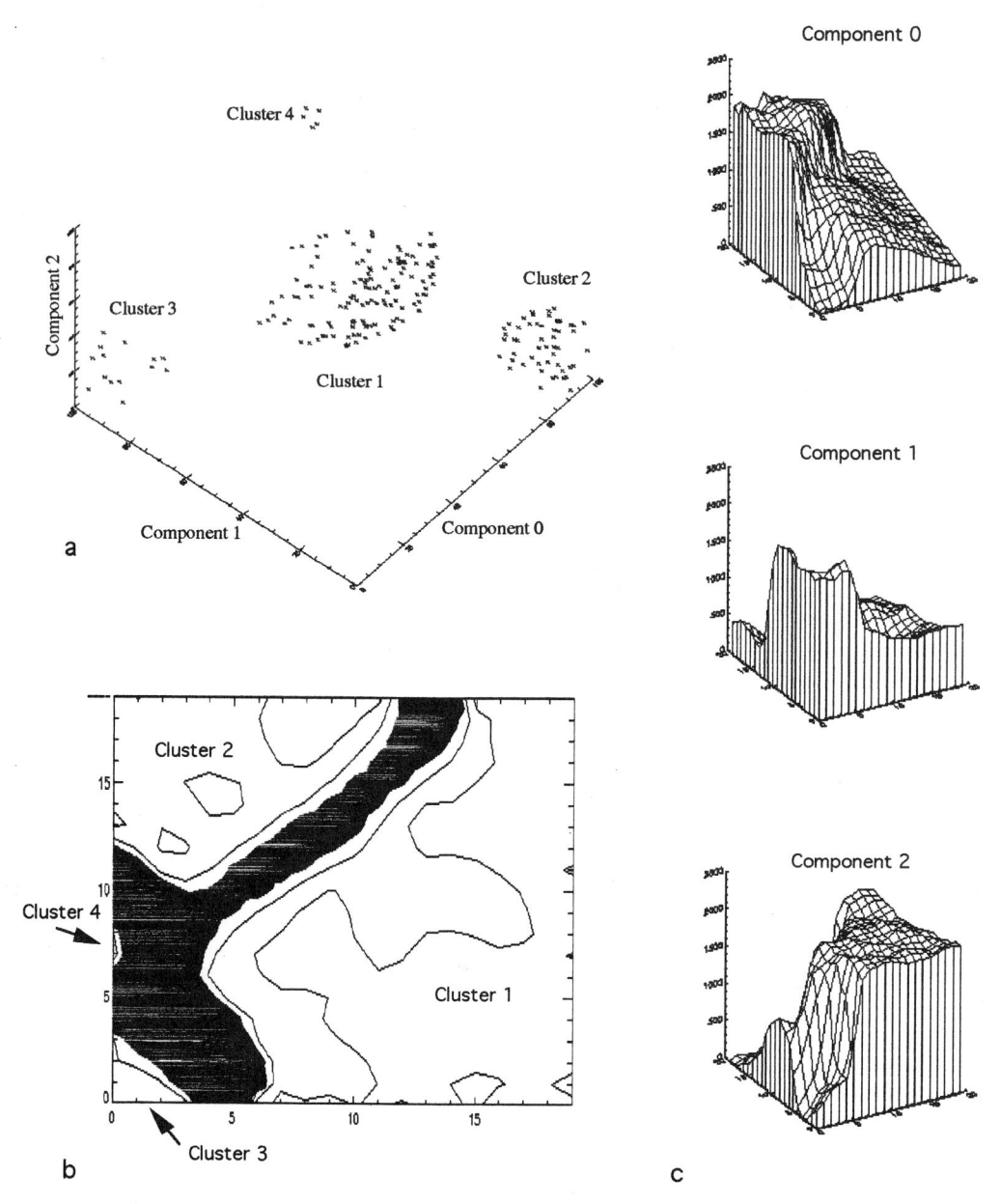

Figure 2 Use of a SOFM for data analysis

However, without knowing the cluster membership of each input vector the clustering as shown in Fig. 2b could not be detected. Consequently, for the application of SOFM an automatic method for visualization of clusters is required. Such a method was developed by Ultsch and Siemon (1989), the so called *unified distance matrix* or short *U-matrix method*. Fig. 2b shows the plot of the resulting U-matrix for the test data displaying its elements as hight over a grid that corresponds to the lattice of the SOFM. This display has valleys (white areas) where the weight vectors in the map are close to each other and hills or walls (black areas) where the weight vectors in the map have large distances. As can be seen, the SOFM is now divided into four different regions, corresponding to the four data clusters. The regions are seperated by a black *wall*, representing a large weight vector to weight vector distance of adjacent neurons and indicating the dissimilarity of the vectors.

In summary, by using the U-matrix method structure in data sets can be detected automatically as classes. These classes represent sets of data that have something in common. Within our project we are now analysing the properties of such methods in comparison to conventional cluster analysis algorithms in more detail.

Rule Extraction

The rule extraction module aims at automatic discovery of the properties of each cluster detected by the neural classifier and its reformulation into a symbolic form. Therefore, the structure learned by the ANN has to be examined and subsequently transformed into PROLOG rules. These rules can then be inspected by a human expert and added to a rule base.

Different approaches are under development within the project at the moment. For example, the well known machine learning algorithm ID3 (Michalski *et al* 1984), an algorithm especially developed for ANN called sig* (Ultsch 1991, Ultsch 1992), and new methods for the generation of fuzzy rules (Ultsch and Höffgen 1991, Surmann *et al* 1992) have been implemented already. All these approaches are based on the clustering result of a SOFM and the resulting U-matrix. By inspecting the individual components of the neurons weight vectors the implemented methods try to discover algorithmically the properties of the clusters. For example, by taking a look at the component cards (Fig. 2c) of the trained SOFM in Fig. 2 it is possible to formulate simple fuzzy rules:

If comp0 is high and comp1 is low and comp2 is low   then vector belongs to cluster2
If comp0 is high and comp1 is high and comp2 is medium then vector belongs to cluster4
....

Obviously, the implementation of the above mentioned intuitive operation *by looking at the component cards* is not straight forward. The proposed algorithms are now tested and evaluated with real life application data in more than three dimensions. First results are reported in Ultsch and Siemon (1989), Ultsch and Höffgen (1991) and Surmann *et al* (1992). Once again, an important aspect is the definition of appropriate evaluation criteria in respect to a quantitative comparison of the different methods.

In summary, the rule extraction module aims at the automatic acquisition of knowledge from a set of examples. Such a module enhances the reasoning capability of classical expert systems with the ability of generalization and the handling of incomplete data.

Neural Associative Memory

One of the application areas, where the computational analysis of the performance of a neural network implementation is relatively advanced is associative memory. Historically, associative memory modelling has been a major focus of neural network research (Kohonen 1977, Kohonen 1984, Palm 1980, Palm 1987, Shaw and Palm 1988). ANNs are well suited for the implementation of neural associative memories (NAMs), at least because the processing elements (artificial neurons) in an ANN operate in a highly parallel way and thus a considerable gain in speed is to be expected (Rückert 1987, Rückert1991). This parallelism is one of the major reasons for investigating new computational models inspired by neurophysiological processing principles.

Another interesting feature of NAMs is that information is stored in a distributed way over many processing units and not anywhere in particular. Hence, each processing unit participates in the encoding of several informations (patterns). The distributed representation of information is a major characteristic of NAMs and appears to be particular appropriate for massively parallel systems.

NAMs are used for two types of tasks: fault tolerant pattern mapping and pattern completion (Kohonen 1977). Obviously, this functionality makes them attractive for the use in our hybrid knowledge processing system. Especially, in regard to noisy or incomplete inputs of subsymbolic and symbolic data sources. Within our system the use of NAMs is investigated for two purposes. The first approach is to use a NAM in the heteroassociative mode in order to enable fault-tolerant logical inferences. In this case, the if-parts of logical rules have to be coded as input patterns and the then-parts as output patterns to the NAM. The second approach is to use a NAM for information-retrieval purposes (Salton and McGill 1983). In this case important or representative features of information units contained in a data base, for example, have to be coded as input patterns and the reference to or the identification key of the stored information units have to be coded as the output patterns. Different methods are under consideration for these applications at the moment.

Sparse Coding

In most ANN models for associative memory the problem of coding of the input and output patterns is not explicitly discussed. Usually, randomly generated patterns are assumed with a certain probability p for a component to be active (1) and a corresponding probability 1-p for a component to be inactive (0). Most of the models only consider $p=1/2$. But it turns out, that for sparsely coded patterns (small p) the storage efficiency of NAMs (number of bits per synapse that can be effectively stored) and the number of patterns that can be stored with low error probability is much larger (e.g. Palm 1980, Palm 1982, Palm 1990). The problem of sparse coding of I/O patterns is one of the basic problems that has to be considered for each prospective application of a NAM.

The investigation of various data analysis techniques (see above) helps in designing sparse codes. For example, a cluster analysis groups similar objects into (usually disjoint) subsets called *cluster*. The membership of a pattern to a cluster can be interpreted as a special property of this pattern which can be used to transform the pattern into a sparsely coded feature vector. In addition to sparseness, such a code can have the further important advantage of being similarity preserving (Palm 1988).

The research group of Palm (Ulm) is actually working on a theory and on formal design strategies for sparse, similarity preserving codes. The research group of Goser/Rückert (Dortmund) explore the use of trained SOFMs for the generation of such codes. Even if the problem of sparse coding has not been solved completely at the moment, such a code is biologically motivated (e.g. Palm 1982, 1990) and has a great deal of potential in associative storage and retrieval of information.

Inference

At present, the inference module consists of a commercially available PROLOG interpreter for symbolic proofs. As mentioned above, it is a subtask of the WINA project to investigate the usage of NAMs for speeding up the inference process. Another extension to the PROLOG interpreter was proposed by Ultsch *et al* (1990). They used a backpropagation network to learn and memorize control knowledge for PROLOG programs. Trained with examples of successful proofs, the network learns a control strategy. Proof strategies are described by certain features like the types and the possible identity of the arguments of a clause. Similarities of these features are used to train the network in order to generalize to similar clause-selection-situations. First results show the ability of the network to learn the presented description of proofs and a speed up for the proofs can be observed (Ultsch *et al*

1990). An important advantage of the system is its ability to learn proof heuristics from examples. No programmer is needed to define explicit control knowledge for the program.

Probabilistic reasoning and the ability to deal with uncertainty and inaccuracy are further important subjects in respect to the inference module. It is intended to test ANN methods for this purpose and to compare them with conventional approaches (e.g. Bayes, Dempster-Shafer, Fuzzy-Sets...).

System Analysis Tools

For the qualitative as well as quantitative evaluation of the properties and performance of different implementations for certain subcomponents it is important to have formal criteria, as already mentioned. In order to get a better understanding of the processes taking place in ANNs, a graphical tool for the display and perhaps animation is greatly appreciated, too. Especially, for large ANNs and data sets automatic tools for system analysis are necessary. In addition, we expect to get important inferences on the theory of ANNs itself by the use of such tools.

To explore the properties of a SOFM we already proposed to interprete the map by means of the U-matrix (Fig. 2). Alternative graphical representations are *spanning trees* or *vector maps* (Kohonen 1984), and *component cards* (Fig. 2c). Numerical parameters for the SOFM, e.g. homogeneity or heterogeneity of the learned clusters, min/max measure or hit rate (Ultsch and Siemon 1989), are examined at the moment.

For NAMs a couple of formal parameters were specified by Palm (1980). For example, the storage efficiency, the expected number of additional ones in the output pattern, the probability of a connection weight (synapse) of being set (1), the number of patterns which can be stored with low error probability, and so on. These parameters can be easily represented graphically.

PROTOTYPE IMPLEMENTATION

Software Implementation

Different versions of the components of the hybrid knowledge processing system (Fig. 1) have been individually implemented and already tested on SUN-workstations. At present, all successful modules are put together to obtain the desired hybrid system as shown in Fig. 3.

Parallel Hardware Support

Simulation of ANNs on conventional (serial) hardware is rather slow, especially for large net sizes. Therefore, parallel hardware support is greatly appreciated. Within our project we have access to three different parallel hardware platforms.

A straigth forward approach is to use the available hardware within a local area network. At the University of Dortmund such a distributed implementation of SOFMs and NAMs have been devoloped (Surmann *et al* 1992). The local area network (thin ethernet) consists of different UNIX-workstations (Sparc, AVIION, HP9000). By using 8 workstations (1 server and 7 clients) a performance of about 3,6 MCUPS (mega connection updates per second) was achieved for the simulation of a 50x50 SOFM.

Siemon and Ultsch (1990) report on an implementation of the SOFM algorithm on a network of transputers (17xT800, 1MB each local memory) using the occam 2 programming language. They chose a simple ring architecture. The network offers a speed of up to 2,7 Mega CUPS. This allows to train even fairly big nets of more than 10,000 units within less than 30 minutes. At present, the number of transputers is increased from 17 to 65 (T800).

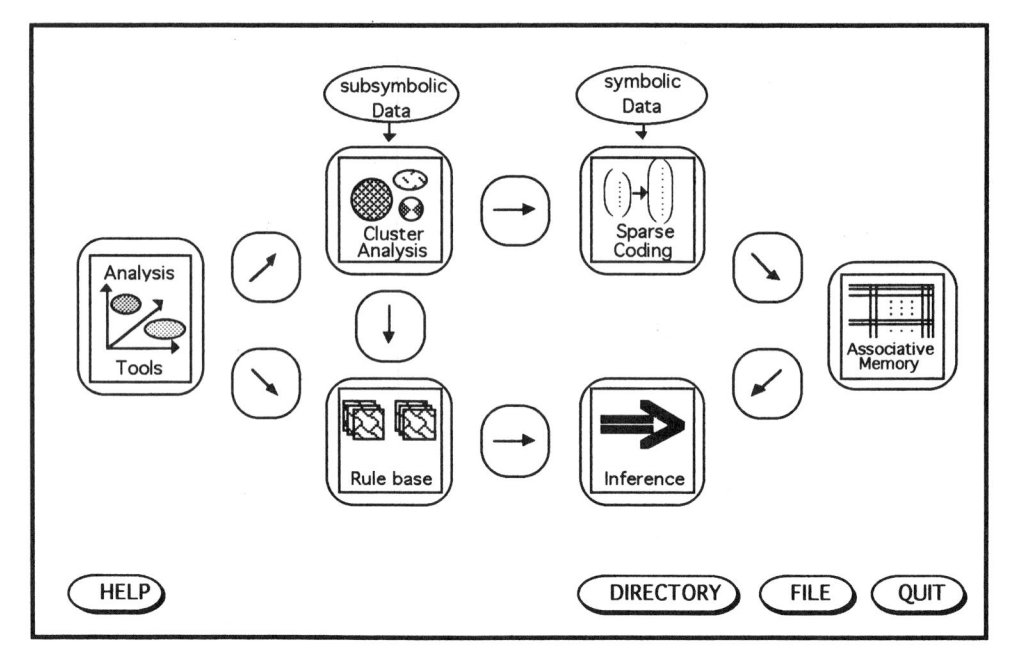

Figure 3 WINA software control window

For NAMs with binary weights and input/output patterns a special purpose SIMD architecure called PAN IV (Parallel Associative Network) was built (Palm and Palm 1991). The first prototype system consists of 144 special purpose ICs (digital ASICs) and 144 MByte memory. On that hardware platform the concept of *virtual networks* builds the fundamental base for multiuser/multitasking operation controlled by the operating system PANOS (Palm and Palm 1991). Several *virtual networks* can be defined and placed arbitrarly into the physical network provided by the PAN IV hardware. The PAN system can be integrated into a local area network. The extentension of the PAN concept to NAMs with multivalued weights and input/output patterns (PAN V) is under devolopment in cooperation with the Univsersity of Dortmund (Goser/Rückert).

In addition to these parallel hardware platforms it is a subgoual of the project to specify task-dedicated VLSI architectures for NAMs and SOFMs as well. At the University of Dortmund (Goser/Rückert) several prototype chips for NAMs based on analog, digital and digital/analog circuit techniques have been successfully realized and tested already (Goser *et al* 1987, Rückert 1991). Based on these experiences, application specific VLSI architectures for NAMs and SOFMs which are optimized in respect to speed, size and testability are under development at the moment.

APPLICATIONS

The three research groups cooperating in the WINA project are investigating different applications of the hybrid knowledge processing system. The work on these applications is done in close cooperation with industrial partners and other faculties at the Universities of Ulm and Dortmund.

The Palm group (Ulm) works on speech recognition (spoken words or syllables, subsymbolic data) and on written text (symbolic data). By using the PAN IV hardware they develop a new associative information-retrieval system which is able to deal with complex subsymbolic and symbolic data structures.

The Goser/Rückert (Dortmund) group applies the hybrid knowledge processing system to process monitoring and quality assurance of their own chip fabrication facilities (Marks and Goser 1988).

The Ultsch group (Dortmund) is engaged with medical (blood) and environmental (local water quality) data for diagnostic purposes. In addition, they apply the hybrid knowledge processing system to quality assurance (mechanical engineering) and process control.

CONCLUSION

The presented national research project *Knowledge Processing in Neural Architecture* (WINA) aims at combining neural information processing and methods known from Artificial Intelligence. Connectionist systems may bridge the gap between the ´subsymbolic´ raw data and symbolic knowledge processing. The hybrid knowledge processing system proposed in this paper is one of the possible ways to combine the advantages of the symbolic and subsymbolic paradigms. Our first results show that the combination of both approaches is not only feasible but also useful.

ACKNOWLEDGEMENT

This work has been supported by the german ministry of research and technology BMFT, contract number 01-IN 103 B/O.

REFERENCES

Goser, K., Hilleringmann, U., Rückert, U., Schumacher, K., *VLSI Technologies for Artificial Neural Networks*, IEEE Micro, Vol. 9, No. 6, pp. 28-44, 1989.

Jain, A.K., Dubes, R.C., *Algorithms for clustering data*, Prentice Hall, Englewood Cliffs, New Jersey, 1988.

Kohonen, T., *Associative Memory: A system-theoretical approach*, Springer Verlag, 1977.

Kohonen, T., *Self-organization and associative memory*, Springer Series in Information Sciences 8, Heidelberg 1984.

Marks, K.M., Goser, K.F., *AI Concepts for VLSI Process Modelling and Monitoring, Proc. Comp. Euro. 87*, IEEE Society Press, pp. 474-477, 1987.

Michalski, R., Carbonell, J.G., Mitchell, T.M., *Machine Learning - An artificial intelligence approach*, Springer Verlag, Berlin 1984.

Oja, E., *Neural netwoks: principal components and subspaces*, Int. J. of Neural Systems, vol. 1, no. 1, pp. 61-68, 1989.

Palm, G., *On Associative Memory*, Biol. Cybern. 36, pp. 19-31, 1980.

Palm, G., *Neural Assemblies: An Alternative Approach to Artificial Intelligence*, Springer-Verlag, Berlin,1982.

Palm, G., *Computing with Neural Networks*, Science, Vol. 235, pp. 1227-1228, 1987.

Palm, G., *Assoziatives Gedächtnis und Gehirntheorie*, (in german) Spektrum der Wissenschaft, pp. 54-64, 1988.

Palm, G., *Local Learning Rules and Sparse Coding in Neural Networks*, in *Advanced Neural Computers*, R. Eckmiller (ed.), North-Holland, 1990.

Palm, G., Palm, M., *Parallel Associative Networks: The PAN-System and the Bacchus-Chip*, in Proc. of the 2nd Int. Conf. *Microelectronics for Neural Networks*, U. Ramacher, U. Rückert, J.A. Nossek (eds.), Kyrill&Method Verlag, München 1991, pp. 411-416.

Pao, Y.H., *Adaptive Pattern Recognition and Neural Networks*, Addison Wesley, 1989.

Reuse, B., *Neurocomputing Initiatives and Perspectives of the German Ministry of research and Technology BMFT*, in *ENNS Newsletter*, May 1992, pp. 11-13, 1992.

Rogers, S.K. (Ed.), *Applications of Artificial Neural Networks*, Proc. SPIE 1294, Washington 1990.

Rückert, U., Kreuzer, I., Goser, K., *A VLSI Concept for an Adaptive Associative Matrix based on Neural Networks*, in: *Proc. Comp. Euro. 87*, IEEE Society Press, pp. 31-34, 1987.

Rückert, U., *An Associative Memory with Neural Architecture and its VLSI Implementation*, Los Alamitos: IEEE Computer Society Press, pp. 212-218, 1991.

Salton, G., McGill, M.J., *Introduction to Modern Information Retrieval*, McGraw-Hill, New York, 1983.

Shaw, G.L., Palm, G., *Brain Theory*, World Scientific, Singapure, 1988.

Surmann, H., Möller, B, Goser, K., *A distributed self-organizing fuzzy rule based system*, in: Proc. of the 5th Int. Conf. on Neural Networks & their Applications, NEURO NIMES 92, pp. 187-194, 1992.

Tou, J.T., Gonzales, R.C., *Pattern Recognition Principles,* Addison-Wesley, Reading, MA, 1974.

Ultsch, A., Siemon, H.P., *Exploratory Data Analysis: Using Kohonen Networks on Transputers*, Research Report No. 329, University of Dortmund, 1989.

Ultsch, A., Siemon, H.P., *Kohonen's Self Organizing Feature Maps for Exploratory Data Analysis*, Proc. of the Int. Neural Network Conf., Kluwer Academic Press, Dordrecht, pp. 305 - 308, 1990.

Ultsch, A., Hannuschka, R., Hartmann, U., Weber, V., *Learning of Control Knowledge for Symbolic Proofs with Backpropagation Networks*, in *Parallel Processing in Neural Systems and Computers*, Eckmiller, R., Hartmann, G., Hauske, G. (eds.), North-Holland 1990.

Ultsch, A., *Konnektionistische Modelle und ihre Integration mit wissensbasierten Systemen*, (in german) Research Report No. 396, University of Dortmund, Dept. of Computer Science, 1991.

Ultsch, A., Höffgen, K.-U., *Automatische Wissensakquisition für Fuzzy-Expertensysteme aus selbstorganisierenden neuronalen Netzen*, Research Report No. 404 (in german), University of Dortmund, Dept. of Computer Science, 1991.

Ultsch, A., *Self-Organizing Neural Networks for Knowledge Acquisition*, in Proc. of the *ECAI 92. 10th European Conference on Artificial Intelligence*, B. Neumann (ed.),Wiley & Sons Ltd, Chichester, pp. 208-210, 1992.

TWO METHODS FOR SOLVING LINEAR EQUATIONS USING NEURAL NETWORKS

M. A. Styblinski and Jill R. Minick

INTRODUCTION

The problem to be considered is to create a circuit able to solve the system of linear equations

$$\mathbf{Dv} = \mathbf{b} \qquad (1)$$

where $\mathbf{D}$ is a nonsingular $n \times n$ matrix of real constant coefficients, $\mathbf{v} \in R^n$ is the vector of variables, and $\mathbf{b} \in R^n$ the vector of real constant coefficients.

Applying Artificial Neural Networks (ANNs) (Vemuri 1988) to the solution of system (1) provides a hardware approach which is faster and more efficient than conventional iterative implementations in software programs. Anticipating technological advancement in the implementation of ANNs with *variable* coefficients, this paper provides the theoretical background for future hardware implementation and checks the feasibility of the proposed solution. In some large real-time problems, where the analysis time is critical, the matrix $\mathbf{D}$ is fixed by principle, and only the right hand side vector $\mathbf{b}$ needs to be changed. In such cases, a special purpose circuit with *fixed* coefficient can be built using the *existing* technology.

Currently, solving a system of linear equations involves different numerical techniques in the form of an algorithm implemented in a computer program. The time needed for the computer to yield an exact solution can grow exponentially with the number of equations for which the solution is desired. With ANNs, a good approximation to the solution can be obtained directly in hardware (an electrical circuit) in a parallel architecture. A possible disadvantage of the ANN approach is that the accuracy of the solution obtained is limited by the accuracy of the ANN element values and the measurements of the solution voltages, since the ANN is an *analog* circuit. Therefore, the proposed method of solving systems of linear equations should be used in applications where speed is more important than the accuracy of the solution.

The proposed approach is to map the process of solving a system of linear equations into a process of solving the Linear Programming Problem (LPP) of Hopfield and Tank (1986), using a neural network implementation. These authors showed that given a linear function to be minimized subject to linear constraints, the solution can be determined using a specially design nonlinear circuit (Hopfield and Tank 1986). The parameters of the function to be minimized (the *cost* function), and the constraint equations are mapped into a corresponding neural network. The network will then converge to the minimum of the corresponding energy surface in time proportional to the network time constants (i.e., very quickly).

Based on the Hopfield and Tank network (modified by Chua and Lin (1984)), two closely related methods are proposed: (1) using an artificially introduced linear cost function and problem mapping into the Linear Programming Problem, (2) a direct, unconstrained approach – obtained from the first one – where a quadratic error function is directly minimized. The second method is simpler, since only linear elements are required.

METHOD 1: MAPPING INTO THE LINEAR PROGRAMMING PROBLEM (LPP)

Solving a system of linear equations requires a variation of the Tank and Hopfield approach. In this case, system (1) corresponds to the *linear constraints* of the LPP. When an artificially induced cost function is minimized subject to those constraints, the system of linear equations is simultaneously solved.

Let the system of linear equations (1) (or the constraint equations) be rewritten as

$$\mathbf{z}(\mathbf{v}) \equiv \mathbf{D} \cdot \mathbf{v} - \mathbf{b} = \mathbf{0} \qquad (2)$$

where $\mathbf{z}(\cdot) \in R^n$ is the vector of residues (errors). Note that (2) represents equality constraints as opposed to the inequality constraints used in the LPP formulated in (Hopfield and Tank 1986). An artificial cost function to be minimized is introduced in the form:

$$Y(\mathbf{v}) = \mathbf{a}^T \cdot \mathbf{v} \qquad (3)$$

where $\mathbf{a} \in R^n$ is the vector of coefficients, and T denotes the transpose. These coefficients are to be selected such that minimization of Y with respect to (w.r.t.) $\mathbf{v}$ gives the solution to $\mathbf{D} \cdot \mathbf{v} = \mathbf{b}$. The jth element of $\mathbf{z}$ is:

$$z_j(\mathbf{v}) = \mathbf{D}_j^T \mathbf{v} - b_i, j = 1, \ldots, n \qquad (4)$$

where $\mathbf{D}_j^T$ is the jth row of $\mathbf{D}$. Let the Lagrange function be defined as:

$$L(\mathbf{v}, \lambda) = Y(\mathbf{v}) - \sum_{j=1}^{n} \lambda_j z_j(\mathbf{v}) \qquad (5)$$

where $\lambda_i \geq 0$, $i = 1, 2, \ldots, n$, are the Lagrange multipliers. Since the number of constraint equations $z_j(\mathbf{v})$ (4) is the same as the space dimension n, and D is assumed nonsingular, there is just *one* point in R^n space where the constraints $z_j(\mathbf{v}) = 0$ are simultaneously fulfilled (as opposed to the general LPP, where there can be many such points). Because of this, the solution to (1) can be recast as the following LPP:

$$\min_{\mathbf{v} \in R^n} \{\mathbf{a}^t \mathbf{v} \mid z_j(\mathbf{v}) = \mathbf{D}_j^T \mathbf{v} - b_j \geq 0, \; j = 1, \ldots, n\}, \qquad (6)$$

where | means "subject to." At the solution $(\mathbf{v}^*, \lambda^*)$, the following necessary Kuhn-Tucker conditions must simultaneously be fulfilled:

$$\nabla L(\mathbf{v}^*, \lambda) = \nabla Y(\mathbf{v}^*) - \sum_{j=1}^{n} \lambda_j^* \nabla z_j(\mathbf{v}^*) = 0 \tag{7a}$$

$$z_j(\mathbf{v}^*) \geq 0, \quad j = 1, \ldots, n \tag{7b}$$

$$\lambda_j^* \geq 0, \quad j = 1, \ldots, n \tag{7c}$$

$$\lambda_j^* z_j(\mathbf{v}^*) = 0, \quad j = 1, \ldots, n, \tag{7d}$$

where ∇ denotes the gradient vector. Therefore, at the solution

$$\nabla Y(\mathbf{v}^*) = \sum_{j=1}^{n} \lambda_j^* \nabla z_j(\mathbf{v}^*). \tag{8}$$

Since $\nabla Y(\mathbf{v}^*) = \mathbf{a}$ and $\nabla z_j(\mathbf{v}^*) = \mathbf{D}_j$, therefore $\mathbf{a} = \sum_{j=1}^{n} \lambda_j^* (\mathbf{D}_j)$ or

$$a_i = \sum_{j=1}^{n} \lambda_j^* D_{ji}, \quad i = 1, \ldots, n. \tag{9}$$

These are the *necessary* conditions (and in our case also sufficient) which must be fulfilled at the solution (otherwise the solution process will be divergent). In our case all $\lambda_j^* \neq 0, j = 1, \ldots, n$, and for *fixed* a_i's, λ_j^*'s can be found at the solution $\mathbf{v}^*$. Instead, we can *assume* their values and find a_i's.

The choice of λ_j^*'s is somehow arbitrary (e.g., λ_j^*'s could be chosen such that the fastest convergence is achieved). If all Lagrange multipliers are chosen to be equal to 1 (which is assumed in what follows), then

$$a_i = \sum_{j=1}^{n} D_{ji}. \tag{10}$$

The circuit implementation for an *ideal* case follows *directly* from conditions (7) and (10) (Chua and Lin 1984 and 1985, Kennedy and Chua 1987) and for a *practical* case is identical with that of Hopfield and Tank (1986), except for the modification to the nonlinearity of the $f(z_j)$ functions (needed for stability and convergence – see below) as derived by Kennedy and Chua (1987). This implementation is shown in Fig. 1 for a system of 2 equations with 2 unknowns (i.e., $n = 2$).

INTERPRETATIONS

Interpretation of Kuhn-Tucker conditions for $n = 2$ is shown in Fig. 2 for the two-variable example presented in a later section. Condition (8) means that $\nabla Y(\mathbf{v}^*)$ is a *linear combination* of the constraint gradients $\nabla z_j(\mathbf{v}^*)$, $j = 1, \ldots, n$, and $\nabla Y(\mathbf{v}^*)$ belongs to the cone spanned by the gradient vectors, as shown in Fig. 2. If this was not the case (i.e., the a_i coefficients were wrongly chosen) then the optimization process would diverge.

The Hopfield and Tank network of Fig. 1 is described by the following system of equations (Hopfield and Tank 1986):

$$C_i \frac{du_i}{dt} = -a_i - \frac{u_i}{R} - G \sum_{j=1}^{n} D_{ji} \tilde{f}(\mathbf{D}_j^T \mathbf{v} - b_j), \quad i = 1, \ldots, n \tag{11}$$

$$v_i = g(u_i) \tag{12}$$

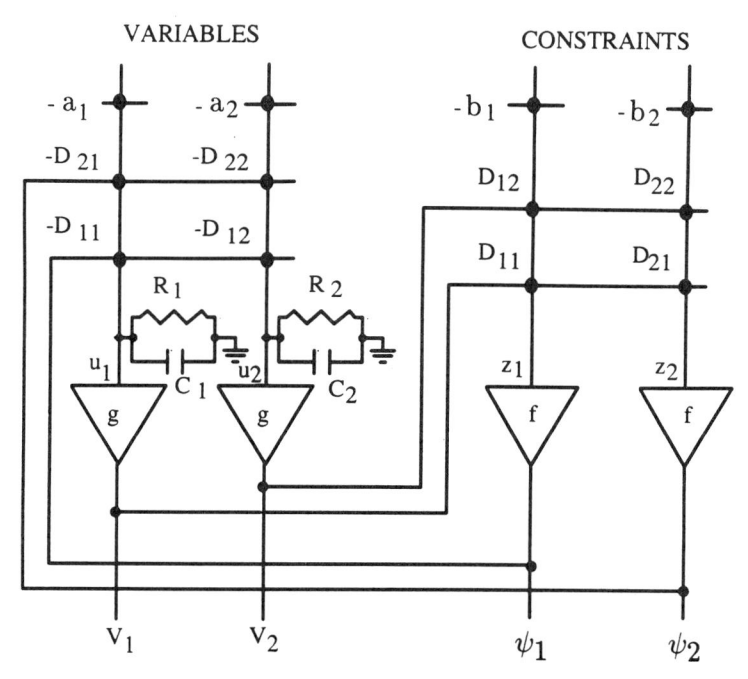

Figure 1. The Hopfield and Tank circuit for the solution of a 2×2 system of equations.

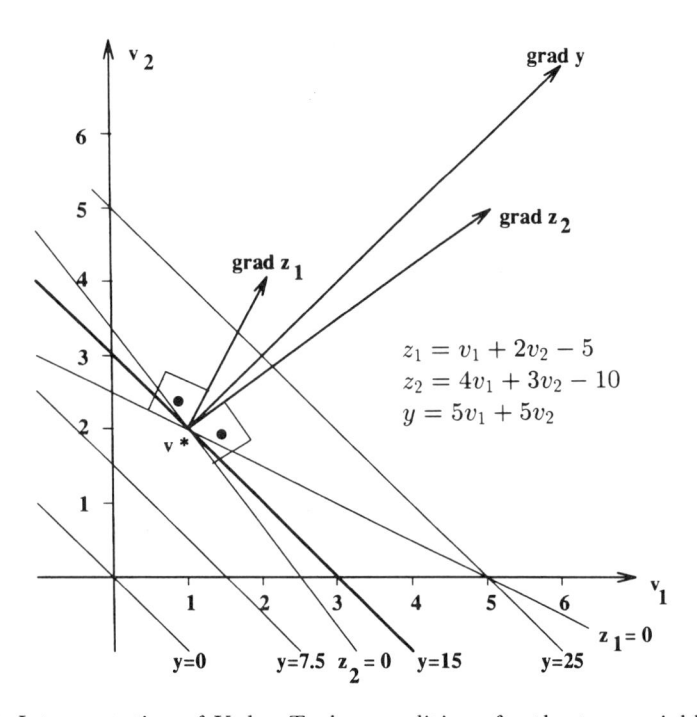

Figure 2. Interpretation of Kuhn–Tucker conditions for the two variable example.

where, without loss of generality it is assumed that: all resistances in parallel with C_i are equal to R, and all $g(\cdot)$ and $\tilde{f}(\cdot)$ functions are identical; moreover, $g(\cdot)$ is a linear increasing function, which in what follows is assumed to be $g(x) = x$, i.e., $v_i = u_i$ (also without loss of generality), and, according to Kennedy and Chua (Chua and Lin 1984, Kennedy and Chua 1987) $f(v)$ should be (for convergence) chosen as

$$f(v) = G \cdot \tilde{f}(v); \quad \tilde{f}(v) = \begin{cases} 0, & \text{if } v > 0; \\ v, & \text{if } v \leq 0 \end{cases} \qquad (13)$$

where $G = 1/r$ is the slope of the $i - v$ characteristics for the corresponding nonlinear element (for $v \leq 0$). This is in contrast with the wrong type of nonlinearity assumed in (Hopfield and Tank 1986), where $f(v) = -v$, for $v \leq 0$.

It is readily shown that under the assumptions above, the system (11) can be written in the following matrix notation

$$\mathbf{C\dot{v}}(t) = -\mathbf{a} - \frac{1}{R}\mathbf{v}(t) - G\,\mathbf{D}^T\tilde{\mathbf{f}}(\mathbf{Dv}(t) - \mathbf{b}) \qquad (14)$$

where $\mathbf{C} \equiv diag(C_1, C_2, \ldots, C_n)$, $\dot{\mathbf{v}}(t)$ is the time derivative of $\mathbf{v}$, $\mathbf{a}$ is the vector of a_i coefficients ($i = 1, \ldots, n$), and $\tilde{\mathbf{f}}(\cdot) \in R^n$ is the vector of $\tilde{f}(\cdot)$ functions defined by (13), i.e., $\tilde{f}_j \equiv \tilde{f}(\mathbf{D}_j^T\mathbf{v} - b_j)$. Representing the time derivative by finite differences:

$$\dot{\mathbf{v}}(t^k) \cong \frac{\mathbf{v}^{k+1} - \mathbf{v}^k}{h} \qquad (15)$$

where h is the time step of the resulting Forward Euler integration formula, the following *iterative process interpretation* of the convergence of the system (14) can be considered

$$\mathbf{v}^{k+1} = \mathbf{v}^k - \mathbf{C}^{-1}h\,[\mathbf{a} + \frac{1}{R}\mathbf{v}^k + G\,\mathbf{D}^T\tilde{\mathbf{f}}(\mathbf{D}\cdot\mathbf{v}^k - \mathbf{b})]. \qquad (16)$$

Now consider the following minimization problem:

$$\min_{\mathbf{v}} \left\{ \phi(\mathbf{v}) = \mathbf{a}^T\mathbf{v} + \frac{1}{2}G\,(\mathbf{Dv} - \mathbf{b})^T\tilde{\mathbf{f}}(\mathbf{Dv} - \mathbf{b}) \right\}, \qquad (17)$$

where the second term realizes the (scaled) Euclidean norm of those elements of the vector $\mathbf{Dv} - \mathbf{b}$ that are less than zero (otherwise, for the satisfied constraints the relevant $\tilde{f}(\cdot)$'s are equal to zero). This is a *quadratic penalty function* approach to minimizing the linear function $\mathbf{a}^T\mathbf{v}$ subject to linear constraints $\mathbf{Dv} - \mathbf{b} \geq 0$, i.e., it provides the solution to the Linear Programming Problem (6)[1]. Due to the properties of $f(z)$, as defined by (13), any time the constraints are violated ($z < 0$) a quadratic penalty is added to the linear term in (17), thus pushing the $\mathbf{v}$ vector into the feasible region. The gradient $\nabla\phi(\mathbf{v})$ of $\phi(\mathbf{v})$ with respect to $\mathbf{v}$ is

$$\nabla\phi(\mathbf{v}) = \mathbf{a} + G\,\mathbf{D}^T\tilde{\mathbf{f}}(\mathbf{Dv} - \mathbf{b}). \qquad (18)$$

Assume now that all C_i's are equal and that the term $\frac{1}{R}\mathbf{v}$ is made arbitrarily small by increasing R. Then, comparing (18) with (16), it is seen that the iterative process (16) implements the *steepest descent* minimization of $\phi(\mathbf{v})$, since the square bracket term of (16) is the approximate gradient of $\phi(\mathbf{v})$, as seen from (18). The term $\mathbf{C}^{-1}h$ controls the step size (if C_i's are different, the minus gradient direction is also modified). $\phi(\mathbf{v})$ is convex and continuously differentiable (even if $\mathbf{D}$ is not positive definite). As

---

[1] It is interesting to notice that the *concontent function* used in (Chua and Lin 1984) and (Kennedy and Chua 1987) to prove convergence is, for $R \to \infty$, *identical* to $\phi(\cdot)$ defined in (17). That function (proportional to the total power dissipated), was, however, developed by the above authors based on reciprocal circuit properties only.

known from the theory of penalty function based optimization, this process leads to the minimum of $\phi(\mathbf{v}^*)$, if $G \to \infty$ as $\mathbf{v} \to \mathbf{v}^*$, at which point $\mathbf{D}\mathbf{v}^* - \mathbf{b} = 0$, i.e., to the solution of the LPP. If $\tilde{f}(z)$ is selected such that $\tilde{f}(z) \equiv f(z) = -z$ if $z \leq 0$ (constraints are violated (Hopfield and Tank 1986)) then the quadratic penalty term is *subtracted* from $\mathbf{a}^T \mathbf{v}$. This means that for increasing v_i's, $\phi(\mathbf{v})$ is unbounded from below. This is yet another justification why the type of nonlinearity assumed in (Hopfield and Tank 1986) was incorrect.

Since, as shown above and formally proved in (Chua and Lin 1984 and 1985, Kennedy and Chua 1987) the circuit described by the system (11) (12) converges to a stable DC solution, then $\dot{\mathbf{v}} = \mathbf{0}$ at $\mathbf{v} = \mathbf{v}^*$, and from (14) we have

$$0 = -\mathbf{a} - \frac{1}{R}\mathbf{v}^* - G\,\mathbf{D}^T\tilde{\mathbf{f}}(\mathbf{D}\cdot\mathbf{v}^* - \mathbf{b}). \tag{19}$$

For an exact solution we must have $R \to \infty$, so the only possibility for a solution to (19) to exist, is that the constraints *are* slightly *violated*, i.e., the third term in (19) is equal to $-G\,\mathbf{D}^T(\mathbf{D}\cdot\mathbf{v}^* - \mathbf{b})$ (in other words, in (19), the function $\tilde{\mathbf{f}}(\mathbf{z}) = \mathbf{z}$). Therefore

$$\mathbf{v}^* = (\frac{1}{R}\mathbf{I} + G\,\mathbf{D}^T\mathbf{D})^{-1}(-\mathbf{a} + G\mathbf{D}^T\mathbf{b}), \tag{20}$$

where $\mathbf{I}$ is the unit matrix. If $R \to \infty$ then

$$\mathbf{v}^* \cong (\mathbf{D}^T\mathbf{D})^{-1}(\frac{-1}{G}\mathbf{a} + \mathbf{D}^T\mathbf{b}). \tag{21}$$

But, $G \to \infty$ (as necessary for obtaining an exact solution), so, with $\mathbf{D}$ nonsingular, we obtain in a limit

$$\lim_{G \to \infty,\, R \to \infty} \mathbf{v}^* = \mathbf{D}^{-1}\mathbf{b}, \tag{22}$$

which is the solution to the system of linear equations of interest. Observe that (21) *does not* indicate that for $G \to \infty$ the correct solution $\mathbf{v}^*$ can be obtained for *any* $\mathbf{a}$. This is due to the fact that (21) and (22) were obtained under the assumption that the optimization process *converges*, so $\mathbf{a}$ still has to be selected according to (19), otherwise the whole process will be divergent, as it was mentioned above and actually confirmed in our experiments (see **Examples** Section). The divergence of this process is due to the form of the $\tilde{f}(z)$ functions, which are zero any time the constraints are fulfilled, which allows $\mathbf{v}$ to freely move inside the feasibility region. To avoid this, it is sufficient to change $\tilde{f}(z)$ to a *linear* function $\tilde{f}(z) = z$. This leads to Method 2 described below.

METHOD 2: A DIRECT APPROACH

By changing $\tilde{f}(z)$ to a linear function $\tilde{f}(z) = z$ in the definition of $\phi(\mathbf{v})$ in (17), a penalty is introduced any time $\mathbf{v} \neq \mathbf{v}^*$, i.e., a *direct* minimization of the quadratic function $(\mathbf{D}\mathbf{v} - \mathbf{b})^T(\mathbf{D}\mathbf{v} - \mathbf{b})$ will take place leading to the solution $\mathbf{v}^*$, provided that the linear term (or actually $\frac{1}{G}\mathbf{a}$ in (21)) is small. So, the best solution is to eliminate $\mathbf{a}$ altogether, i.e., making $\mathbf{a} = \mathbf{0}$. This leads to a simpler and rather obvious formulation

$$\min_{\mathbf{v}}\{\tilde{\phi}(\mathbf{v}) = \frac{1}{2}(\mathbf{D}\mathbf{v} - \mathbf{b})^T(\mathbf{D}\mathbf{v} - \mathbf{b}) = \frac{1}{2}\|\mathbf{D}\mathbf{v} - \mathbf{b}\|^2\}, \tag{23}$$

where, $\|\mathbf{p}\|$ denotes the Euclidean norm of $\mathbf{p}$ and, in comparison with (17), $G = 1$. The whole process of constructing the relevant optimization circuit could now be repeated, using the necessary condition

$$\nabla \tilde{\phi}(\mathbf{v}) = \mathbf{D}^T(\mathbf{D}\mathbf{v} - \mathbf{b}) = \mathbf{0}, \tag{24}$$

and applying, e.g., the method of Chua and Lin (1984 and 1985). However, this is not necessary, since it is obvious that all the conditions are the same as in formulation (17), except for $\mathbf{a} = \mathbf{0}$ and $\tilde{f}(z) = z$. Therefore, the same circuit as the one shown in Fig. 1 can be used, just setting a_i's to zero and realizing $\tilde{f}(z)$'s by linear noninverting amplifiers. This circuit is much simpler than the neural network resulting from using Method 1.

As before, the method can be interpreted by the iterative process (16), in which $\mathbf{a} = \mathbf{0}$, and all the functions $\tilde{f}$ are linear. Since in (23) $\tilde{\phi}(\mathbf{v})$ is a convex quadratic form, then the iterative process (16) is convergent to the solution $\mathbf{v}^*$, provided that, as before, $R \to \infty$. To make this point clear, let us notice that the concontent function used in (Kennedy and Chua 1987) (Eq. (28)), is expressed now as (using our notation, $z_j = \mathbf{D}_j^T \mathbf{v} - b_j$ (Eq. (4)), and $f(z) = G\tilde{f}(z) = Gz$)

$$\bar{G}(\mathbf{v}) = \sum_{i=1}^{n} \frac{v_i^2}{2R} + \sum_{j=1}^{n} \int_0^{z_j(\mathbf{v})} Gz\, dz =$$

$$= \sum_{i=1}^{n} \frac{v_i^2}{2R} + \frac{1}{2} G \sum_{j=1}^{n} (\mathbf{D}_j^T \mathbf{v} - b_j)^2 =$$

$$= \frac{1}{2R} \| \mathbf{v} \| + \frac{1}{2} G \| \mathbf{D}\mathbf{v} - \mathbf{b} \|^2, \tag{25}$$

where $G = 1$. With $R \to \infty$, $\bar{G}(\mathbf{v}) \to \tilde{\phi}(\mathbf{v})$ used in (23). Since each solution of the Hopfield and Tank network is a stationary point of (25) (Chua and Lin 1984 and 1985, Chua 1973, Kennedy and Chua 1987) and (25) is bounded from below by zero (which is also its stationary point with $R \to \infty$), therefore, if the system of equations has a solution, the circuit has also a solution. Referring now to (20)-(22), with $R \to \infty$, $G = 1$, $\mathbf{a} = \mathbf{0}$, we see clearly that $\mathbf{v}^* \to \mathbf{D}^{-1}\mathbf{b}$, as it should. Thus, the only error in the solution is that due to the parasitic amplifier resistance R.

EXAMPLES

Two Variable Example: Controlled Source Neuron Implementation

Let us consider the following system $\mathbf{D}\mathbf{v} = \mathbf{b}$, involving two variables v_1, v_2:

$$\begin{bmatrix} 1 & 2 \\ 4 & 3 \end{bmatrix} \begin{bmatrix} v_1 \\ v_2 \end{bmatrix} = \begin{bmatrix} 5 \\ 10 \end{bmatrix} \tag{26}$$

Method 1.

Since, according to (10) $a_1 = D_{11} + D_{21} = 1 + 4 = 5$, and $a_2 = D_{12} + D_{22} = 2 + 3 = 5$, the cost function (3) $Y\mathbf{v} = \mathbf{a}^T\mathbf{v}$ is defined as

$$\Phi(v_1, v_2) = 5 v_1 + 5 v_2. \tag{27}$$

Mapping into the Hopfield and Tank Linear Programming Problem (Hopfield and Tank 1986) was performed (Minick 1990) (see Fig. 3) and the resulting neural network was simulated using the SPICE circuit simulation program in order to check the validity of the proposed approach. In the simulations, ideal elements were applied first to confirm the theoretical predictions. Voltage-controlled current sources (VCCS's) were used as connective elements (corresponding to resistors in the Hopfield and Tank approach), ideal diodes were used as the nonlinear elements implementing the f functions and current-controlled voltage sources (CCVS's) were used for the g amplifiers.

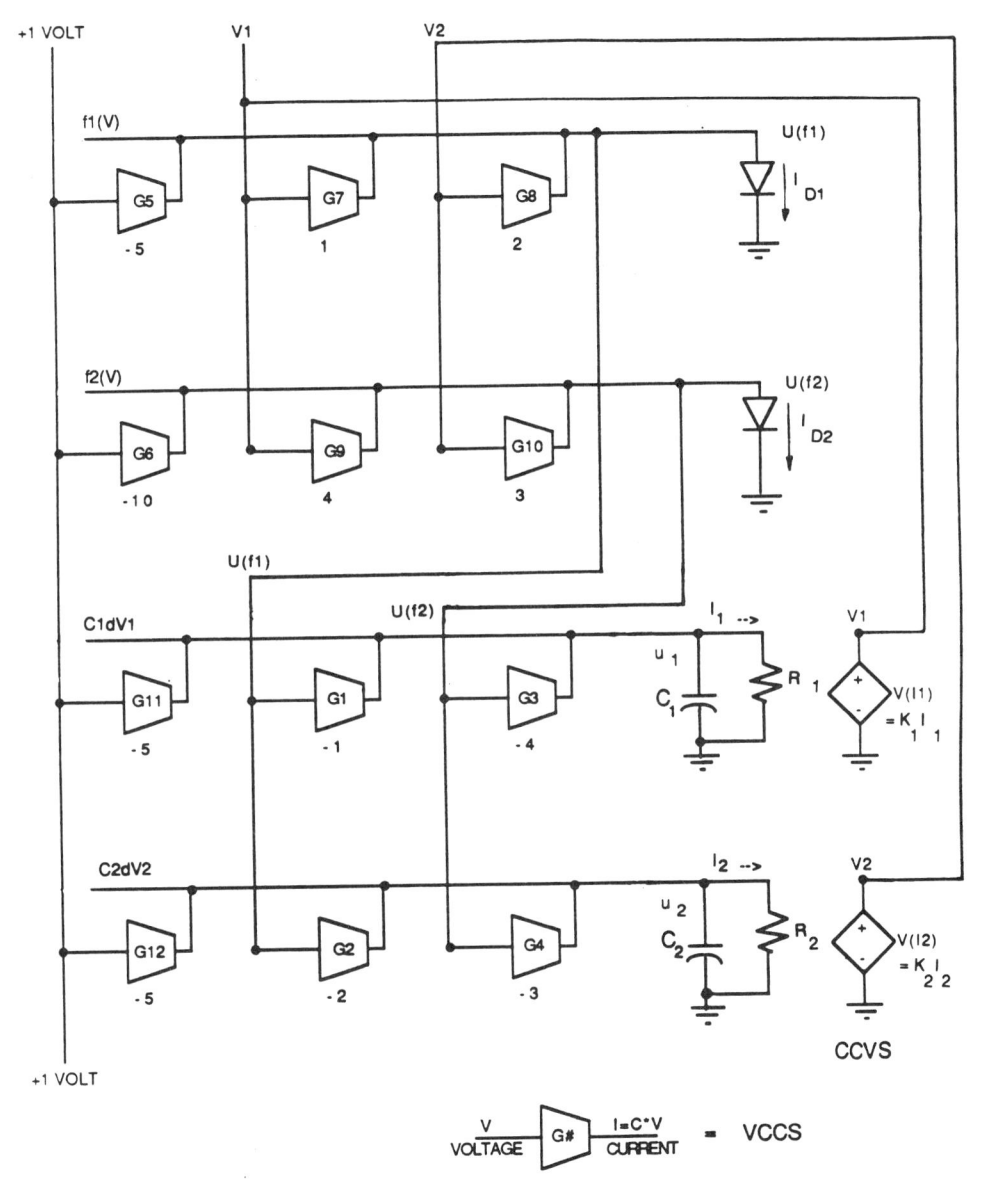

Figure 3. Control source implementation of the neural network for the two variable example.

The working of the circuit shown in Fig. 3 is as follows: the Gx VCCS's drain currents *out* of the corresponding lines if their coefficients are *negative*, and they *put* currents *into* the lines if their coefficients are positive. The currents in the lines labeled as $f1(V)$ and $f2(V)$ correspond to $z_1 = v_1 + 2v_2 - 5$ and $z_2 = 4v_1 + 3v_2 - 10$ constraint equations. If the sum of currents injected by $G7$ and $G8$ (or $G9$ and $G10$) is larger than that drained by $G5$ (or $G8$), then the constraints are satisfied and the current I_{D1}, I_{D2} flow *into* the diodes, thus making them conduct. As a result, the voltages $U(f1) \cong 0$ and $U(f2) \cong 0$, and there are no signals carried to the lines labeled as $C1dV1$ and $C2dV2$. On the other hand, if a constraint is violated and the corresponding line current flows *out* of the diode (e.g., $I_{D1} < 0$), the diode is shut off, which creates a very large reverse resistance R_{REV}, equivalent to the inverse of the parameter G in (13). Writing nodal equations for the circuit of Fig. 3, the following identities are readily shown: $D_{11} = G_7 = -G_1$; $D_{12} = G_8 = -G_2$; $D_{21} = G_9 = -G_3$: $D_{22} = G_{10} = -G_4$; $-b_1 = G_5$; $-b_2 = G_6$; $-a_1 = G_{11}$; $-a_2 = G_{12}$; $z_1 = I_{D1}$; $z_2 = I_{D2}$, where G_i are the coefficients of the VCCS's; the resistances R are actually R_1 and R_2; the gain functions $v_i = g(v_i)$ are actually $v_1 = K_1/R_1$, $v_2 = K_2/R_2$, where K_1, K_2 are the gain coefficients of the two CCVS's; $U(f1) = I_{D1}R_{D1}$, $U(f2) = I_{D2}R_{D2}$, where $R_{D1} = R_{D2} = R_{REV}$ if $I_{D1}, I_{D2} < 0$, and zero otherwise; $G = 1/R_{REV}$, therefore $f(z) \equiv \tilde{f}(z)/R_{REV}$ (where $z = z_1$ or $z = z_2$ and $\tilde{f}(z)$ is defined by (13)). For this system the solution vector is $v_1 = 1$ and $v_2 = 2$.

It was found from SPICE simulations that the circuit was converging to the exact solution (several significant figures) practically instantaneously, with the delay time related only to the slope of the input $1V$ pulse (piece-wise linear with the input ramp extending from 0 to 1-10 ns). The solution was obtained practically at the same time (within about 1 ns) as the input ramp was reaching $1V$, and was almost independent of the C_1, C_2 capacitor values (for larger C_1 and C_2, SPICE did not converge). This was attributed to a very strong negative feedback present in the circuit. To investigate the importance of the choice of a_i coefficients, several cases were studied. If formula (9) was used, perfect solutions were obtained. On the other hand, changing the sign of a_1 to negative, i.e., using $a_1 = -6$, $a_2 = 7$ led to the divergence down the $z_1 = 0$ line in Fig. 2, as expected.

Method 2. In this case, $a_1 = a_2 = 0$ was set and the two diodes in Fig. 3 were replaced by linear resistors. Perfect solutions were again obtained, but the convergence was now more dependent on the capacitor values (due, probably, to the reduced feedback strength, since instead of the very large R_{REV} value, a moderate resistances replacing the diodes in the range of 100-10,000 Ω were used). It was realized at this point that the speed of convergence problem requires more investigation, which was left as a subject of future research.

The influence of finite values of $a_1 = -6$, $a_2 = 7$ (for which the system was divergent using Method 1) on the solution was also studied. Replacing the diodes by $R_{D1} = R_{D2} = 1000\Omega$ resistors led to convergence, but with much poorer accuracy (about 1% - 5% errors). Increasing these resistors to 10,000 Ω (due to the increased loop gain), again led to full accuracy (several significant digits).

System of 20 equations (Minick 1990)

Method 1 was also used to solve a system $n = 20$ linear equations with 20 unknowns, using the same circuit structure as described above. Full accuracy was again obtained. The convergence delays with respect to the time length of the input ramp signal were only about $1ns$. The parallel nature of the solution process was thus clearly seen: the size of the system had very little influence on the convergence time.

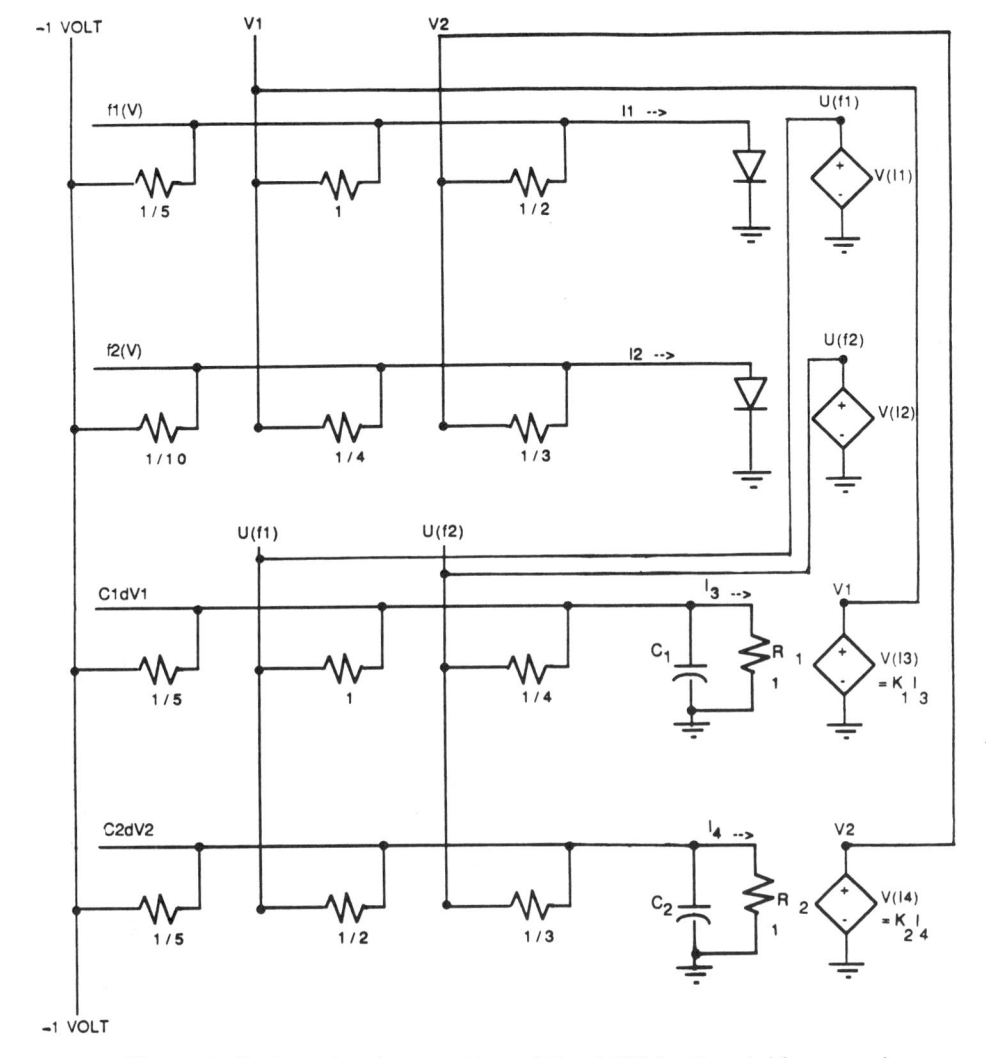

Figure 4. Resistor implementation of the ANN for 2 variable example.

Two Variable Example: Resistor Neuron ANN Implementation (Minick 1990)

The implementation using VCCS's for connection weight coefficients was altered by using resistors in place of VCCS's to simplify the circuit components. Using the same 2×2 system of equations as before, a resistor neuron implementation was developed as shown in Fig. 4.

This implementation had an error of approximately .1%. The VCCS implementation achieved an exact solution because it was simulated with *ideal* voltage-controlled current sources with gains equal to the values of the coefficients and ideal current summation. On the other hand, the resistor neuron implementation led to poorer accuracy due to the non-ideal current summation using strictly resistive adders. However, the convergence for the resistor neuron implementation was achieved within the same time as for the VCCS implementation.

The influence of resistor inaccuracies on the solution accuracy was also studied. When two resistors were altered by 1%, the solutions developed errors around 5%.

When six resistors were altered by 1% the errors of the solutions varied by 13% for v_1 and 5.8% for v_2. The influence of those changes on the solution is, obviously, related to the conditioning of the D matrix (ill-conditioned matrices will produce large solution errors).

CONCLUSION

Two methods for solving systems of linear equations using Artificial Neural Networks were proposed, stimulated by the Linear Programming Problem solved by Hopfield and Tank. The resulting Artificial Neural Networks are highly interconnected circuits composed of simple processing elements which can be resistors or voltage-controlled current source amplifiers (or other amplifier types). Several interesting theoretical results were obtained and interpretations for the solution process given.

The resulting implications are that the hardware implementation of the ANN circuitry has a potential of providing a very efficient solution of linear equations. The relevant ANN circuitry could become a section of a hybrid digital/neural computer.

The ANN implementation is problem specific (i.e., a distinct ANN is required for each system of linear equations). Technical progress in the implementation of reprogrammable connections, in which the weights of the processing elements can be altered, will expand the capabilities of the ANN to solve different systems of equations.

One of the implementation problems is that inaccuracies in components such as voltage supplies, resistors, amplifiers, etc. contribute to a less accurate solution. Some improvements can be obtained by using a switched capacitor implementation of the proposed architecture, where resistor values are implemented as capacitor *ratios* which can be implemented with great accuracy (about .01 to .1%) in the existing VLSI technology. Even if the errors are present in the solution, the proposed approach can still be used for those cases were the accuracy of the solution is not critical, e.g., in solving certain classification and decision making problems.

An interesting open research problem is how to map the proposed architecture into the *existing digital* parallel multiprocessor computers (digital emulation of a neural network) and how competitive would that approach be in comparison with the existing parallel schemes for the solution of linear equations. Another problem is how to solve very large systems of equations, for which the number of ANN processing elements (neurons, weights, and interconnections) is not sufficient for the representation of the whole system. Partial *optimal* mapping and coordination of the partial results so obtained, are some of the most intriguing problems for future research.

References

Chua, L.O., "Stationary Principles and Potential Functions for Nonlinear Networks", *J. Franklin Inst.*, vol. 296, no. 2, pp. 91-114, Aug. 1973.

Chua, L.O. and Lin, G.-N., "Nonlinear Programming without Computation", *IEEE Trans. on Circuits Syst.*, vol. CAS-31, pp. 182-188, Feb. 1984.

Chua, L.O. and Lin, G.-N., "Errata to 'Nonlinear Programming without Computation'", *IEEE Trans. on Circuits Syst.*, vol. CAS-32, p. 736, July 1985.

Hopfield, J.J. and Tank, D.W., "Simple 'Neural' Optimization Networks: An A/D Converter, Signal Decision Circuit and a Linear Programming Circuit.", *IEEE Trans. on Circuits Syst.*, vol. CAS-33, no. 5, pp. 533-541, May 1986.

Kennedy, M.P. and Chua, L.O., "Unifying the Tank and Hopfield Linear Programming Circuit and the Canonical Nonlinear Programming Circuit of Chua and Lin", *IEEE Trans. on Circuits Syst.*, vol. CAS-34, no. 2, pp. 210-214, Feb. 1987.

Minick, J.R., "Application of Neural Networks to Computer-Aided Design of Electronic Circuits: Solution of Linear Equations", Undergraduate Fellow Honor Thesis, Department of Electrical Engineering, Texas A&M University, College Station, TX, April 1990.

Vemuri, V., "Artificial Neural Networks: An Introduction", in *Artificial Neural Networks: Theoretical Concepts*, Computer Society Press Technology Series, pp. 1-7, 1988.

HARDWARE SUPPORT FOR DATA PARALLELISM IN PRODUCTION SYSTEMS

S. H. Lavington, C. J. Wang, N. Kasabov and S. Lin

INTRODUCTION

Production systems, such as OPS5 [1] and CLIPS [2], have been widely used to implement expert systems and other AI problem solvers. The knowledge representation formalism in a form of *if-then* rules and the computational paradigm that incorporates an event-driven control mechanism provide a natural platform for realizing knowledge based systems. However, production systems are normally computation intensive and run slowly on conventional sequential computers. This severely limits the applicability of production systems in problem domains where rapid response is crucial, such as real-time control or financial trader data-feeds.

A production system consists of three components: a set of facts known as Working Memory (WM), a set of rules or productions known as Production Memory (PM), and an inference engine. The facts are represented as tuples of attribute-value pairs and each tuple is called a Working Memory Element (WME). The rules are expressed in a form of *if-then* structure. The *if* part, known as Left-Hand Side (LHS), declares the preconditions of the rule as a pattern of facts or a conjunction of Condition Elements (CEs) some of which may be negated; the *then* part, known as Right-Hand Side (RHS), specifies actions to take if the LHS of a rule is satisfied by facts. The actions include adding or retracting some WMEs so as to update the information known to the system. The inference engine controls the production system execution in cycles of three steps: (1) Match facts against rules; (2) Select an instantiation of the rules to fire; and (3) Execute the selected instantiation (firing), which may result in adding and/or deleting some WMEs. When WMEs change, this triggers the next execution cycle.

For efficient implementation of general production systems, Forgy [3] developed a state-saving RETE algorithm which trades off memory space for speed and has been adopted in OPS5 and CLIPS. Gupta's analysis of six real production systems based on the RETE algorithm, running on a VAX-11/780, showed that around 90% of execution time is due to the match step [4]. With a view to exploiting massively parallel processing, Miranker [5] proposed a non-state-saving Treat algorithm which simply recomputes the matching state as and when necessary. Since the target machine for the Treat algorithm, DADO [6], was to have 1024 8-bit processors (organized in a binary tree structure), it was thought that the recomputation would not normally slow down the match step. However, as Gupta concluded from his experiments, the potential parallelism inherent in production systems that may be exploited by parallel processing architectures is well below 100 and this applies to the Treat algorithm as well [4, 7 & 8].

It should be pointed out that the type of parallelism investigated by Gupta and Miranker as mentioned above was conditional on, and restricted by, von Neumann type architectures - even when parallelized with shared- or local-memories. For these architectures, processing localities, and overheads of memory contentions and communications, will restrict significantly the effectiveness of the actual available parallel processability.

The parallelisms in production systems investigated by Gupta [4] were *thorough and extensive with regard to the presumed framework of machine architecture*, PSM [4] and DADO [6 & 7], as well as the six sampled production systems. For instance, the rule level parallelism, node level and intra-node parallelism (matching facts against CEs and variable binding consistence check between CE instantiations), and action level parallelism were all taken into account. Since the conflict resolution and RHS evaluation each only took 5% of the total computation time and these operations can always largely overlap with the match step, the potential parallelisms in them were not considered worth pursuing unless they began to take significant proportions of the run time. Finally, *data parallelism*, which was loosely defined as *finer grain parallelism* that may be exploited in checking consistence of variable binding, was considered and rejected on the ground that the overhead of scheduling and synchronizing these very fine grained tasks in a shared-memory multiprocessing system would nullify its advantage. Set in the above framework, Gupta et al achieved an average overall speed up of 21.5-fold for two Soar programs and an average overall speed up of 13.6-fold for four OPS5 programs, using 32 processors in PSM [8].

DATA PARALLELISM IN PRODUCTION SYSTEMS

The data-parallelism neglected by Gupta et al due to the machine architectural limitations is, perhaps, the richest source of parallelism in real-life production systems in which the volumes of facts and production rules may be much larger than those adopted in Gupta's experiments. This had been exploited by Stolfo and Miranker et al to certain extend using a *copy-and-constrain* strategy on DADO [9]. This technique divides a *hot spot* rule into a few sub-domain bound new rules so as to maximize the rule level parallelism that can be exploited by the architecture of multiprocessor with distributed-memory. However, it is not clear how effective this technique would be when we take into account the overheads of dividing and copying new sub-rules and sub-domains across the DADO binary tree structured massively parallel architecture.

Data parallelism has also been investigated on NON-VON, a massively parallel tree-structured architecture [10], with tens of Large Processing Elements (32 bit commercial microprocessors) running multiple instruction streams to control a tree of tens of thousands of Small Processing Elements (SPEs, simple 8 bit processors). The SPEs together function as an associative memory. However, the estimated performance of NON-VON did not justify its cost [8].

Kogge et al [11] investigated the feasibility of using Content Addressable Memory (CAM) technology to support rule-based programming. They concluded that, using their enhanced CAM Chips to implement production systems based on the RETE algorithm, a factor of 50 over conventional computers would be achievable. Two other recent projects exploiting data parallelism using CAM devices should be mentioned.

Pratibha and Dasiewicz [12] propose a CAM-based machine architecture CAMPS for implementing a RETE-like match algorithm and claim that it can speed up some production systems by up to 3 orders of magnitude compared with OPS5 on a VAX 11/785. This approach disregards all the parallelisms that Gupta exploited and simply exploits the data parallelism as the control flow traverses the CAMPS network. However, this approach may suffer from the cross-product effect that can quickly saturate the CAM token pool and prolong the matching step if applied to real-life systems.

Fujita et al [13] attempted to enhance the performance of Gupta's PSM architecture by using CAM devices as local memory in order to exploit data parallelism, which they called search parallelism. A potential speed up of 2 orders of magnitude, attributable to the use of CAM, was indicated.

In summary, there are two points worth noting:

(i) The *data parallelism* in production systems may offer greater potential speed up than the *control parallelism* investigated by Gupta. This is at least intuitively true in applications where the volume of data is overwhelmingly larger than the size of machine code.

(ii) The von Neumann type of parallel processing architectures do not seem to be able to exploit the data parallelism effectively. Content-Addressable (i.e. associative) memory, in contrast, is based upon an ability to search data in parallel; architectures incorporating CAM have shown promising speed improvements in production systems.

In the next section we describe an active memory system based on CAM and designed to handle large volumes of data. The system, known as the IFS/2, is intended to support (deductive) databases, smart information systems, and general AI applications. We show in the rest of this paper how this CAM-based architecture may be used with advantage to exploit the inherent data parallelism in production systems.

IFS/2 - AN ACTIVE MEMORY SYSTEM FOR AI APPLICATIONS

It is well-known that pattern-directed search is a fundamental operation in all information systems, and is central to most AI problem-solving paradigms. It is therefore a prime candidate for some form of hardware support. Content-addressable (ie associative) memory has, in principle, the right physical properties needed for pattern-directed search. Apart from obvious questions of size, speed and cost-effectiveness, the main architectural issues in applying CAM to AI systems in general, and to Production Systems in particular, concern the following problems:

(a) devising a semantics-free convention for low-level information representation and memory-management within an associative memory;

(b) integrating the memory-dominant and processing-dominant phases of the total problem-solving cycle including, where appropriate, the exploitation of any natural parallelism.

Based on experience gained from constructing several Mbytes of semiconductor content-addressable memory for a knowledge-base server known as the IFS/1[14], we have designed a successor known as the IFS/2 which uses the same associative technology both for the storage and processing of information. To solve problem (a) above, the IFS/2 uses a tuple-based format for each CAM entry as follows:

$$T_i = <A_{i1}, A_{i2}, A_{i3}, ..., A_{im}>.$$

As explained more fully in [15], the ith tuple, T_i, consists of m atomic objects (ie fields), each of which can be either a ground atom (ie constant) or a named wild card (an abstraction of the variable of logic programming) or the un-named wild card. The scope of a wild card is the tuple and its extensions. Wild cards can be used to encapsulate constraints. As detailed in [15], this suggests five possible modes of associative matching, corresponding to five cases of pattern-directed search. In particular, the IFS/2 offers two variants of one-way matching, and unifiability matching.

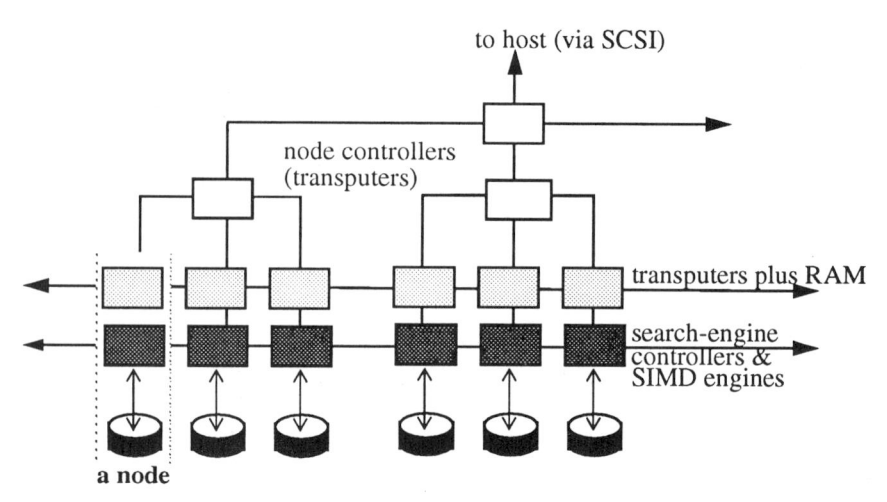

Figure 1 IFS/2 System architecture

Tuples are arranged in classes, via the *make.tuple.set* constructor. Each tuple class has a user-specified format of atomic objects which is fixed for that set of tuples. Tuples can, in principle, have any number of fields, though the IFS/2 prototype has an upper limit of 128 fields. The prototype has provision for up to 2^{16} classes, with a very large number of tuples (many megabytes, see below) per class. Memory management (ie 'paging' and protection) within the IFS/2 is handled by a scheme known as semantic caching [16].

In addressing problem (b) above, the IFS/2 has been designed on what may be called the *active memory* principle. That is to say, the unit responds to whole-structure commands on named tuple-sets. The IFS/2 may be used both to store and process structures such as relations, sets and graphs, and indeed any ADTs that are embraced by the super-type *relation*. The repertoire of IFS/2 commands includes such primitive operations as *search, join* and *transitive closure* on persistent data which is held within the active memory unit.

The IFS/2 is constructed as an add-on device, connected by a standard SCSI interface to a host Workstation - see Figure 1. The design, which is modularly extensible, consists of transputers for controlling nodes of SIMD search engines and associatively-accessed discs [17]. The SIMD search engines perform two roles: they act as a pseudo-associative cache to the discs, and they carry out the data-element comparisons which lie at the heart of primitive operations such as *join* and *transitive closure*. They exploit naturally-occurring data parallelism in a manner which is invisible to the host programmer, who accesses the IFS/2 via a simple C procedural interface [15].

The prototype IFS/2 at Essex consists of three nodes, each having 9 Mbytes of SIMD associative cache backed by a 720Mbyte associatively-accessed SCSI disc. Typical operating times for the IFS/2 prototype, when storing 276,480 simple 3-field tuples within the 27 Mbytes of semiconductor associative memory, are:

insert a tuple:	~ 339 microsecs.
delete a tuple:	~ 113 microsecs.
member operation:	~ 117 microsecs.
search with one wild card:	~ 688 microsecs. (no responders).

join measurements on simple 3-field relations of varying cardinalities were carried out to compare the IFS/2 with software *joins* written in C, Quintus Prolog, Kyoto Common Lisp, and various database systems, all running on a 24MHz, 16Mbyte Sun Sparc. The results, more fully described in [18], show that the IFS/2 is between 5 and 5,000 times faster than software for relation cardinalities greater than about 1000. For cardinalities below about 100, the IFS/2 is generally slower than software. Apart from its speed capabilities, the IFS/2 has a functional interface that is closer to the requirements of the higher-level problem- solver than is the case for conventional computing platforms. Use of the IFS/2 therefore has the potential to reduce software complexity.

In the next section we show how a production system may be mapped conveniently onto the IFS/2's tuple-based representation, so as to take advantage of hardware support for pattern-directed searching during the central WME/WM matching phase.

IMPLEMENTING PRODUCTION SYSTEMS ON THE IFS/2

With the IFS/2 native data structure, WMEs and rules are naturally represented as tuples of constants and named wildcards. A WME tuple contains a *fact-id* followed by a fact-type and a number of constant fields:

<fact_id, fact_type, const1, const2, ..., constk>

For example, the fact "the father of Tom is John" may be represented as:

<f_1, father_of, Tom, John>.

Logically, a rule is represented as a flattened tuple, with a rule_id followed by CEs each taking a few consecutive fields as required by its type.

In this format, the IFS/2's local unification search operation will ensure consistent variable bindings while matching a CE. In the current experimental work, this logical rule tuple is actually stored as a group of IFS/2 tuples. The CEs in the LHS of a rule is stored as tuples of the same format as that for a WME except that the fact_id is replace with a rule_id and a CE ordering number. Variables in CEs are represented as named wildcards and the RHS of a rule is tagged by its rule_id.

The inference engine is implemented as a sequence of IFS/2 operations with the following two control data structures for efficient system performance. The first control data structure is the Binding Constraint Matrix (BCM, one for each rule) which indicates the positions of a shared variable in each CE of a rule. The BCM, which is produced by the compiler, helps reduce the cross product effect in the match phase. The second control data structure is the Constant Test Vector (CTV, one for each rule), which indicates constant patterns that are necessary to satisfy the rule. A Constant Test Interrogand Vector (CTIV) is constructed which is essentially a superimposed codeword for the constant patterns that are supported by WMEs in the system. As WMEs are inserted/deleted, the CTIV is modified accordingly (for details, cf. [20]).

The match cycle starts with matching the new WMEs against CEs which will produce a set of partially affected rules, say S1. At the same time, the CTIV is matched against CTVs, stored in an IFS/2 associative table - Constant Table (CT), which will produce a set of possibly satisfiable rules, say S2. If the intersection of S1 and S2, on the rule_id field, is not empty, then the elements of the resultant set, say S3, identify the potentially instantiable rules. One of the elements of S3 is then selected for further instantiation. The selection of the next CE to instantiate is guided by the BCM so as to reduce the cross product effect, and the IFS/2 internal unification search is used to extend variable bindings across CEs at run time. The first response of the production system as implemented in this approach features a lazy evaluation analogous to [21].

The principle of lazy evaluation of production systems is to integrate the conflict resolution strategy in the match phase. A conflict resolution strategy similar to LEX, which is used in CLIPS and by default in OPS5, is implemented in our system, since the rule selected to instantiate first is affected by the most recent WME and specificity can be examined when selecting a rule from S3. The match algorithm is shown in Figure 2 which

Figure 2 Flow of control in the prototype of IFS-CLIPS

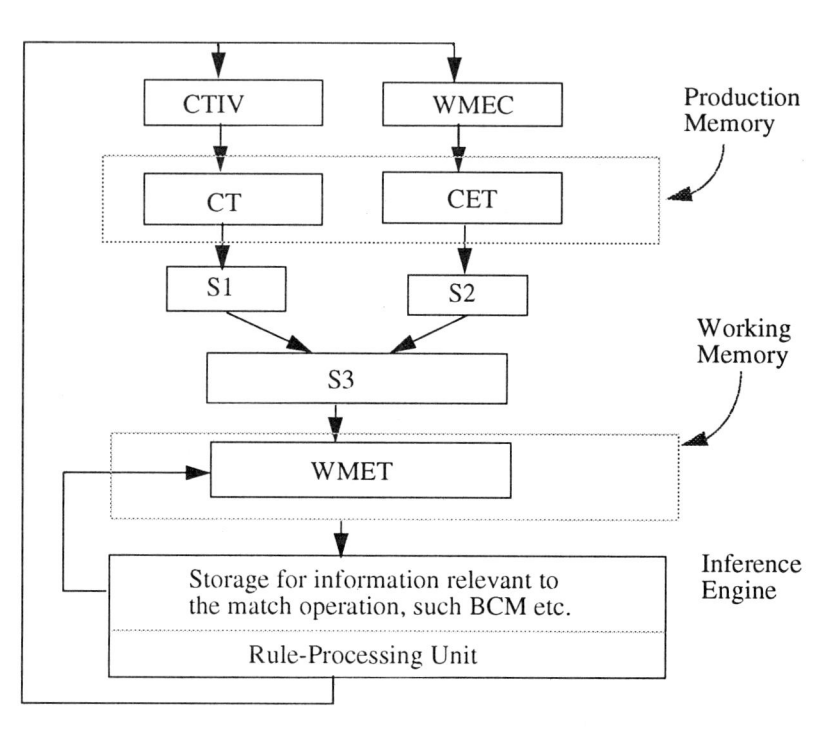

Key:

CT	- constant table;	CET	- condition element table;
WMET	- working memory element table	WMEC	- working memory element change
CTIV	- constant test interrogand vector	S1	- set of possibly satisfiable rules;
S2	- set of partially affected rules	S3	- set of active rules in the present cycle

is self-explanatory. Note that tables CT, CET, and WMET are held as tuple-sets in the IFS/2. The rationale of adopting lazy evaluation is as follows.

Most production systems, such as OPS5 and CLIPS, only fire one rule in most of their execution cycles. This is due to the LEX conflict resolution strategy adopted in these systems that requires the most recent WME change, which happens in almost every execution cycle, to be responded to promptly [21].

Given the above observation, it may be appreciated that it is the system response time that is crucial in real applications, and a faster system response may be realized by a best-first instantiation, i.e. integrating the conflict resolution strategy in the match phase.

There are several functionalities required for implementing the above match algorithm. Firstly, the associative hardware should be able to deal with variant length tuples since CEs, WMEs, and CTs may be of different lengths. Secondly, it should be capable of expressing variables that may appear in CEs. Thirdly, it should be able to extend variable bindings consistently across CEs in the same rule. All these functionalities are met by the IFS/2 operations as described in the previous section.

A prototype of our approach has been implemented based on a front-end processor (Sun work station) connected to the parallel associative hardware unit, the IFS/2. The target production system is CLIPS - C Language Integrated Production System, produced by NASA - Johnson Space Research Centre. Our prototype of production system is called IFS-CLIPS.

The forward-chaining problem-solving strategy is employed in the CLIPS production system. It is implemented using the RETE match algorithm [3] as the internal matcher. An example of a CLIPS rule is shown below.

```
(defrule ancestor                --- define the rule name
    (father_of  ?x  is  ?y)      --- the first condition element
    (father_of  ?y  is  ?z)      --- the second condition element
=>
(assert (ancestor_of  ?x  is  ?z)   --- the action
```

This means, in English, that if ?y is the father of ?x and ?z is the father of ?y then a new fact "?z is an ancestor of ?x", can be asserted into the working memory.

IFS-CLIPS keeps the same syntax as the original CLIPS. However, the main components of the production system - working memory and production memory - are implemented as associative tables which are kept in the parallel associative hardware. The inference engine, a set of C subroutines which controls the overall operations of the production system, resides in the Sun workstation which is connected to the IFS/2. As mentioned above, the pattern-matching is carried out by looking up associative tables.

EVALUATION OF THE PROTOTYPE

The evaluation of our prototype was carried out by comparing the match time of CLIPS and IFS-CLIPS running the same program with the same initial state. The average time of the match phase in each execution cycle of both systems was worked out in the following way:

The average match time per cycle of the standard CLIPS was regarded as 90% of each execution cycle time. Each cycle time was calculated as the whole execution time divided by the number of execution cycles. The average match time per execution cycle of the IFS-CLIPS was assessed using the number and type of CAM-operations performed in the match phase per cycle. The timing of each CAM operation varies with interrogand pattern (i.e. number of wild cards) in a manner illustrated in [17]. Sample times have been given in section 2.

Different production system programs sometimes have very different properties, which bring about different behaviour of a production system interpreter. In order to make a fair evaluation of a production system, several test production system programs must be used to arrive at an average and objective conclusion.

Table 1. surface characteristics of two test program

	No. of CEs	No. of Attr.	No. S. Var.	No. U. Var.	No. Rules
P. 1	3	3	2	2	100
P. 2	3	6	2	2	100

No. of CEs ::= Number of condition elements; No. of Attr. ::= Number of attributes;
No. S. Var. ::= Number of shared variables; No. Rules ::= Number of rules;
No. of U. Var. ::= Number of unshared variables.

Synthetic production systems are used to evaluate our prototype. The idea is that if a synthetic production system reflecting the average behaviour of several production systems is used as the test program, an assessment of average performance can be obtained directly.

Two synthetic production system programs were constructed according to the statistics given in [19]. They have the following surface characteristics:

The number of rules in each program is not necessarily critical to the performance of a production system interpreter because, at a given time, only a subset of rules is relevant to handling the situation at that time. However, the number of rules will affect the size of the condition element table (CET). However, CET is an IFS/2 associative table. Therefore, the size (i.e. length) of the associative table CET will not be significant to the performance of the IFS-CLIPS production system prototype.

The dynamic properties of the synthetic production system programs were carefully controlled. In each cycle, the number of affected rules was around 28. However, the distribution of the working memory elements to individual condition elements was done in a random manner. The total number of working memory elements formed the control variable for our performance tests (see below).

A sample of the rules in our artificially generated benchmark program is as follows:

```
(defrule r-1
    ?class   <-   (class1   ?x   A   ?q)      ---- CE-1
                  (class2   ?z   ?x  B)       ---- CE-2
                  (class3   C    ?z  ?p)      ---- CE-3
    =>
    (printout  t   "r-1")
    (retract   ?class)
    (assert    (classn  B  0)))
```

The n in the last line indicates that we actually provide two versions of the rule:

Version V1: n = 1; this yields a WMEC matching CE-1 which contains only one shared variable;

Version V2: n = 2; this yields a WMEC matching CE-2 which contains two shared variables.

These two versions allow us to investigate effects dependent upon the number of shared variables in matching condition element.

RESULTS

We have assessed the performance of synthetic programs P1 and P2 (see table 1) on two systems:

a) standard CLIPS on a Sun Sparc (24MHz);

b) IFS-CLIPS on the same Sun host plus the IFS/2 active memory unit.

We used two versions of each program, as explained above, so that the four tests were denoted as P1V1, P1V2, P2V1 and P2V2.

Figure 3 shows the results of these tests when performed for various numbers of initial facts (i.e. working memory elements, WME). The vertical axis is a measure of speed-up capability, calculated as the ratio of match times per execution cycle as defined in the previous section. It is seen from Figure 3 that use of the IFS/2 may be expected to speed CLIPS execution times by three orders of magnitude when there are several tens of thousand initial facts. Conversely, use of the IFS/2 actually slows down performance for production systems having fewer than about 100 initial facts.

Figure 4 shows how the match time varies with number of WMEs (initial facts) for test P1V1. Match times for standard CLIPS increase rapidly as facts increase, whereas IFS-CLIPS match times remain reasonably stable. This is an illustration of the advantage of a CAM-based architecture. Another advantage is that IFS-CLIPS is not at all sensitive to the position (ordering) of individual condition elements. In contrast, standard CLIPS can be very sensitive as is shown in Figure 5. The results in Figure 5 were obtained by swapping CE-2 and CE-3 in the sample rule given earlier for test P1V1. After swapping these two condition elements, a cross-product affect occurs since there are no shared variables between CE-1 and CE-3.

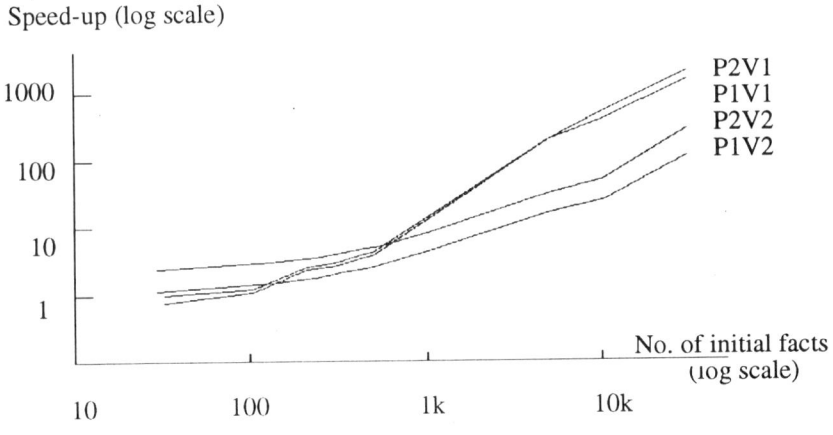

Figure 3 Speed-up of IFS-CLIPS versus standard CLIPS

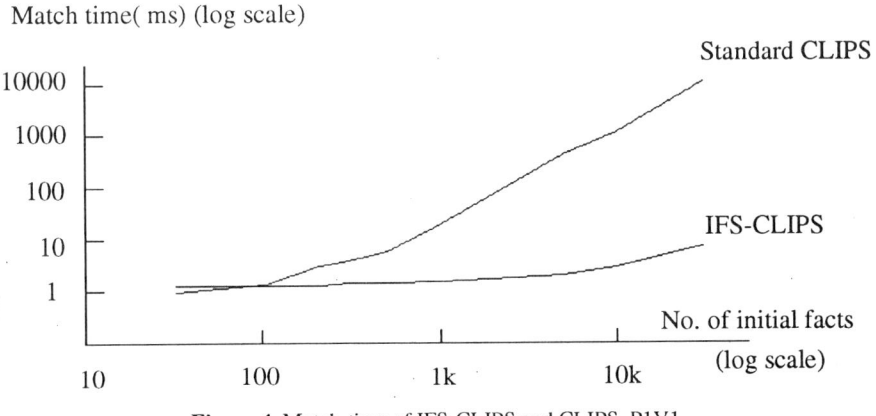

Figure 4 Match-time of IFS-CLIPS and CLIPS, P1V1

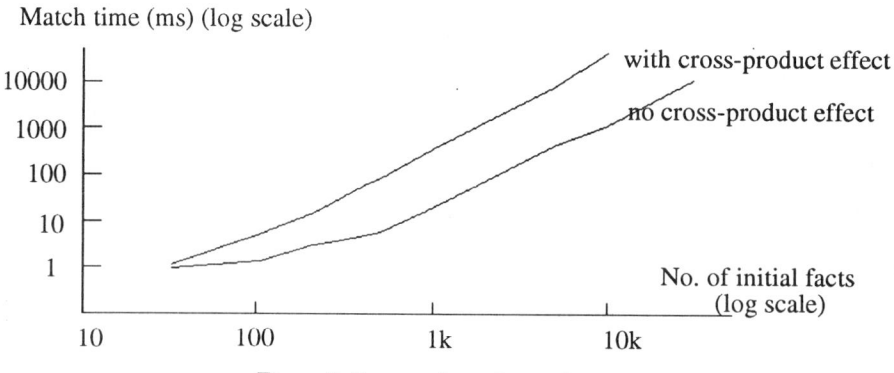

Figure 5 Cross-product effect in CLIPS

Many investigators measure the speed of a production system in terms of the maximum number of WME changes per second. For example, the figure for NON-VON [10] has been estimated to be 2,000. For the IFS/2, an estimate of the maximum number of WME changes per second is given by the reciprocal of:

$$T_m + (T_i + T_d) / 2$$

Where: T_m = time of member operation = 117 microsecs;

T_i = insertion time = 339 microsecs;

T_d = deletion time = 113 microsecs.

(taking figures which apply to a reasonable volume of tuples, equivalent to over 4 Mbytes of data) This gives an IFS/2 figure of 2,915 WME changes per second. This estimate assumes that the firing of a rule will cause one WME change, with an equal probability of either inserting or deleting a WME.

CONCLUSIONS

There is much inherent data parallelism in production systems which are of realistic size. By using an associative (i.e. CAM-based) architecture, this data parallelism can be exploited to produce significant improvements in run times. We have described an add-on, SIMD-parallel unit called the IFS/2 which uses associative techniques both to store and process sets of tuples. The IFS/2 is designed to be scalable, i.e. modularly extensible. Our prototype has 27 Mbytes of semiconductor CAM, backed by 2.2 Gbytes of associatively-assessed disc.

The WMES and CEs of the CLIPS production system have been represented as tuples and stored within the IFS/2. A lazy evaluation strategy has been adopted. For large production systems, use of the IFS/2 can speed up the CLIPS matching phase by about three orders of magnitude.

ACKNOWLEDGEMENTS

It is a pleasure to acknowledge the contribution of other members of the IFS team at Essex. In particular, K. C. Chen assisted with the synthetic benchmark tests. The work described in this paper has been supported by SERC grants GR/F/06319, GR/F/61028 and GR/G/30867.

REFERENCES

[1] Brownston L., Farrel R., Kant E., and Martin N., "Programming Expert Systems in OPS5: An Introduction to Rule-Based Programming", Addison-Wesley, 1985.

[2] Giarratano J. and Riley G., "Expert Systems - Principles and Programming", PWS - KENT Publishing Company, Boston, 1989.

[3] Forgy C. L., "Rete: A Fast Algorithm for the Many Pattern/Many Object Pattern Match Problem", Artificial Intelligence, Vol. 19, September, 1982, Page 17 - 37.

[4] Gupta A., "Parallelism in Production systems", PhD thesis Carnegie-Mellon University, Pittsburgh, March 1986.

[5] Miranker D. P., "Treat: A Better Match Algorithm for AI Production Systems", In National Conference on Artificial Intelligence, AAAI-1987. AAAI proc. Vol.1 1987, Page 449 - 457.

[6] Stolfo S. J., Shaw D. E., "DADO: A Tree-Structured Machine Architecture for Production Systems", Proc. of National Conference on Artificial Intelligence, AAAI-1982, Pittsburgh, Pa., Aug. 1982, Page 242-246.

[7] Gupta A., "Implementing OPS5 Production Systems on DADO", Computer Architecture, 1984, Page 83 - 91.

[8] Gupta A., "High-Speed Implementations of Rule-Based Systems", ACM Trans. on Computer Systems, Vol. 7, No. 2, May, 1989, Page 119 - 146.

[9] Stolfo S., Miranker D. P. & Mill R., "A Simple Processing Scheme to Extract and Load Balance Implicit Parallelism in the Concurrent Match of Production Rules", In Proc. of the AFIPS Symposium on Fifth Generation Computing, 1985.

[10] Hillyer B. and Shaw D., "Execution of OPS5 Production System on a Massively Parallel Machine", J. of Parallel and Distributed Computing, Vol. 3 No. 2, June, 1986.

[11] Kogge P. M., Oldfield J. V., Brule M. R. & Stormon C. D., "VLSI and Rule-Based Systems", in "VLSI for Artificial Intelligence", ed. Delgado-Frias & Moore, Kluwer, 1989, Pages 95 - 108.

[12] Pratibha and Dasiewicz P., "A CAM Based Architecture for Production System Matching", VLSI for Artificial Intelligence and Neural Networks, ed. Delgado-Frias & Moore, Kluwer, 1991, Pages 57 - 66.

[13] Fujita S., Yamashita M. and Ae T., "Search Level Parallel Processing of Production Systems", PARLE '91, Volume 2: Parallel Languages, LNC 506. Pages 471 - 488.

[14] Lavington S. H., "A Technical Overview of the Intelligent File Store", Knowledge-based Systems, Vol. 1, No. 3, June 1988, Page 166-172.

[15] Lavington S. H., Waite W. E., Robinson J. and Dewhurst N. E. J., "Exploiting parallelism on primitive operations on bulk data types". Proceedings of PARLE-92, the Conference on Parallel Architectures and Languages Europe, Paris, June 1992. Published by Springer-Verlag as LNCS 605, pages 893 - 908.

[16] Lavington S. H, Standring M., Jiang Y. J., Wang C. J. and Waite M. E., "Hardware Memory Management for Large Knowledge Bases". Proceedings of PARLE, the conference on Parallel Architectures and Languages Europe, Eindhoven, June 1987, Pages 226 - 241. (Published by Springer-Verlag as Lecture Notes in Computer Science, Nos. 258 & 259).

[17] Lavington S. H., Emby J., Marsh A., James E., Lear M., "A Modularly Extensible Scheme for Exploiting Data Parallelism", Presented at the 3rd Intl. Conf. on Transputer Applications, Glasgow, Aug. 1991, Published in applications of transputers 3, IOS Press 1991, Pages 620 - 625.

[18] Lavington S. H.,"A novel architecture for handling persistent objects". To appear in the Special Issue of Microprocessore and Microsystems devoted to Persistent System Architectures, autumn 1992.

[19] Gupta A, and Forgy C., "Measurements on Production Systems", Report CMU-CS-83-167, Department of Computer Science, Carnegie-Mellon University, 1983.

[20] Lin S., "An Investigation of Search Parallelism in Production Systems Using Parallel Associative Hardware", M.Phil Thesis, Department of Computer Science, University of Essex, 1991.

[21] Miranker D., Brant D., Lofaso B. and Gadbois D.,"On the Performance of Lazy Matching in Production Systems", Proceedings of AAAI, 1990, Page 685 - 692.

SPACE: SYMBOLIC PROCESSING IN ASSOCIATIVE COMPUTING ELEMENTS

Denis B. Howe and Krste Asanović

INTRODUCTION

Many AI tasks require extensive searching and processing within large data structures. Two example applications are semantic network processing [Higuchi et al., 1991], and maintaining hypothesis blackboards in a multi-agent knowledge-based system for speech understanding [Asanović and Chapman, 1988]. Associative processors promise significant improvements in cost/performance for these data parallel AI applications [Foster, 1976, Lea, 1977, Kohonen, 1980]. SPACE is an associative processor architecture designed to allow experimentation with such applications.

A large SPACE array has been built as part of the PADMAVATI project [Guichard-Jary, 1990]. The core of PADMAVATI is a MIMD transputer array, where each processor has a small amount of fast on-chip SRAM and a large bank of slower external DRAM. Each transputer acts as controller for a local SPACE array.

In this paper we first present the architecture of SPACE, then describe its implementation within the PADMAVATI prototype. We also present performance figures for a range of primitive operations before concluding.

INSTRUCTION SET ARCHITECTURE

Programming model

Figure 1 presents the programming model for SPACE. Data is stored and processed in an ordered array of 36b associative memory words. Two 36b control registers, the Write Enable Register (**wr**) and the Mask Register (**mr**), modify the effect of **write** and **search** operations as described below. The 36b data width was chosen to match current 32b microprocessors, allowing a full 32 bits of address or data to be stored along with a few tag bits in each word.

Each word w of the associative memory has a single flag bit, $f[w]$. The flag bits are used to select the words that will participate in an instruction and can be conditionally set or reset as a side effect of most instructions. Each word w has connections to the flags of the word before, $f[w-1]$, and the word after, $f[w+1]$, and these flag values can be used in place of $f[w]$ to select the words that will be active during a given instruction. This flag chain connects the words in a linear array. Multi-word operations to be constructed through sequences of primitive instructions. This simple linear communications topology performs well for algorithms of interest and is easily

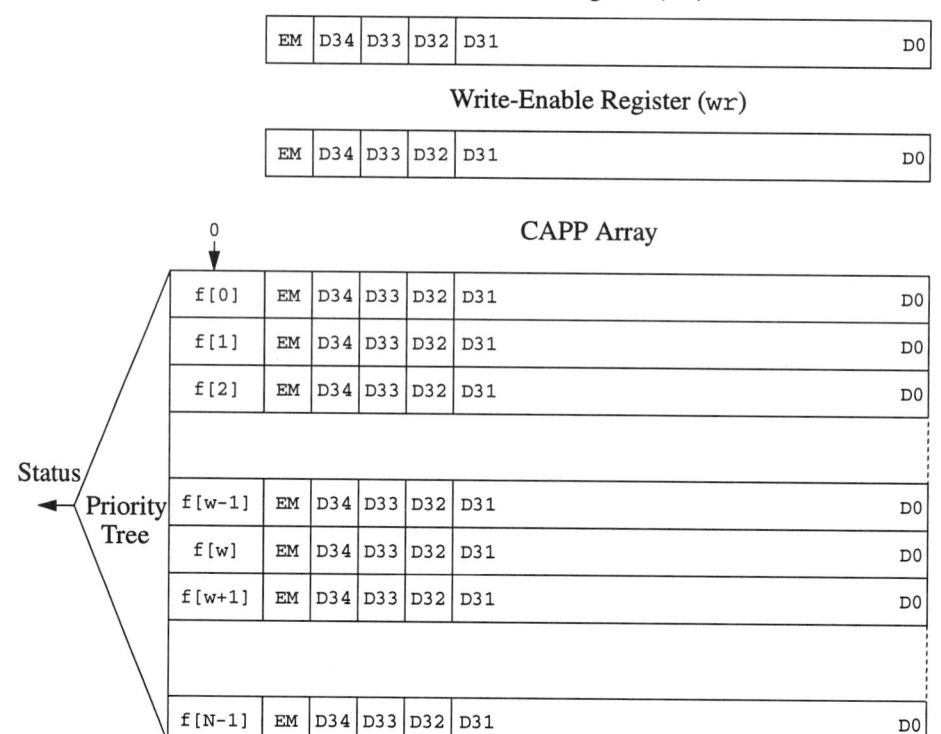

Figure 1. SPACE Programming Model.

scaled to large numbers of processors. The first word in the array (word 0) has its flag input from the previous word (f[-1]) set to 0.

The architecture includes a priority resolution tree that can activate only the *first* (lowest numbered) selected word for an instruction. Alternatively the resolution tree can be used to update the flags of all words after the first matching word during a search operation. Data words in SPACE are addressed only by their contents, or by the contents of their neighbours.

Instruction format

SPACE instructions are encoded in seven bits as an orthogonal combination of an opcode, a select mode and a flag update mode. The opcode occupies four bits and is encoded as shown in Table 1.

Two further bits in the instruction encode the select mode. These are the Adjacent/Own Flag bit, **AOF** and the Previous/Next Flag bit, **PNF**. Select modes are SPACE's equivalent of the addressing modes found in conventional processors and determine the flag value used to select participating words as shown in Table 2.

New Flag, **NF**, is the final bit in the instruction format and this encodes the flag update mode. The two modes are set ($NF = 1$) and clear ($NF = 0$). In the assembler syntax either s or c is appended to the opcode plus select mode. For read and write

Table 1. Opcode encoding, x indicates don't care.

Assembler Syntax	Operation	CD	RW	TS	SA
wwr	Write Write-Enable Register	1	0	0	0
wmr	Write Mask Register	1	0	0	1
wbr	Write both Registers	1	0	1	x
rwr	Read Write-Enable Register	1	1	x	0
rmr	Read Mask Register	1	1	x	1
wfi	Write first selected	0	0	1	1
wal	Write all selected	0	0	1	0
rfi	Read first	0	1	1	x
rst	Read status	0	1	0	x
smo	Search for match only	0	0	0	0
smf	Search for match & following	0	0	0	1

Table 2. Select Mode Encoding

AOF	PNF	words selected	assembler syntax	flag used
0	0	all	*	1
0	1	flagged	@	$f[w]$
1	0	before flagged	-	$f[w+1]$
1	1	after flagged	+	$f[w-1]$

operations, the flags of active words are updated to hold **NF**. For search operations with **NF** = 1, flags of search *hits* are set and flags of all non-hits are cleared. When **NF** = 0, a search operation clears the flags of hits and leaves other flags unaffected.

Instructions can be divided by the **CD** bit into those that act on the control registers and those that act on the array. Array instructions can be further divided into three major categories, search, read and write.

Control register instructions

An early decision in the SPACE design was to specify search operand don't cares and write-enabled bit columns through separately loaded control registers rather than by adding a second 36b operand to search and write instructions. In many of the applications we have investigated, the same mask and write-enable values are reused over a number of instructions. By adding programmer visible registers for these values we reduce off-chip operand bandwidth requirements substantially.

The wwr, wmr, rwr and rmr instructions allow a 36b value to be written to and read from a control register. The wbr instruction allows both registers to be written with the same 36b value in one cycle.

Search instructions

Search instructions take a single 36b operand, the search key, and modify the values of flag bits. Don't cares in the search key are specified with mr. Any bit i, for which $mr[i] = 0$, is a don't care that always matches.

As well as allowing don't care values to be specified in the search key, SPACE also allows don't cares to be stored with the data. One of our applications is Prolog pre-unification, which relies heavily on stored don't cares to represent unification variables in database clause heads [Asanović and Howe, 1989]. To reduce the cost of implementing stored don't cares, we adopted a compromise whereby a data word can be masked in units of a byte rather than bit by bit. Software can be used to simulate stored don't cares, but this simulation is expensive. Simulating the four maskable fields per word in software would take 49 SPACE instructions.

The 36b data word is grouped into four 8b data bytes, **D0–D31**, three tag data bits, **D32–D34** and an Exact/Masked bit, **EM**. The value of the **EM** bit stored in a word affects the way the word responds to searches. If the **EM** bit of a stored data word is set, all 36 bits must match the corresponding masked search key bits. This is an *Exact* word with no stored don't cares; all 35 data bits are available for data storage. If **EM** = 0, the word is *Masked* and byte-wise masking is enabled. The most-significant bit of each of the four data bytes controls whether that byte is a don't care. All four bytes and the four tag bits must match the search key to produce a hit. The 8b size of the maskable field provides sufficient granularity for Prolog pre-unification and is convenient for storing and searching 7b ASCII text, while minimising don't care storage overhead.

There are two search instructions: smo is search for match only, smf is search for match and following. Smo compares all selected words with the masked search key and all matching words are considered *hits*. Smf compares all selected words with the masked search key and the first matching word and *all* words following are considered *hits*. The smf instruction is typically used to flag blocks of words, first flagging all words following a block header word, then clearing all flags following a block trailer word.

If **NF** = 1, then the flags of hits are set and the flags of all non-hits are cleared. If **NF** = 0, then the flags of hits are cleared and the flags of all non-hits are unchanged. This behaviour differs from that of read and writes, which simply set the flags of active words to **NF**. The scheme used for searches allows the results of successive searches to be AND-ed together, whereas employing the read/write flag update semantics would have OR-ed results instead. We introduced this asymmetry on the basis of our experience in writing various associative algorithms. Especially when coding multi-word operations, it is more common to AND together the results of search operations. If required, search OR-ing can be readily accomplished with a simple series of instructions using a data bit as temporary flag storage.

Write instructions

Write instructions take a single 36b operand and update the value of array data words, modifying flag bits as a side effect. The Write-Enable Register determines which bits of array words are updated in a write instruction. Only those bit positions, i, for which wr[i] = 1 are written; others are unaffected. This makes it possible to address specific bit columns as well as specific words. This ability to selectively write individual bit columns, together with the ability to write multiple selected words in parallel, distinguishes a content addressable parallel processor (CAPP) such as SPACE from a less powerful content addressable memory (CAM) [Foster, 1976]. There are two write instructions: wal writes to all selected words, wfi writes only to the first selected word (if any). The flags of the written words are updated to the value of **NF**.

Read instructions

There is a single read instruction `rfi` which returns the 36b value of the first selected word in the array. The read word's flag bit is assigned the value of **NF**. A value of all 1s is returned if there are no selected words. In many cases, this allows the readout of the last in a sequence of flagged words to be determined without separate explicit checks of flag status.

The read status instruction `rst` returns a single bit indicating if any flagged words would be selected by a given select mode.

SPACE IMPLEMENTATION

We have constructed a large SPACE system as part of the PADMAVATI project [Guichard-Jary, 1990]. Full custom VLSI SPACE chips are packaged using Tape Automated Bonding onto compact SPACE modules. These modules are attached to conventional circuit boards, which are then connected by a backplane to host processor boards within a MIMD parallel computer system. The following sections describe each level in the packaging hierarchy.

SPACE chip

The SPACE chip implements a small version of the SPACE programming model described in Figure 1. Each chip contains an array of 5328 static CAPP cells arranged as 148 words of 36 bits, a 148-bit flag chain, 36-bit Mask and Write-Enable registers and a 148-input priority resolution tree. The array is never coordinate addressed, so there is no incentive to make the number of words per die a power of two. A conservative 16 transistor static CAPP cell design was adopted to simplify design and ensure robustness. Priority resolution was implemented using a high-radix tree to reduce the latency of selection operations.

The SPACE chip has 56 pins: 7 instruction inputs, a 36b tri-state data bus, 4 pins for cascading chips (pins **PRF**, **NXF**, **REQ** and **PRQ**), 3 timing and control inputs (pins **CS**, **CE** and **PCH**), and 6 power supply connections.

PRF and **NXF** are tri-state pins used to connect the flag chain of a chip to the previous and next chips' flag chains respectively. **REQ** and **PRQ** are used to connect the internal priority resolution tree as a leaf node of an external priority resolution tree. A chip asserts **REQ** if it contains selected words, the external tree asserts **PRQ** if any preceding chip asserted **REQ**. Each SPACE chip in a large system maintains its own copy of the Mask and Write-Enable registers. For write register operations all registers are updated in parallel. For read register operations, the priority resolution tree is used to ensure that only the first chip in the array responds with the values stored in these registers.

The **CS** input is used to partition a large array into independent banks. When low, it disables all chip operations and outputs. Multiple selected chips must be adjacent if flag chain continuity is to be preserved. This feature was added to allow a large SPACE array to be partitioned amongst a number of different applications. For example, on the PADMAVATI machine the run-time system may use one bank to speed global address translation for interprocess communication while the rest of the array is made available for user applications.

SPACE chip timing is controlled by the **CE** input. The **PCH** pin controls the internal prechargers. It can be tied high in which case the chip is fully static but burns

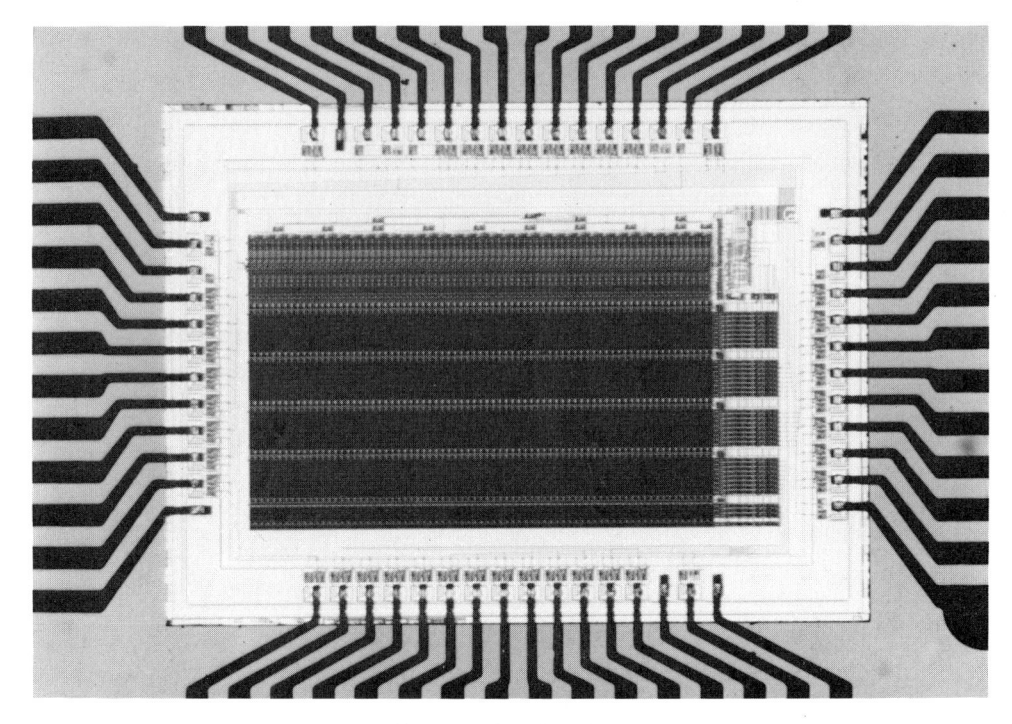

Figure 2. SPACE Chip.

more power, or can be pulsed high between supplying the data and strobing **CE**.

The SPACE chips were fabricated in a 1.2 μm two-level-metal n-well CMOS technology from a fully-custom layout. The die measures $5.8 \times 7.9\,\text{mm}^2$ and contains over 80,000 transistors. The minimum chip cycle time is 125ns.

Figure 2 is a photomicrograph of the chip. Samples were received in June 1989 and passed all functional and speed tests first time, thanks to extensive functional and electrical simulation prior to fabrication. Over 1,200 working parts have since been fabricated for the PADMAVATI project. A more comprehensive data sheet is available [Howe and Asanović, 1990].

SPACE modules

A SPACE module contains twelve SPACE chips, buffering and one stage of external priority resolution, and is functionally equivalent to a 1776-word SPACE chip. There are numerous advantages to mounting the chips on modules rather than directly onto a large circuit board. Fabrication, testing and repair are simplified, and the modules can be reused in different target systems.

Each module measures $100 \times 60\,\text{mm}^2$ and holds twelve 56-pin SPACE chips plus seven 20–28 pin SSI/MSI components. Tape Automated Bonding (TAB) is used to bond the SPACE chips to the module PCB; the remaining components are surface mounted. With TAB, each SPACE chip has a PCB footprint of only $12 \times 10\,\text{mm}^2$. The module PCB has power and ground planes, and 4 signal layers with a trace pitch of 0.5 mm. The pin-out of the SPACE chip was designed to allow devices to be tightly tessellated on the module PCB. Each face of the die has the connections for one data byte. Since all four data bytes within a word are equivalent, the north facing data

Figure 3. SPACE Board.

byte of one die can be directly connected to the south facing data byte of another die, minimising the area needed for routing and vias on the module PCB.

SPACE in PADMAVATI

Each PADMAVATI processor node occupies a standard 280×233 mm^2 6u board and includes a 25MHz Inmos T800 transputer with 16MB of 200ns DRAM and a 32b asynchronous bus interface. SPACE boards have the same form factor and connect to the processor through the bus interface. Figure 3 is a photograph of a SPACE board. Each board holds six SPACE modules, and each operation can select any subset of the six modules. The selected subset then acts like a single SPACE chip with up to 10656 words.

The transputer acts as the microcode sequencer for the SPACE array, sending instructions and data over the inter-board bus. The SPACE array is memory mapped, with instructions encoded on the bus address lines. SPACE board write and search operations are pipelined, and have a minimum cycle time of 320ns. SPACE read operations require a wait for the returned value, and have a minimum cycle time of 480ns. The transputer limits the rate at which instructions are issued to the board. Measurements on the PADMAVATI prototype give an average instruction cycle time of around 1250ns. A more tightly coupled, dedicated microcontroller could reduce the board cycle time for each SPACE instruction to around 160ns with the existing modules.

The final level of packaging yields a complete PADMAVATI system. Sixteen transputer/SPACE nodes are fully connected through a custom VLSI dynamic routing switch that obeys the transputer link protocol [Rabet et al., 1990]. By partitioning the SPACE array amongst the transputers in this manner, we gain higher array I/O bandwidth and

Table 3. Measured SPACE performance with 170496 words in PADMAVATI.

Operation	Cycles	Operations/s
Scalar-Vector		
36b search	1	136×10^9
1b AND/OR	3	45.5×10^9
1b XOR	8	17.0×10^9
1b half add	3–8	
1b full add	5	27.3×10^9
16b add	83	1.64×10^9
8b×8b multiply	3–539	
32b =	8	17.0×10^9
16b <	68	2.01×10^9
Vector-Vector		
1b AND/OR	3	45.5×10^9
1b XOR	8	17.0×10^9
1b half add	8	17.0×10^9
1b full add	9	15.2×10^9
16b add	144	947×10^6
8b×8b multiply	539	253×10^6
16b < and =	84	1.62×10^9
Vector Reduction		
Find 16b Max/Min	48	2.84×10^9

the flexibility of MIMD control of SIMD subarrays. The total associative storage capacity is 170496×36b words, or around 750KBytes. The complete system is attached as a compute server to a host Sun workstation.

PERFORMANCE

Table 3 lists the number of cycles taken to perform some primitive operations in SPACE, and the resulting measured peak performance for the PADMAVATI prototype. The reader is referred to [Foster, 1976] for a detailed exposition of CAPP algorithms. All these routines are executed conditionally, taking effect only in those words where a user defined tag field in each word matches a user defined tag value.

In the PADMAVATI prototype, execution speed is limited by the transputer used to issue instructions. The performance of the existing modules could be increased by nearly a factor of 8 by using a dedicated microcontroller operating over a synchronous bus. In this case, the peak performance of the array would be over 1×10^{12} comparisons per second.

SUMMARY

SPACE, an associative processor architecture, has been designed to allow experimentation with AI applications. The instruction set is small and highly orthogonal. All data bits in a word support parallel, maskable writes, allowing parallel logical and arithmetic transformations on stored data. Stored data bytes can be individually masked as don't cares. Processors can communicate with their neighbours in a bidirectional linear array to support multi-word data objects, and to allow parallel communication between neighbouring objects.

An implementation of SPACE containing 170496 processors has been completed. This implementation used aggressive packaging techniques to reduce the relative cost of associative storage. Performance has been measured for a range of primitive operations.

ACKNOWLEDGEMENTS

The authors gratefully acknowledge the vital contributions to the success of this work made by all those involved, especially Bruce Cameron, John Birch, John Robson and Virendra Patel. This work was partially funded by ESPRIT Project 1219 (967) "PADMAVATI" (Parallel and Associative Development Machine As a Vehicle for ArTificial Intelligence). This project ran from January 1986 to April 1991 and involved GEC Hirst Research Centre in Great Britain, Thomson-CSF in France, CSELT in Italy and FIRST International in Greece [Guichard-Jary, 1990]. Krste Asanović thanks the International Computer Science Institute for its support.

REFERENCES

[Asanović and Chapman, 1988] Asanović, K. and Chapman, J. R. Spoken Natural Language as a Parallel Application. In *Proc. CONPAR 88*, volume B. BCS Parallel Processing Specialist Group, 1988.

[Asanović and Howe, 1989] Asanović, K. and Howe, D. B. Simulation of CAM Pre-unification. PADMAVATI project report, GEC Hirst Research Centre, East Lane, Wembley Middlesex HA9 7PP, Great Britain, January 1989.

[Foster, 1976] Foster, C. C. *Content Addressable Parallel Processors*. Computer Science Series. Van Nostrand Reinhold, 1976.

[Guichard-Jary, 1990] Guichard-Jary, P. PADMAVATI. In Commission of the European Communities, DG XIII: Telecommunications, Information Industries and Innovation, editor, *Proc. Annual ESPRIT Conference '90*, page 227. ESPRIT, Kluwer Academic Publishers, 1990.

[Higuchi et al., 1991] Higuchi, T., Furuya, T., Handa, K., Takahashi, N., Nishiyama, H., and Kokubu, A. IXM2 : A Parallel Associative Processor. In *Proc. 18th Int. Symp. on Computer Architecture*, pages 22–31, 1991.

[Howe and Asanović, 1990] Howe, D. and Asanović, K. PADMAVATI CAM Functional Specification and Data Sheet. PADMAVATI project report, GEC Hirst Research Centre, East Lane, Wembley Middlesex HA9 7PP, Great Britain, April 1990.

[Kohonen, 1980] Kohonen, T. *Content Addressable Memories*. Springer Series in Information Sciences. Springer-Verlag, 1980.

[Lea, 1977] Lea, R. M. Associative processing of non-numerical information. In Reidel, D., editor, *Computer architecture: NATO Advanced Study Institute, St Raphael, France, September 1976*, volume C-32 of *NATO ASI series*, pages 171–215, Holland, 1977. Dordrecht.

[Rabet et al., 1990] Rabet, N. M., Guichard-Jary, P., and Deschatres, M. Padmavati delta network. Technical Report DEL-04, Thomson-CSF DOI, December 1990.

PALM: A LOGIC PROGRAMMING SYSTEM ON A HIGHLY PARALLEL ARCHITECTURE

Mario Cannataro, Giandomenico Spezzano and Domenico Talia

INTRODUCTION

Since Kowalski (1974) defined the framework for the procedural interpretation of Horn clauses, many models and techniques have been designed for the implementation of logic programming languages. These techniques are mainly concerned with the implementation of logic programming systems on sequential Von Neumann architectures. The major problem of these implementation derives from their low efficiency compared to imperative languages. The main reason for this is that the high level of the logic programming paradigm requires more than a translation to generate the machine code of a logic program. Thus, some overhead is added to bridge the gap between the logic programming level and the machine level. On the other hand, imperative languages are closer than logic languages to the machine level because they embody many concepts of the Von Neumann architecture.

Parallel computers (Seitz 1984) may constitute an alternative to traditional systems to support the execution of logic programming languages. The parallel execution of logic programs may allow to overcome the low performance problems of the traditional implementations of logic programming systems on sequential Von Neumann computers. Recently, many models and architectures for the parallel execution of logic programs using different process grain sizes and memory models have been proposed. These models exploit the various kinds of parallelism of logic programs: AND parallelism, OR parallelism, search parallelism, etc.

This paper describes a parallel logic programming system running on a massively parallel VLSI architecture. The system is called PALM (Parallel Logic Machine). PALM has been defined for a message-passing parallel architecture. It is based on an abstract model that allows the exploitation of OR and Restricted-AND parallelism by the automatic decomposition of the logic program in a network of concurrent processes.

We have implemented the system on a parallel computer which consists of 32 Transputers (Inmos 1989). The processes of the PALM run-time support are implemented as concurrent Occam (Inmos 1988) processes mapped onto the Transputer nodes. Techniques to explicitly control the grain size of computation, parallelism degree, and dynamic task allocation are used. Furthermore, a virtual machine level offering a logical fully connected network of dynamic AND-OR processes has been developed. Experimental performance results of the PALM system on some test programs have been obtained.

The paper discusses the policies, techniques and tools utilized in PALM to solve problems such as process mapping, distributed scheduling, routing of messages through the network, and distributed termination. The rest of the paper is organized as follows. After a brief description of the abstract model and the architecture of PALM, a deep analysis of the problems solved in the parallel implementation of PALM on a highly parallel architecture is presented. Finally, some experiments are shown and discussed.

THE ABSTRACT MODEL

The PALM system is the implementation of an abstract model for the parallel execution of logic programs based on the use of concurrent processes cooperating by message-passing. The goal of this model is the exploitation of OR and Restricted-AND parallelism. Restricted-AND parallelism exploits the AND-parallelism among subgoals which do not share variables (DeGroot 1984).

According to the abstract model, a logic program is solved by a network of concurrent processes. A process is associated to each node of the AND/OR tree of a logic program. In the abstract model of PALM, there are two kinds of processes: *AND-OR processes* and *Clause Manager processes*. An AND-OR process can be executed in AND mode (*AND Process*) or in OR mode (*OR Process*), depending on the parameters of the *start* message that the parent process sends it at the activation time. An AND process is created to solve a goal statement which consists of a conjunction of literals (subgoals). An OR process is created by an AND process to solve one of those literals. The OR Processes cooperate with the Clause Manager processes to ask for and to receive the unification of literals. The message exchanged between the processes are:

- *start, redo, cancel,* sent from an AND-OR process to its immediate descendants;
- *success, fail,* sent from an AND-OR process to its parent;
- *unifyrequest,* sent from an OR Process to a Clause Manager;
- *unifyresult,* sent from a Clause Manager to an OR Process;

Two models for the parallel execution of logic programs which are similar to the PALM model are the AND/OR Process Model (Conery 1987) and the REDUCE-OR Model (Kalé 1991). Compared to these models, the main differences are:

- O Unification algorithm is performed into the Clause Manager not in the OR Process. This also allows a kind of *search parallelism* by partitioning of clauses of a procedure and storing each partition into a different Clause Manager.
- O Restricted-AND parallelism is implemented instead of complete AND parallelism which requires a computation overhead that may abate the benefits deriving from parallel execution. The same choice has been done in the REDUCE-OR Model.
- O In the resolution of goals which unify with clauses whose body is composed of only one literal (e.g., *p(X) :- q(X).*), an OR Process instead of an AND Process is activated to solve the body. In this way a considerable reduction in the number of the activated processes is achieved.

PALM

The architecture of PALM consists of a set of modules or processes mapped on each node of a parallel computer (Figure 1). On each node there is a set of *AND-OR processes* (AO) that implement the clause resolution, a *Clause Manager process* (CM) which handles one portion of clauses, a process which carries out the loading of the database clauses called *Loader* (LD), and a *Communication Server* (CS) that performs the routing of messages to and from the network nodes.

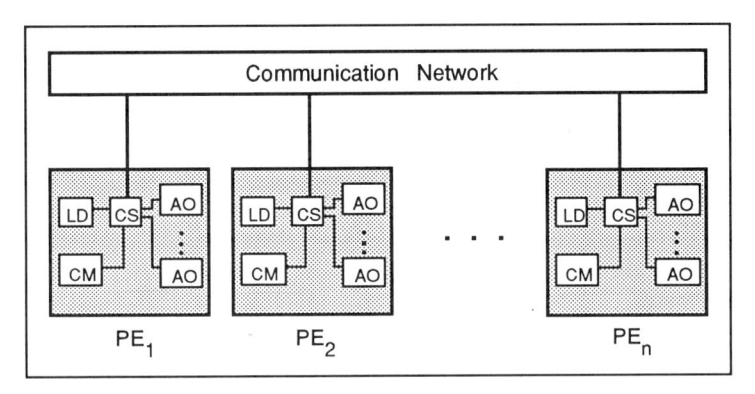

Figure 1 Architecture of PALM

The AND-OR processes implement the parallel resolution of the goal statement according to the algorithms provided by the abstract model. The AND-OR Processes are created statically and activated dynamically. The Clause Manager process performs the management of the local partition of the clauses database. Furthermore, it performs the unification of a literal with a clause of its partition.

The parallel computer on which PALM has been implemented is a *distributed memory* MIMD machine and consists of a network of T800 Inmos Transputers connected in a toroidal mesh topology. A source logic program is translated into a network of cooperating processes written in the concurrent "target" language *Occam 2*, then these processes are mapped onto the Transputer network. An important capability of Transputer is related to the ability to execute many processes at the same time (from tens to thousands) and to create new processes easily allowing communication between processes within a single Transputer and between processes running on different Transputers. The PALM prototype has been evaluated on a network of up to 32 Transputers, but its implementation does not limit the use on larger networks.

IMPLEMENTATION TECHNIQUES

Most parallel logic programming systems have been implemented on shared-memory multiprocessors. Differently, PALM has been designed to be implemented on message-passing multicomputers. It is based on a completely decentralized model where the memory and process scheduling are highly distributed.

In this section the main problems faced in a distributed implementation of PALM are outlined. For each problem the chosen solution is discussed. The problems discussed are related to the following topics: dynamic processes and reusability, process scheduling, communication handling, control of parallelism degree, clause distribution and variable binding, and distributed termination of a parallel logic programming support. These issues are typical of parallel implementations of a logic language on a distributed architecture.

Dynamic processes

In the PALM model a logic program is implemented by a set of dynamic processes which cooperate by message passing. For the implementation of the model some mechanisms for the activation/termination of processes and for their remote execution are needed. The Occam language does not allow dynamic process creation. Therefore a mechanism which simulates dynamic process creation must be implemented on top of Occam using the mechanisms offered by the language. Three main ways may be followed to do this.

1. Static creation of a very large number of processes equal to the number of nodes of the computation graph of the parallel program.
2. Process migration between different processors.
3. Static creation of a limited number of processes and a mechanism for their reuse. According to this technique, a small number of processes are re-executed many times on a different set of data.

Although the first approach looks like very simple and immediate, it poses some problems. The high number of processes require a large amount of memory on each processor of the parallel system. Further, this requires the use of dynamic channels to support the communication between two generic processes allowing the dynamic configuration of the parallel program as the execution proceeds. The second approach has proved to be not practical on current distributed parallel architectures due to its high implementation overhead.

The third approach has been used in the implementation of PALM. This solution limits the amount of memory used and supports an object-based approach to software development. This technique is useful in the parallel implementation of logic languages as well as in the implementation of object-oriented languages and divide-and-conquer algorithms.

In PALM the AND-OR Processes are created statically and activated dynamically. They are designed in such a way to allow, after a virtual termination, the process "wake up" and initialization to carry out a new task, by receiving a special message (*start*). The Communication Server supports the activation of a process on each processor and its communication with other processes in a dynamic way.

The process termination management and its inclusion in the idle processes pool allows to achieve a high reusability of processes, especially when their execution time is short. Thus, using a small number of processes is possible to solve a logic query with a large AND/OR tree. For example, to solve the *6-queens* problem, the AND/OR tree is composed of about 6000 nodes, whereas the system only uses 128 AND-OR processes which are re-activated about 50 times.

Process scheduling

Process scheduling and load balancing algorithms provide a distribution of tasks (processes) to the computing nodes of a parallel computer attempting to balance the load on all nodes such that the whole throughput of the system is maximized. In PALM the problem to be solved is related to the dynamic activation of AND-OR processes on the nodes of the multicomputer.

The implementation of load balancing strategies on multicomputers is very critical due to the large number of computing nodes and the overhead resulting from communication necessary to accomplish the application load balancing. Most of the proposed load-balancing strategies use centralized or fully distributed models for taking the task assignment decision. In a centralized model, a special node collects the global information about the state of the system and assigns tasks to individual nodes. In a fully distributed model, each node executes a scheduling algorithm by exchanging information with all the other nodes. For multicomputers consisting of a large number of nodes, centralized and fully distributed load-balancing strategies are not very suitable for many reasons.

In a multicomputer system a central scheduler becomes a bottleneck and may abate the performance. Moreover, the memory requirements for storing the state information about all the nodes may become excessively high. On the other hand, fully distributed strategies incur large communication overhead due to the information exchange among all the nodes of a multicomputer. Due to these reasons, a load-balancing algorithm for a multicomputer should take into account these requirements using a locality-based or semi-distributed scheduling strategy.

In PALM we used a dynamic process scheduling and load balancing strategy called Probabilistic strategy with Neighbourhood Synchronization (PNS) (Cannataro *et al* 1992b) The PNS strategy uses only information from neighbours and takes into account the effect of information lags in multicomputer systems for estimating the system load and for task assignment decision. This effect means that dynamic information about state of nodes among which we choose one for processing a new task is not immediately available, because an information measured at some time t on a node i will reach node j at time $t+t'$, where t' is the communication delay between i and j. Thus, the information held by a node producing the assignment decision will be obsolete. Then if the flow of task arrivals has no regularity, decisions taken on the basis of currently available information will be nonoptimal.

To overcome this problem PNS uses a probabilistic strategy based on two combined ideas. First, a new task should be sent with a higher probability to a processor with a short task queue and with a lower probability to a processor with a long queue. Second, a new task should be sent with a lower probability to a processor about which we have more obsolete state information and with a higher probability to a processor about which we have more recent information.

Communication handling

A major problem faced in the implementation of PALM was the implementation of network communication between AND-OR processes and Clause Managers. In PALM this remote communication level is offered by a *Communication Server* (CS) which is implemented by a set of processes, *Multiplexer, Demultiplexer, Router,* and *Process Table*. The main functions of the CS are:

- receiving of every message sent from PALM processes,
- message compression,
- naming of AND-OR processes,
- message through-routing,
- break up and decompression of messages on the target node, and
- message delivering to the destination process.

Furthermore, the CS manages some other operations such as activation, suspending and termination of AND-OR processes.

Multiplexer and Demultiplexer implement an interface between the Router and the processes of the system. The Router process routes the messages through the network nodes. It uses a deadlock-free routing algorithm which is adaptive, guarantees message arrival, and automatic local congestion reduction (Cannataro et al 1992a). Finally, the ProcessTable process manages the status of AND-OR processes and stores the logic links of an AND-OR process with its parent and its descendants in a transparent way as regards their location on the nodes of the network.

To describe how the CS works, we shortly discuss the local and remote activation of an AND-OR process. For a local activation the Multiplexer process books a process from the pool of idle AND-OR processes. As result of this operation, the identifier of the process is sent from ProcessTable to the Multiplexer. Then the *start* message is delivered to the AND-OR process. ProcessTable updates the status of the activated process and adds the process identifier to the list of descendants of the parent AND-OR process.

When a remote activation (from node *A* to node *B*) must be carried out, the operations are performed on different processors. In particular, MultiplexerA communicates the start of the remote activation to ProcessTableA. After that the activation message reaches the node chosen by the load balancing algorithm (node *B*), DemultiplexerB asks to ProcessTableB the activation of a process. When the activation is completed, ProcessTableB sends back the process identifier, then the list of the descendant processes of the parent process is updated by ProcessTableA.

Degree of parallelism

The degree of parallelism of a program indicates the number of program operations which can be executed in parallel. Controlling the degree of parallelism in a program means to specify which parallel activities are actually executed in parallel depending on the available resources (e.g., the number of processors).

In the parallel execution of logic programs, the main objective is the exploiting of the highest degree of parallelism. On the other hand, supporting a high degree of parallelism does not automatically improves the throughput. High parallelism introduces some overheads due to process creation, communication costs, and task switching that can annihilate the benefits of parallelism. Therefore a trade-off between the parallelism degree we want to exploit and the overhead it can introduce must be found.

A solution which can be used in parallel logic programming systems such as PALM should be based on the dynamic control of the degree of parallelism during the logic program execution. In PALM, the degree of exploited parallelism can be controlled at run time by the use of communications among the AND-OR processes and by the size of data structures. For example, the activation of a larger number of parallel processes can depend on the time when the request of a solution is sent to a descendant process or on the size of the queue where the partial solutions are stored.

In the PALM implementation of OR parallelism, an OR process sends a *redo* message asking for a new solution to each AND process which already sent a *success* message. To control the degree of OR parallelism, the time at which the *redo* message is sent may be varied. If the message is sent immediately after the reception of a *success* the highest degree of parallelism is exploited. On the other hand, OR parallelism can be largely reduced if a *redo* message is sent only when a new solution is necessary as occurs in the system described in (Biswas et al 1988).

A simple technique which can be used in PALM for the control of the degree of parallelism is based on the use of buffers with a limited size to store the solutions arriving in the *success* messages. Thus, the exploited parallelism depends on the buffer size because an OR process can send a *redo* message only if at least one free buffer element is available. If the buffer is full the *redo* is delayed until one buffer element will be free. This example and the previous one show how the degree of parallelism can be controlled at run time in PALM without the overhead of particular data or control structures.

Clause distribution and variable binding

There are two possible ways to store the clauses of a logic program on a highly parallel architecture:

1. *Clause replication*: the whole set of clauses of the logic program must be replicated on every node of the parallel system;
2. *Clause partition*: the clauses are partitioned, so that on every node is allocated a single partition with the information about the complete allocation of the other clauses on the other nodes.

The choice of clause replication on every node allows the system to have only local accesses (i.e., on the same node), but on the other hand it raises memory capacity problems. On the contrary, the clause partitioning is suitable for a highly parallel architectures. This solution is better from the scalability point of view. Indeed, when the number of nodes (n) increases the total amount of memory (M) does not change, whereas the amount of memory of a single partition is reduced ($m=M/n$). On the basis of these motivations we have chosen to partition the clauses entrusting the management of the local partition to one Clause Manager per node.

In the parallel implementation of a logic language, one of the problems to be solved is the management of multiple bindings of variables or *binding environments*. A binding environment consists of a set of frames which contain the values of the variables of a clause for a given goal call. An interpreter can produce different values for a variable due to the parallel execution of subgoals of a conjunction. In PALM we used a modified version of a method for variable binding management in a distributed environment known as *Closed Environments* (Conery 1988). This method follows the copying approach. The major feature which makes it different from other methods is that a process accesses only its local environment where the variables of its subgoal are stored. Thus, the processes cooperate using messages for the exchange of frames.

A closed environment E is defined as a set of frames such that no variable in the frames of E contains a reference to a frame that is not in E. Generally, an environment contains two frames named *top* and *bottom*. The top frame holds the terms sent to the process from its parent. The bottom frame contains terms that will be sent to its descendents. Frames are exchanged among processes by means of *start* and *success* messages.

During the execution of the unification algorithm it is necessary to perform some transformations on the frames. After a transformation, one frame will be closed, that is all external references will be solved. The closure of the top frame with respect to the bottom frame occurs after the unification with the head of a unit clause (assertion). In this case the OR Process passes the top back to the parent AND Process in the arguments of the *success* message. For every unification with the head of a non-unit clause (implication), the bottom frame is closed and is sent to the AND Process which must solve the body of the clause. In the former, the closure is used to import the bindings back in the AND/OR tree. In the latter, the closure is used to pass to the descendant process the terms of the body which must be solved.

This technique can be implemented on a distributed memory architecture. In fact, if on the one hand the Closed Environments technique presents an overhead due to the intensive use of copies, on the other hand there is locality of reference, and the allocation of memory required by a process is small since every process only needs to store the goals that it must solve. In (Cannataro *et al* 1991) is presented a detailed description of the implementation of this technique in PALM.

Distributed termination

A typical issue in concurrent programming is to ensure the correct distributed termination of processes which compose a concurrent program. A concurrent program is terminated when every process is passive and there are no messages in transit. The detection of the termination of a concurrent program is a non-trivial problem and it is essential for the program correctness (Francez 1980).

A decentralized algorithm has been implemented in PALM for the termination detection of processes which support the parallel execution of a logic program. When every AND-OR process is terminated and no more solutions can be found, the termination algorithm forces the termination of the other processes of the PALM run-time support. The algorithm uses a virtual ring to propagate the termination message. No central controller exists and the algorithm is symmetric, that is no process has the special role of initiating the detection of termination. Furthermore, no delaying of the basic application (freezing) is necessary.

According to the termination algorithm, when on a node every AND-OR process is terminated and no more solutions can be found, the Router process sends a token through a channel on the virtual ring. A simple procedure can assure that only one process will begin the termination operation. We shortly describe this procedure for dynamic selection of the process which begins the distributed termination. Let n be the number of processes. When a process is ready to terminate it generates a token ($t=1$). The token must be routed among the processes until all have finished their tasks.

When a process ends its task, it increases the value of the token ($t=t+1$) and checks if $t=n$. When all processes are ready to terminate the token will be $t=n$. The last process to finish its task, after increasing the value of the token will find it equal to n. This one is the process which begins the distributed termination operation sending the termination message and terminating itself. Every process performs the same operation until all the processes are terminated.

EXPERIMENTS

This section shortly discusses some performance figures of PALM on two test programs. The first one is a simple database program which for a given person gives its *parents*, *siblings*, and *cousins* (144 solutions). The program contains three rules and twenty facts. The second one is the *n-queens* program with $n=6$ (4 solutions).

Figure 2 shows the speedup of the two programs with respect to the number of Transputers. We can see that on 32 Transputers the speedup which has been obtained is about 5. In this case the speed up is limited by the low utilization of the parallel system due to the small size of the test programs. Using more complex programs a greater speed up might be obtained.

In general, the major limit to the program speed-up is due to the communication overhead. An improvement may derive from the optimization and reducing the messages exchanged between the AND-OR processes.

Figure 2 Speed-up

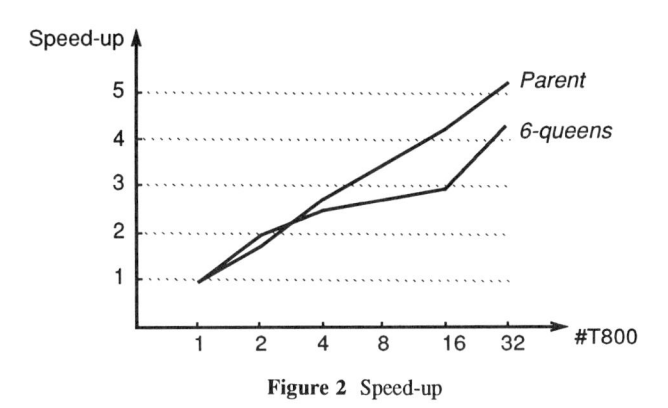

Tables 1 and 2 summarize the values of communications in the two examples. Notice that the average journey (number of crossed links) of remote communications is about 1. This means that a *high locality degree* is present in the system.

Table 1

6-queens	Total	Local	Remote
Comm.no.	17157	346	16811
Latency (ms)	1.95	1.43	1.97
Size (bytes)	132	250	130
Journey (link)	-	-	1.00

Table 2

Parent	Total	Local	Remote
Comm.no.	16183	12728	3455
Latency (ms)	1.35	1.25	1.70
Size (bytes)	61	68	33
Journey (link)	-	-	1.02

Table 3 shows the number of activated AND-OR processes with respect to the available (created) processes. It shows the high reusability of AND-OR processes. Thus, with a relatively small number of processes is possible to solve a logic query with a large AND/OR tree. Notice that in the *6-queens* program an OR Process instead of an AND Process is activated to solve the body of clauses whose body is composed of only one literal. In this way the number of the activated processes and the elapsed time have been reduced by half.

Table 3

	Active Procs	Availab. Procs	Reuse Degree
6-queens	6140	128	48
Parent	5298	128	41

Figure 3 shows the lifetime of AND-OR processes of the *6-queens* (a) and *parent* (b) programs. These figures indicates that PALM is a very dynamic environment. In fact, more than 95% of processes have a lifetime lower that 100 milliseconds. This means that the process grain size is very fine and the task switching is high.

Considering that the number of communications is high and the medium latency of a message is 1.95 milliseconds, we can deduce that the communication cost is a major overhead in the parallel execution of logic programs in PALM. A special attention must be paid to reduce this overhead.

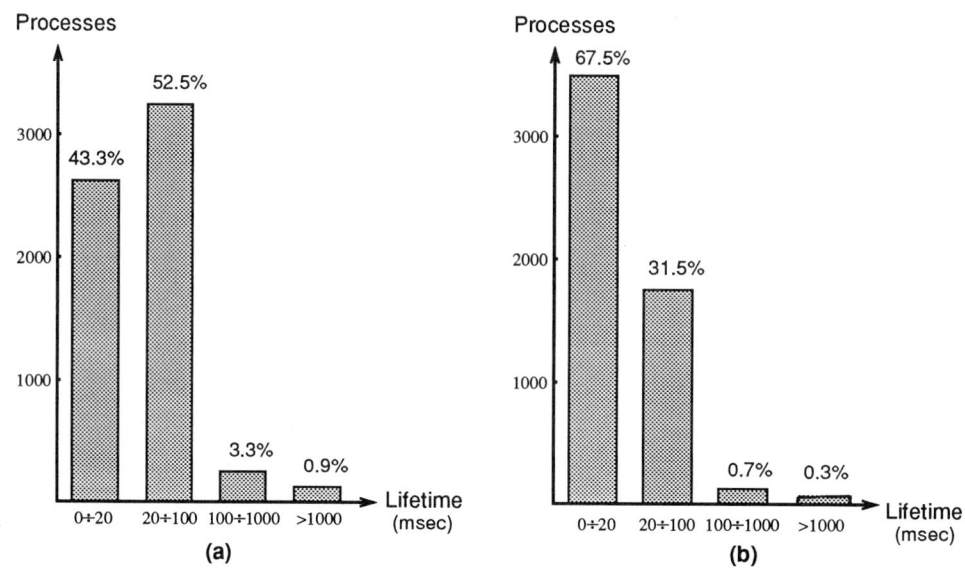

Figure 3 Process lifetime

CONCLUSIONS

The availability of a logic programming environment on a parallel VLSI architecture makes possible to reduce the software design overhead and to achieve high performance. PALM provides tools to decompose automatically the problem (program) in many concurrent tasks and to allocate these tasks on the network processing elements. We implemented the initial version of the model on a parallel computer which consists of 32 Transputers.

The main results of the work are the integration of the expressive power of logic programming with the computational power of highly parallel VLSI systems, and the implementation of a logic programming language on a fully distributed parallel architecture. Currently we are evaluating the prototype. These experiments will drive the future work.

Acknowledgements

This work has been partially supported by "Progetto Finalizzato Sistemi Informatici e Calcolo Parallelo" of CNR under grant 91.00986.69.

References

Biswas P., Su S.C., and Yun D.Y.Y., "A scalable abstract machine to support limited-OR/restricted-AND parallelism in logic programs," *Proc. Fifth Int. Conference and Symposium on Logic Programming,* MIT Press, pp. 1160-1179, 1988.

Cannataro M., Spezzano G., and Talia D., "A parallel logic system on a multicomputer architecture", *Future Generation Computer Systems,* North-Holland, vol. 6, no. 4, pp. 317-331, 1991.

Cannataro M., Gallizzi E., Spezzano G., and Talia D., "Design, implementation and evaluation of a deadlock-free routing algorithm for concurrent computers", *Concurrency: Practice and Experience,* vol. 4, no. 2, pp. 143-161, April 1992a.

Cannataro M., Sergeyev Ya., Spezzano G., and Talia D., "A probabilistic load balancing strategy with neighbourhood synchronization for multicomputer systems," submitted for publication, 1992b.

Conery J.S., *Parallel Execution of Logic Programs*, Kluwer Academic Publisher, 1987.

Conery J.S., "Bindings environments for parallel logic programs in non-shared memory multiprocessor", *Technical Report*, Univ. of Oregon, 1988.

DeGroot D., "Restricted AND-parallelism", *Proc. Int. Conf. on Fifth Generation Computer Systems 1984*, pp. 471-478, ICOT, Tokyo, 1984.

Francez N., "Distributed termination," *ACM Trans. on Programming Languages and Systems*, vol. 2, no. 1, pp. 42-55, 1980.

Kalé L. V., "The REDUCE-OR process model for parallel execution of logic programs", *The Journal of Logic Programming*, North-Holland, vol. 11, pp. 55-84, 1991.

Inmos, *Occam 2 Reference Manual*, Prentice Hall, England, 1988.

Inmos, *Transputer Databook*, Inmos Ltd., England, 1989.

Kowalski R., "Predicate logic as programming language", *IFIP Information Processing*, North-Holland, pp. 569-574, 1974.

Seitz C., "Concurrent VLSI architectures", *IEEE Trans. on Computers,* vol. C-33, pp. 1247-1265, 1984.

A DISTRIBUTED PARALLEL ASSOCIATIVE PROCESSOR (DPAP) FOR THE EXECUTION OF LOGIC PROGRAMS

Darren Rodohan and Raymond Glover

INTRODUCTION

The majority of parallel architectures for Prolog exploit the inherent AND/OR parallelism available within the language, whilst still maintaining the depth first search strategy. Alternative control strategies are usually implemented with a meta-interpreter in Prolog but Guiliano (1991) found them to be inefficient. This inefficiency has lead us to investigate a distributed architecture that exploits search and data parallelism in Prolog rather than the AND/OR parallelism of other systems. State space search and in particular the A* algorithm (Nilsson 1980) provide a possible solution.

The A* algorithm is an optimal state space search technique used to navigate through a weighted graph of nodes. Each node contains the estimated cost of a solution, Fn; the cost of the path so far, Gn; the search depth, Dn; the current state, S, and a pointer to the previous state, P. The algorithm can be parameterised as shown below (Pohl 1969) so that breadth and depth first searches can also be easily obtained.

$$Fn = (1-W).Gn + W.Hn, \qquad \text{where } 0 \leq W \leq 1$$

If the weighting value, W, is set to 0 then a breadth first search is obtained, with W = 0.5 an A* search is obtained and using a value of 1 gives a pseudo depth first search. The depth or breadth first strategy can be made to depend on a domain specific value or the search depth. A depth first search dependent on search depth implements the same strategy used in Prolog.

A hierarchically structured, pipelined architecture can be derived from the division of the A* algorithm into three constituent parts: generation, evaluation and selection. The generation stage can be considered an examination of a database and the application of system constraints, resulting in a set of successor nodes. The generated successor nodes are fed into a pool of processors, organised as a broadcast network, for evaluation. These processors operate as a sequential abstract machine, independently processing sequential instructions and calculating the heuristic function, Fn, of the nodes allocated to them. The evaluated nodes are then passed to a selection network whose sort key is determined by the search strategy adopted. Rodohan and Glover (1991) showed that the expected reduction in execution time compared with Warren Abstract Machine (Warren 1983) implementations of Prolog is approximately proportional to 0.67 times the number of pool processors.

In the following sections we will focus our attention on set processing as this area has been greatly enhanced by advances in VLSI and WSI techniques that allow many processing elements to be fabricated on the same chip.

A SET BASED PROLOG

Previous attempts to exploit data parallelism in logic programs have concentrated on unification with content addressable memory, CAM, for interpreted systems (Kogge et al 1989, Ng et al 1989 and Naganuma et al 1988). These schemes are all aimed at accelerating the sequential execution of Prolog. Kacsuk (Kacsuk and Bale 1987) showed how a set based derivative of Prolog could be executed in parallel on a Distributed Array Processor, DAP, removing the need for shallow backtracking. The model allows Prolog to obtain a set of instantiations for a variable during parallel execution, as long as data consistency is maintained. The model relies on the pattern matching, arithmetic and relational capabilities of the DAP. However, the physical size of the DAP, 4096 processing elements, is a severe limitation of the performance.

We will now describe a mapping of a set based Prolog to a scalable architecture - the Associative String Processor, ASP, (Lea 1988) which can provide a much larger number of processing elements than the DAP with comparable performance per processing element. The ASP is a SIMD machine consisting of a large number of associative processing elements, APEs, connected together as a string. Each APE contains a one bit arithmetic unit and 64 bits of data memory, Data Register, which it can process bit serially. The Data Register is made up of CAM which along with six activity bits in the Activity Register can be addressed associatively in Bit, Byte and Word (32 Bits) modes. When data in an APE matches a search operation the match reply line, MR, goes high to indicate that at least one APE is tagged in one of the tag registers, M or D. An activation register, A, is used as a mask to include or remove an APE from the current operation. The structure of an APE is shown in figure 1.

The ASP supports both synchronous bit serial and asynchronous word parallel communication. Bit serial communication allows N APEs to shift data via a tag register to APEs a distance D away, thus providing many to many communication. Word parallel communication is possible by marking source and destination APEs, then reading the source APE data and then broadcasting it to the destination APE(s), thus providing one to many communication.

In the VLSI implementation of an ASP chip, VASP, there are currently 64 APEs. These chips are then connected together to form strings of the desired length. The Trax machine, currently being commissioned at Brunel University, contains 16K APEs in 256 chips. A wafer scale version of the ASP chip, WASP, will contain 15,360 APEs within a 6 cm X 6 cm device using 1 μm CMOS (double-layer metal). The device contains defect/fault-tolerant circuitry allowing harvest targets of at least 8,192 APEs on 50% of wafers passing parametric probe tests (Lea 1991). Thus the WASP device will make systems containing a very large number of APEs, $O(10^6)$, feasible.

In the following sections the method used to implement a set based Prolog are discussed. All timing results were obtained using a hardware simulator for the VASP chips running at a 20 MHz clock frequency.

In order to implement a set based Prolog on the ASP, two stack structures are required. One stack, the Set Stack, contains the relationship between different sets with the current search depth and the other, the Trail Stack, contains the set name, search depth and rule name when a choice point is reached. The relationship between the sets are divided into five separate types :- base, associated, sibling, restricted and derived. A new set that is only related to the universal set is known as a base set. A sibling set is created during unification when two Prolog clauses contain the same variable name. For instance, given the Prolog fragment 'parent(X, Y), parent(Y, Z)' two separate sets would be created for Y during unification. This effectively provides a link between the two clauses although they are semantically the same set. A restricted set is created if during unification or a conditional expression elements are removed from an existing set. A derived set is created to indicate that a new set was created that is connected to a previous set by the 'is' operator. For example, given the Prolog statement 'X is f(Y)', X would be a derived set of Y.

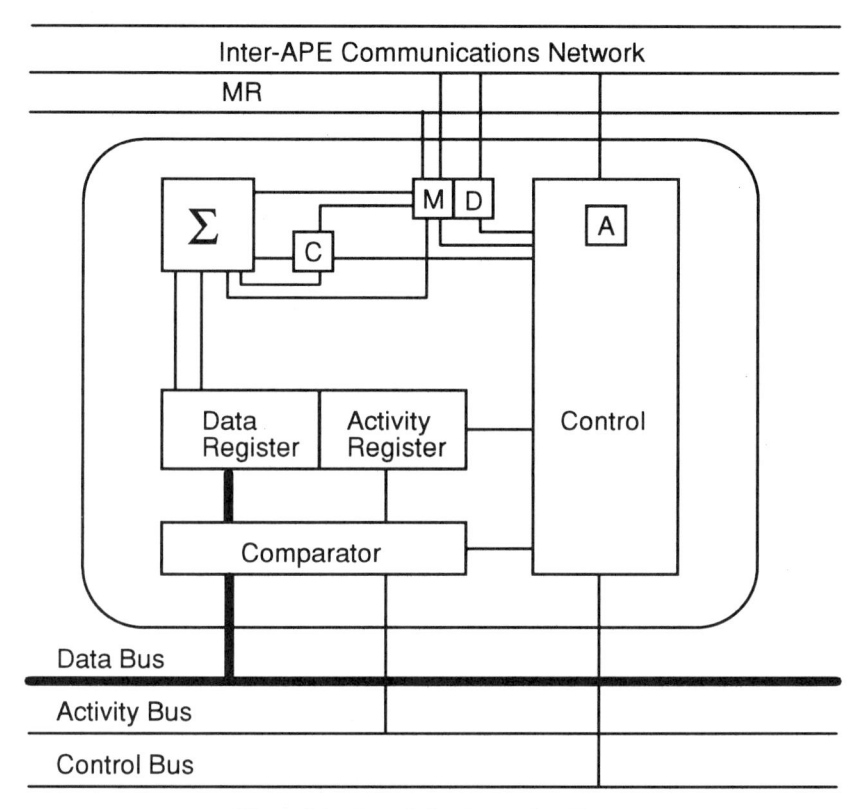

Figure 1 An Associative Processing Element

Execution Model

The execution model for a set based Prolog can be divided into four distinct blocks; unify, data process, restrict and backtrack, as shown in figure 2. In a typical processing cycle either a unification or data process operation is performed. If the operation was successful a restrict operation is required to ensure data consistency, otherwise a backtracking operation is started so that the program state at the last choice point is restored.

Unification. Unification can be split into two categories, those amongst variables in the same clause head known as associated unifications and those amongst variables in different clause heads, known as independent unifications. Associated unifications make use of the ASP's synchronous communication network allowing all APEs that require the unification operation to concurrently process their data. For example, if you wanted to locate all points on a grid at a forty five degree angle from the origin, (0, 0), this would require that both the X and co-ordinate are identical. This would be expressed in Prolog as shown below.
   grid(1, 1). grid(2, 3).

grid(3, 3). grid(7, 6).
diagonal_45(X) :- grid(X, X).

Set Prolog would return X = {1, 3} in one step, whereas Prolog would return the same solution set on backtracking. The time taken to perform the unification is independent of the number of clauses involved, although there is a dependency on the number of elements in the structure being unified.

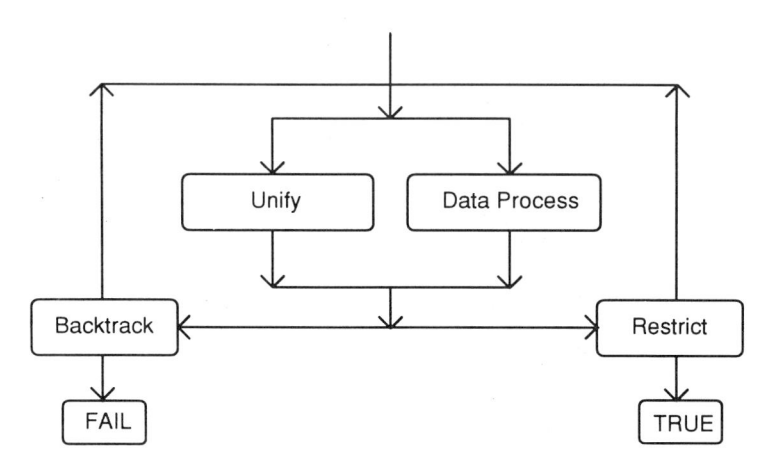

Figure 2 Execution Model

Independent set unification is a one to many operation. This is essentially a join between two databases using a common key value and must therefore use the less efficient word parallel communication. For instance, if we have two different databases of facts, one with the location of university colleges and another with the location of rivers, we can write a relation in Prolog to find which rivers are close to the colleges.

```
univ(jesus, oxford).         univ(imperial, london).      univ(sussex, brighton).
river(oxford, isis).         river(oxford, cherwell).     river(london, thames).
relation(College, River) :-
            univ(College, Town), river(Town, River).
```

This would return the following variable bindings for College/River = {jesus/isis, jesus/cherwell, imperial/thames}. This is achieved by creating a sibling set for Town in the river relation, Town', and then performing a word parallel comparison between an element in the Town set and all of the Town' set for each element in the Town set. The time taken for the operation is therefore dependent on the number of elements in the Town set. Although the time taken for independent unification is much longer than for associated unifications it should be noted that it is a less frequently performed operation and is a similar to problem to that faced in AND parallelism.

Data Process. The ASP supports a wide range of arithmetic and relational operations which are executed bit serially by an APE. The operations can be executed concurrently by all APEs that have matched a given search criterion, thus giving high overall performance. Lea

(1991) estimates peak performance for a WASP device with 40 MHz clock as 39.7 Giga-OPS for 32 bit fixed point addition/subtraction. This was based on the assumption that all APEs (65, 536) were operating concurrently but this is somewhat misleading since it is rare for all processors to be active simultaneously. However, if only 1000 processors were active a performance of 605.7 MOPS is achievable. Similar performance is given for relational operations such as 'less than' and 'greater than' searches. If a conditional expression has been used, a check needs to be made for any elements of the set that did not satisfy the condition. These elements and all related elements in other sets must be restricted to maintain data consistency.

Set Restriction. Restriction is a fundamental operation that ensures consistent data is maintained between sets. It is required when a unification or condition results in elements of a set being removed. The corresponding elements must then be removed from all associated, derived and sibling sets. For example, given the query

$$:-? \ a(X, Y), Y > 4.$$

where $X = \{1, 2, 3, 4\}$ and $Y = \{4, 5, 6, 7\}$, the condition $Y > 4$ requires the deletion of '4' from set Y and the corresponding element '1' from set X. The restriction on set X is known as implicit restriction, because X is associated with Y. When a set is restricted its related sets must be restricted too; this must continue, with restriction of related sets, until a base or restricted set has been reached. This results in a chain of restrictions, where the length of the chain depends on the number of sets unified since the last restriction or base set unification. An algorithm to perform the restrict operation is now given, where S1 is the set to be restricted.

1. Find related set S3 of set S1 on the set stack. If the relation, T1, between S1 and S3 is base or restrict then exit.
2. Create a new set, S4, on the set stack related to set S1 via relation T1.
3. If relation T1 is associated or derived then perform associative restrict operation, otherwise perform sibling restrict operation.
4. S1 = S3 and then go back to 1.

An associative restrict operation takes advantage of the relative locations of associated or derived sets. The individual members of the sets are physically close to one another so that the time taken to perform this restrict operation is independent of the number of elements to be restricted. The associative restrict operation can be performed in 675 nS, using the method outlined below. Let S1 denote the set that has just been processed, S3 the set related to S1 on the set stack and S4 be the new name for the restricted version of S3.

1. Find those APEs with set mark S3 and tag them with tag register D.
2. Find those APEs with set mark S1 and tag them with tag register M.
3. For every APE tagged in M write activity of 'Not These', in the subsequent APE above it, tagged in D.
4. Find those APEs not marked with activity of 'Not These' and then write the new set mark, S4, to them.
5. Clear all activity marks.

The restriction of a sibling set is more complex as the data is no longer local and so the time taken is dependent on the number of elements, N, in the sibling set. The time taken to perform a sibling restrict operation was found to 925 nS + N x 675 nS using the algorithm outlined below; where N is the number of elements in the set that has just been restricted, S1. Let S3 denote the sibling set of S1 and S4 be the new name for the restricted version of S3.

1. Find set S3 and mark with activity of 'Candidate'.
2. Find set S1 and mark with activity of 'Restricted'.
3. Find those APEs marked with 'Restricted', if no matches, then go to 7.
4. Read top match to Data and remove activity mark 'Restricted' from it.
5. Find Data in those APEs marked with 'Candidate' and remove their activity mark.
6. Go back to 3.
7. Find those APEs still marked with 'Candidate' and then write new set mark, S4, to them.
8. Clear all activity bits.

The time taken to perform a complete restriction cycle can now be estimated from the following expression.

Let
N_a be the number of associative restrict operations,
N_s be the number of sibling restrict operations,
T_f be the time taken to find a related set on the set stack,
T_c be the time taken to create a new set on the set stack,
T_a be the time for an associative restrict operation,
T_s be the time taken for a sibling restrict operation, this depends on the number of elements to be restricted, and
T_r be the total restriction time.

Then
$$T_r = (N_s + N_a).(T_f + T_c) + N_a.T_a + N_s.T_s$$

where $N_s + N_a$ is the total number of sets bound since either the latest restrict operation or a base set was pushed onto the set stack.

Backtracking. When a failure occurs, a previous program state must be regained by backtracking. In order that backtracking can be accomplished, a Trail stack must be maintained. Whenever a rule or data choice point is reached, the following information is pushed onto the Trail stack : the current set name, the current search depth and the name of the rule being executed. If the ASP is being run with a strict depth first, left-to-right search strategy, only one data item is partitioned into a new set at a time. After this partition a check is made to see if any other data that has already been processed, using data parallelism, would have satisfied the current rule. If another solution was available, a choice point is established by pushing data onto the Trail stack. The backtracking algorithm employed can be summarised :

1. Clear all temporary storage.
2. Pop Trail stack to obtain set name, depth and rule name where the last choice point occurred.
3. Find the set name on the ASP and mark with an activation called Fail.
4. Sequentially pop the set stack changing set marks back to their previous values until the set name and depth obtained from above is reached.
5. Change the current set name and depth value to those found above.
6. Restart execution at the rule name found above.

The activation called Fail is required to prevent the choice that caused failure and all previous choices for that set variable from being included when the rule is re-satisfied. The time taken to backtrack to the previous choice point can be estimated as follows :

$$Time = (D_c - D).B.T_s + T_f$$

where D_c is the current search depth, D is the depth of the last choice point, B is the average number of bindings at any depth, T_s is the time taken (875 nS) to reset a set marking to its previous value and T_f is the time taken (1000 nS) to flush temporary storage.

RESULTS

The results given in this section were obtained using a hardware simulator for the ASP running at a clock rate of 20 MHz. The ASP code was hand written using a library of routines to execute compiled code. The results for SB-Prolog were obtained using compiled code on a SUN 3/60. Results are given for two benchmark programs, the Nqueens problem (Bratko 1990) and a program that performs joins between database relations (Rodohan 1992).

The Nqueens problem demonstrates the ability to backtrack. In Table 1 the time taken to perform the Nqueens by the ASP and SB-Prolog are given. The improvement factor given was calculated by dividing the time taken for SB-Prolog by the ASP time. The exploitation of data parallelism on the ASP gives performance increases in the region of 200. The time taken to perform the backtrack operation was found to be in the region of 10% of the total processing time, no restrict operations are required by this program.

Table 1 Results for the Nqueens problem

No. of Queens	SB-Prolog (mS)	ASP (mS)	Improvement Factor	No. of ASP backtracks	No. of sets reset	ASP Backtracking Time (µS)
8	920	5.438	169.2	42	672	630
10	1,160	5.023	230.9	35	560	525
12	3,960	13.23	299.3	100	800	800

In Table 2 the time taken for SB-Prolog and the ASP to perform the following code fragment is given for varying numbers of clauses in the last term. The number of clauses in the first and second clause of the program was kept constant at 225.

```
test(Town, Travel, Accom, Price) :-
    travel(Town, Travel),
    stay(Town, Accom),
    hotel(Accom, Price).
```

The time was measured for all solutions to be found using the findall predicate in SB-Prolog and the set execution mode on the ASP. An improvement of more than 400 times was noted for the ASP code which can again be accounted for by the exploitation of data parallelism. This benchmark measures both the independent set unification and set restriction operations of the ASP.

Table 2 Results for joining database relations

Number of Clauses	SB-Prolog (mS)	ASP (mS)	Improvement Factor
0	-	6.788	-
50	4,340	9.153	474.2
100	5,640	11.58	487.0
150	6,920	14.54	475.9
200	8.180	17.83	458.8

SUMMARY

We have demonstrated the feasibility of mapping a version of Prolog based on sets to the ASP architecture. This mapping is based on the ability of the ASP to provide data parallelism on a massive scale through the exploitation of VLSI and WSI techniques. Performance estimates indicate that execution time can be reduced by $O(10^2)$ over sequential implementations.

Future Work

This work is to be continued under SERC contract number GR/H46893 entitled "An Investigation of a Distributed Parallel Associative Processor for the Execution of Logic Programs". The aim of this research is to develop a Prolog compiler for the ASP using the library routines already implemented. The compiler will then be evaluated for use in various application areas such as image understanding and natural language processing.

Acknowledgement

This work is supported by the Science and Engineering Research Council award reference number 89315132.

References

Bratko, I. *Prolog Programming for Artificial Intelligence*, Addison-Wesley, pp. 120-122, 1990.

Guiliano, M.E., "The control and execution of parallel logic programs", *PhD Thesis*, Dept. of Computer Science, University of Maryland, USA, 1991.

Kacsuk, P. and Bale, A., "DAP Prolog: A Set-oriented Approach to Prolog" in *The Computer Journal*, vol. 30, no. 5, pp. 393-403, 1987.

Kogge, P.M., Oldfield, J.V., Brule, M.R. and Stormon, C.D., "VLSI and Rule-Based Systems" in *VLSI for Artificial Intelligence*, J. Delgado-Frias and W. Moore (ed), Kluwer Academic Press, 1989.

Lea, R.M., "The ASP: A cost-effective parallel microcomputer", in *IEEE Micro*, pp. 10-29, 1988.

Lea, R.M., "WASP: A WSI Associative String Processor" in *Journal of VLSI Signal Processing*, vol. 2, no. 4, pp. 271-285, May 1991.

Naganuma, J., Ogura, T., Yamada, S.I. and Kimura, T., "High-Speed CAM-Based Architecture for a Prolog Machine (ASCA)" in *IEEE Trans. on Computers*, vol. 37, no. 11, pp. 1375-1383, Nov. 1988.

Ng, Y.H., Glover, R.J. and Chng, C.L., "Unify with Active Memory", in *VLSI for Artificial Intelligence*, J. Delgado-Frias and W. Moore (ed), Kluwer Academic Press, 1989.

Nilsson, N.J., *Principles of Artificial Intelligence,* Morgan Kaufmann, 1980.

Pohl, I., "First Results on the Effect of Error in Heuristic Search", in *Machine Intelligence 5*, B. Meltzer and D. Michie (ed), Edinburgh University Press, pp. 219-236, 1969.

Rodohan, D.P. and Glover, R.J., "An Overview of the A* Architecture for Optimisation Problems in a Logic Programming Environment" in *Computer Architecture News*, vol. 19, no. 4, pp. 124-131, June 1991.

Rodohan, D.P. "Associative Evaluation of Logic Programs using Artificial Intelligence Search Techniques", *PhD Thesis*, Dept. of Electrical Engineering, Brunel University, UK, in preparation.

Warren, D.H.D, "An abstract Prolog instruction set", *Tech. Note 309*, Artificial Intelligence Centre, SRI Int., Oct. 1983.

PERFORMANCE ANALYSIS OF A PARALLEL VLSI ARCHITECTURE FOR PROLOG

Alessandro De Gloria, Paolo Faraboschi and Mauro Olivieri

INTRODUCTION

The novel trends in high performance Prolog processors suggest the implementation of RISC-based architectures committed to Prolog only through the adoption of a few basic dedicated features, like the Berkeley Abstract Machine (BAM) (Holmer et al. 1990) architecture.

In fact it is possible to exploit the determinism in Prolog programs and to eliminate all the redundancies of Warren Code (Warren 1983) by applying data-flow analysis and specialized unification (Van Roy 1990). This leads to an impressive speedup with respect to previous Warren Machine based implementations, as it is shown in recent research results (De Gloria and Faraboschi 1992).

By using a sophisticated compiler and a simple instruction set, our project (SYMBOL) applies global compaction techniques and VLIW design philosophy to the static exploitation of instruction-level parallelism in Prolog. In this way, the approach benefits from the fine-grain parallelism of code and at the same time does not compromise general-purpose computation capabilities.

A first result of the SYMBOL project is a VLIW incremental architecture and compiler, intended to work as a Hardware Accelerator attached to a host workstation (De Gloria et al. 1991).

This paper presents performance analysis results, regarding two kinds of aspects:

- *Relative Performance Analysis*, in terms of:

    - ideal maximum speedup versus a sequential implementation that the characteristics of the code allow;
    - real speedup versus a sequential implementation that the developed compiler and architecture obtain.

    These results show the effectiveness of the VLIW approach in terms of theoretical speedup and prediction of dynamic performance;

- *Absolute Performance Analysis*, in terms of absolute execution time on a set of standard benchmarks. In this context we show the values obtained from an instruction emulator, where the instruction timing have been validated by a test of the single ASIC on a test board. The obtained values show the competitiveness of the architecture with the best available Prolog machines.

ARCHITECTURE FEATURES

The need of managing tagged data in Prolog imposes heavy limits on general–purpose machine performance. For this reason, our processor incorporates hardware support both for tags and for efficient branching, including multi–way branches and direct branching on tags. Figure 1 shows the architecture of the single VLSI processor.

We can note that a single processor is self sufficient, contains registers, an ALU and a sequencer, and is itself a VLIW machine (64 bits of instruction), since it allows some instruction level parallelism.

- The instruction is organized in two formats, one for ALU operations and one for control operations, because of pinout limitations, as we show in figure 2.

- Memory access to code and data space has been organized in a three–cycle pipeline. This does not compromise the peak rate of one access per cycle, but causes two–cycles delayed branches and a longer completion time of data memory operations. In addition, in this way, three memory operations and three control instructions can be in progress simultaneously.

- A memory operation can be executed in parallel with an ALU operation (two–cycle pipeline) or a control operation and always with a local register movement in all data fields. This gives additional instruction level capabilities to the single processor.

We have built the processor by using a hierarchical standard cell environment. The obtained chip has an area of 12.5×13.5 mm^2 and is composed of about 100,000 transistors packaged in a 180 pins PGA. The measured operating frequency is 30 MHz.

We have designed the processor to be integrated in a synchronous multi–processor architecture, and we have developed a first prototype board to be attached to a host workstation, with three ASICs and a simple ISA bus interface. Figure 3 shows the organization of the prototype board.

When more sequencers are used for multi–way branches, synchronization among them is achieved by means of an external logic network. Since multiple branch instructions can be issued in the same cycle, the compiler includes bits in the instructions to specify the priority of the branch operations. At each branch, a controller decides which processor Program Counter will be enabled in the next cycle.

The multi–processor architecture is still organized as a VLIW machine and is completely exposed to the compiler, that schedules instructions and statically organizes communication avoiding the need of dynamic management. This results in a remarkable hardware simplicity and speed, at the expense of a longer compilation time and some compromises in the exploitable parallelism as not all information are exact during compilation.

CODE ANALYSIS

Due to the peculiarity of the language, which does not allow to disambiguate memory accesses (absence of arrays and loops) we have limited our research scope to shared–memory architectures. Figure 4 shows the instruction profile of the execution of the considered benchmarks, for the considered class of operation: memory accesses, *value* data field movement, *tag* data field movement, ALU operations, and control operations (branches). The hardest problem in finding parallelism in Prolog code comes from

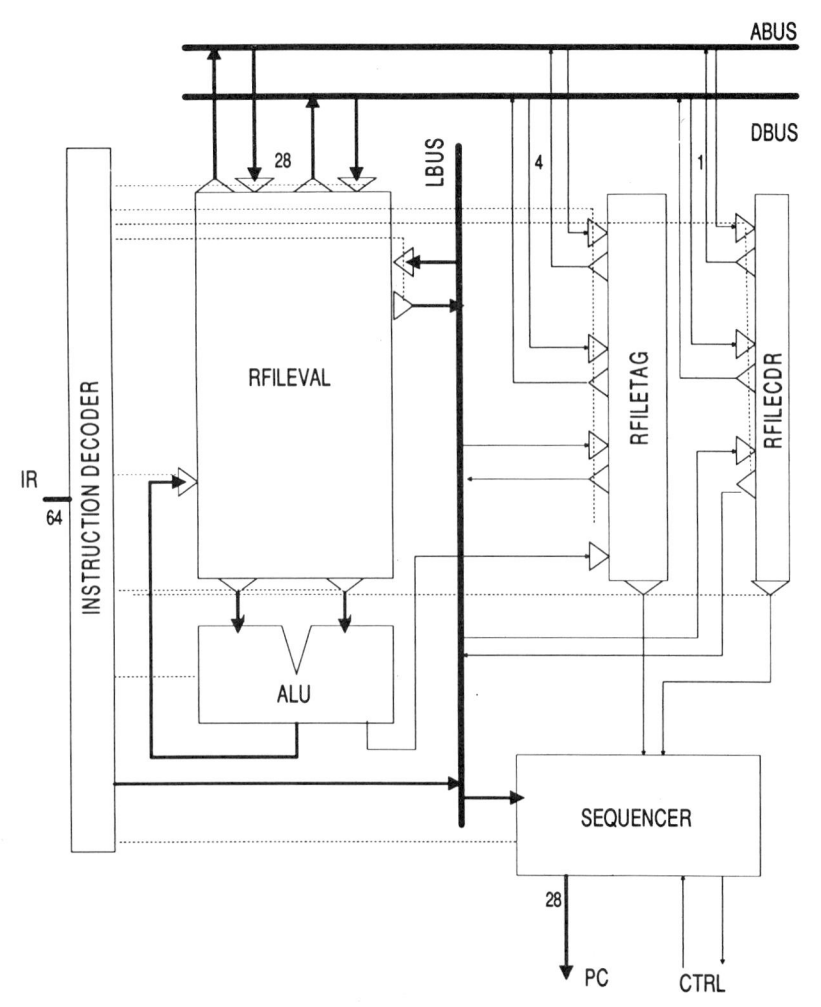

Figure 1. The microarchitecture of the VLSI processor

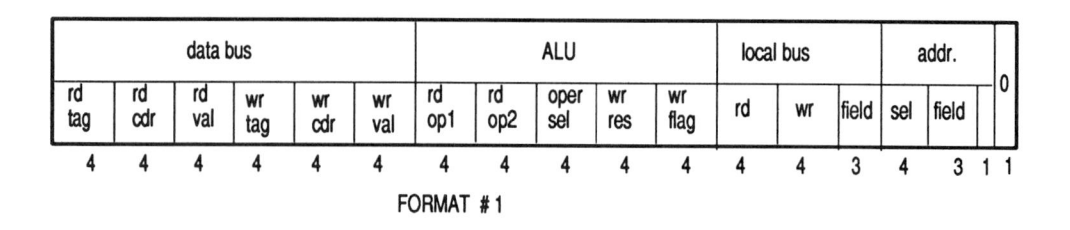

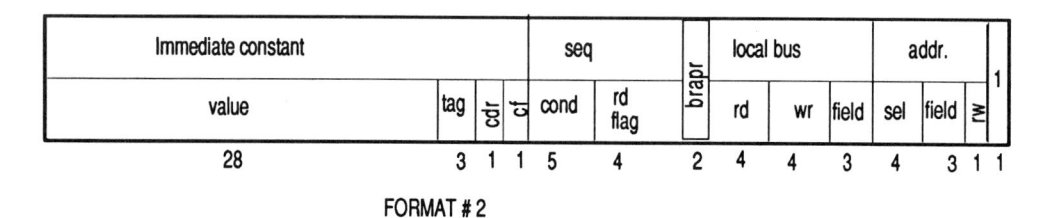

data bus						ALU					local bus			addr.		0
rd tag	rd cdr	rd val	wr tag	wr cdr	wr val	rd op1	rd op2	oper sel	wr res	wr flag	rd	wr	field	sel	field	
4	4	4	4	4	4	4	4	4	4	4	4	4	3	4	3	1 1

FORMAT #1

Immediate constant				seq		brapr	local bus			addr.		1
value	tag	cdr	cf	cond	rd flag		rd	wr	field	sel	field	rw
28	3	1	1	5	4	2	4	4	3	4	3	1 1

FORMAT #2

Figure 2. The instruction formats of the VLIW processor

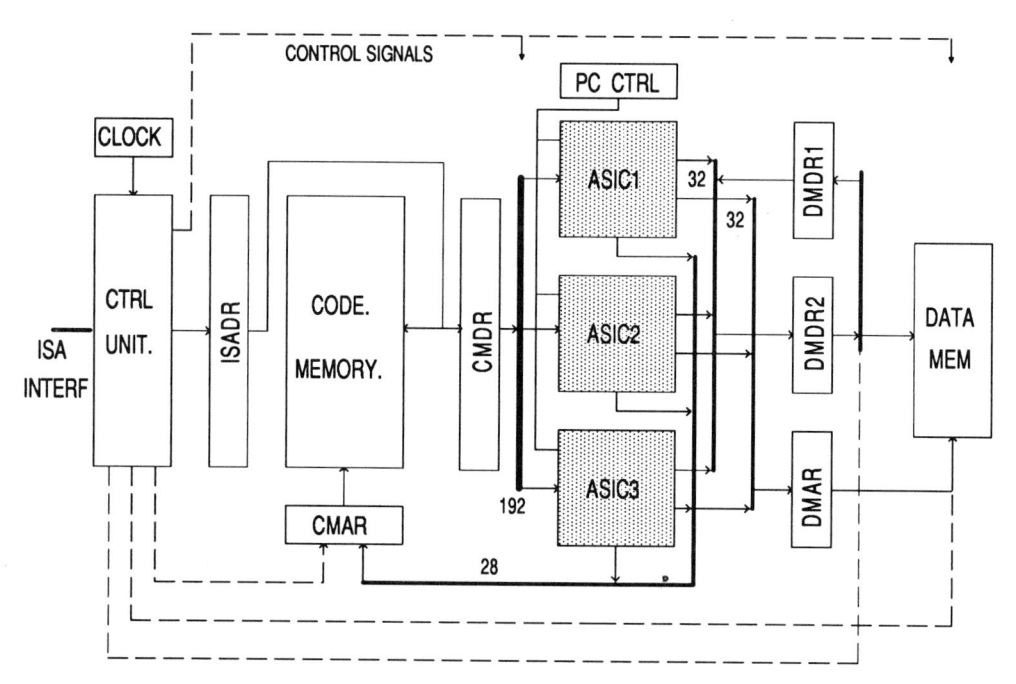

Figure 3. The organization of the three-processor prototype board

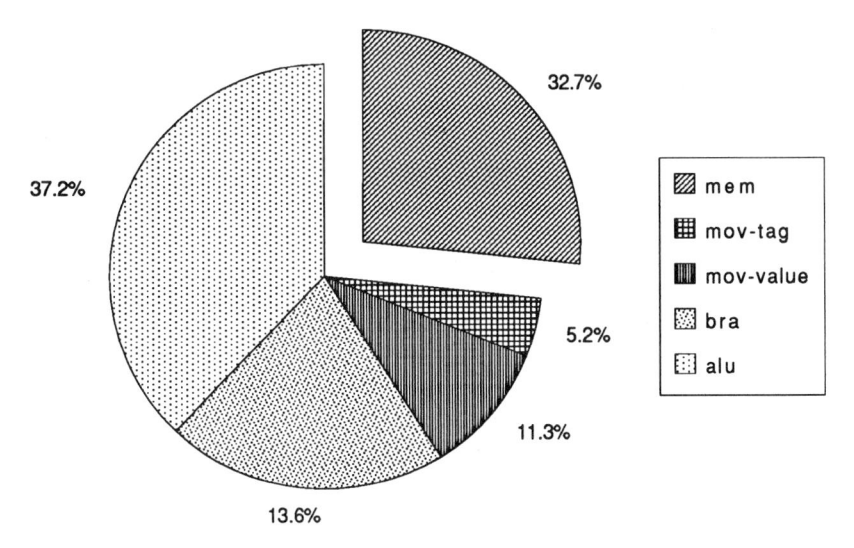

Figure 4. Instruction execution profile

the high percentage of branch operations (more than 15%) which limit the degree of instruction concurrency inside each basic block (composed of not more than 6–7 instructions).

To overcome this limitation, we have applied a Global Compaction technique which extend its optimizations beyond basic blocks and consequently allow a higher possibility of code motion. Measurements (De Gloria and Faraboschi 1992) have shown that a global optimization technique increases the block size to about 15 instructions, and performs roughly 30% faster than simple basic–blocks optimizations.

Global Compaction approaches are based on statistic data on the execution flow of the program. The Trace Scheduling algorithm, in particular, operates compaction by choosing the most probable path at each branch instruction. Measuring the actual distribution of branch probability in the benchmarks, we have noticed that the average probability of a faulty prediction of a branch direction is less than 15%, value which guarantees a low performance decay due to run–time unpredictable execution flow. Figure 5 shows the distribution of a faulty prediction of a branch direction in the considered benchmarks. It is evident that almost 90% of all executed branch instructions are predictable with an error of less than 0.1.

RELATIVE PERFORMANCE ANALYSIS

We have performed our analysis on a set of small and medium size programs extracted from the Aquarius Benchmark Suite (Haygood 1989). We have obtained the code for the architecture by means of the following steps:

- We compile Prolog into the BAM abstract machine model by using the BAM compiler (Van Roy 1990).

- We translate BAM code into machine–independent Intermediate Code Instruction (IntCode). In this step we avoid all optimizations which are delayed to the back–end compiler. The IntCode is composed of simple instructions expressing functionalities which are directly executed by a hardware resource.

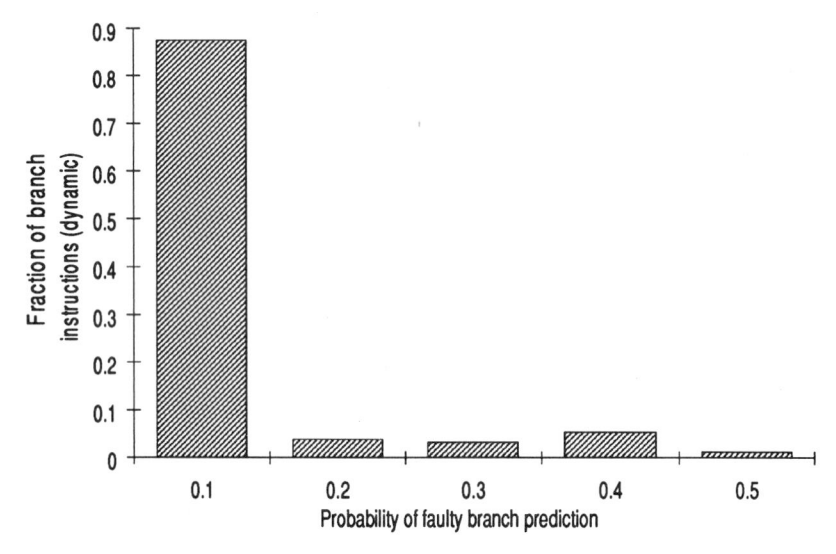

Figure 5. Distribution of probability of faulty branch prediction

- We parallelize the IntCode by means of the back–end compiler that maps the instructions onto the resources of the target architecture. The back–end compiler is based on a global parallelizing technique derived from the Trace Scheduling approach (Fisher 1981), that operates compaction beyond basic blocks of code. Trace choice is based on statistical information about execution frequency extracted by preliminary simulation. The Code generator is implemented by means of a variation on the Bottom–Up–Greedy algorithm described in (Ellis 1985).

Speedup

The first measurement we have computed is the speedup relative to a sequential execution of the same code.

In order to analyze the different factors that limit performance, we have also computed some comparison figures, and in particular:

1. The maximum ideal speedup allowed by Amdahl's law (Amdahl 1967), considering the instruction execution profile, a single memory access per cycle, and the possibility to overlap memory operations with ALU and control operations.

2. The speedup obtained by an ideal VLIW architecture with the same functional units of the real machine, but with no constraints on the instruction format, and a unique register set.

We have experimented architectures composed of different units, to find the degree of scalability of the architecture.

Since the register set is physically distributed on three separated processors, it is important to determine the initial locations of the environment variables of the abstract machine. A statistical analysis on the relationships among the variables has provided the data to find the best allocation.

By applying the obtained register partitions, we have measured the performance of real architectures composed of one, two and three processors, which reach respectively a speedup of 1.5, 2.0 and 2.1 with respect to a sequential implementation.

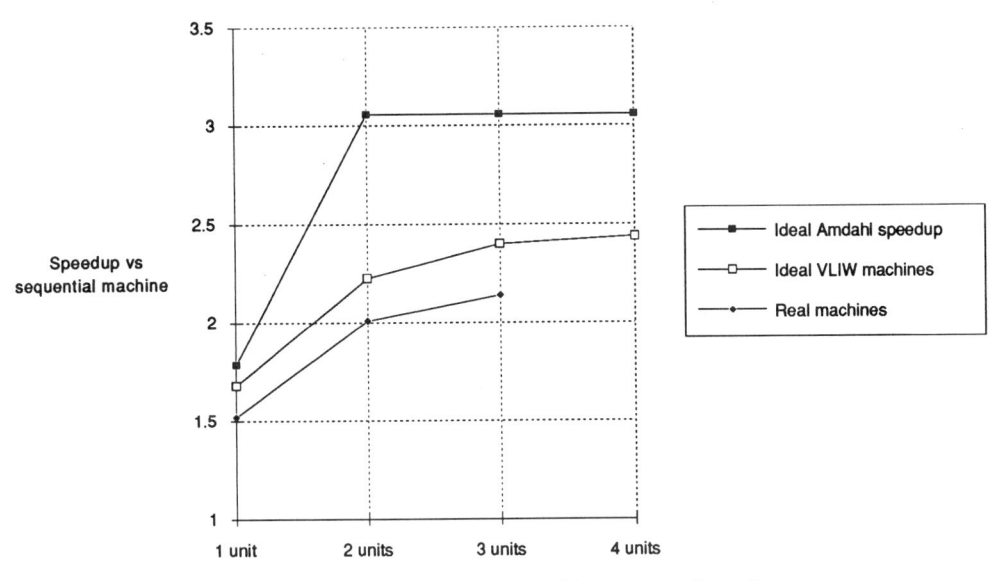

Figure 6. Speedup of different architecture configurations

Figure 6 shows the results for the different configurations together with the ideal speedup (Amdahl Law), and the performance of a single register set ideal VLIW machine. We can draw some conclusions from the analysis of the graph:

- The compiler is far from reaching the maximum ideal speedup (around 3.0) theoretically allowed by Amdahl law, with the single memory access constraint.

- The limitation of resources necessary to build a machine at reasonable VLSI costs (two instruction formats, separate register banks), does not cause a heavy degradation of performance. This is due to the goodness of the scheduler that succeeds in exploiting the resources efficiently.

Resource usage

To compute if the compiler succeeds in exploiting the architecture, we have measured the dynamic distribution of the resource usage.

Figure 7 shows the results for a three processor architecture on the considered benchmarks. In the graph, we can see the different types of operations generated by the compiler. We note that we have separated the *no-operations* caused by data pipeline stalls (data memory and ALU) from the ones derived from control pipeline stalls.

The consideration that emerges from an analysis of the statistic is that resource usage is rather low. This is due to different factors:

1. There are not enough instructions in a trace. The number of instruction compacted by the scheduler is typically between 15 and 20, and includes 2–3 basic blocks. In this situation, there are a few chances to keep the data path fully operative, as the machine features different degrees of instruction level parallelism (and pipelining).

2. The compiler is unable to disambiguate memory accesses. As Prolog manages data dynamically and through pointers, the compiler cannot find memory aliases

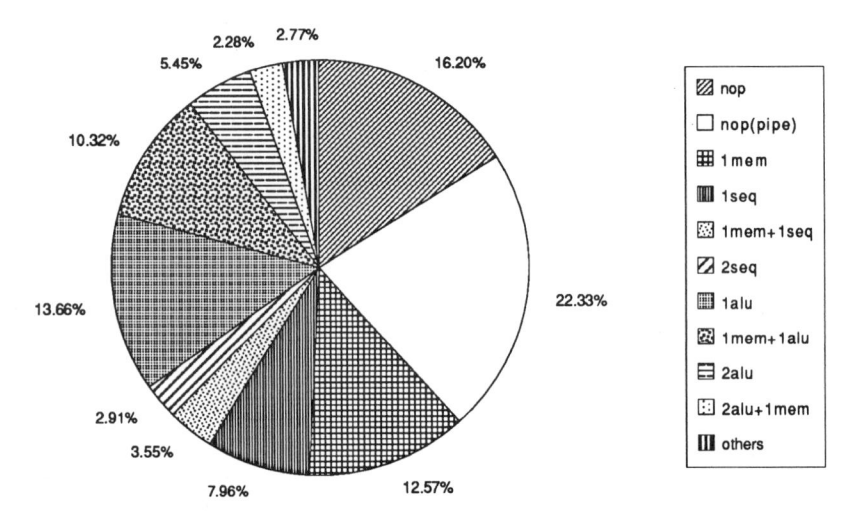

Figure 7. Resource usage in a three-processor architecture

and is forced to take conservative decisions, keeping memory operations strictly in–order.

These considerations give way to further compiler optimizations that we plan to implement in the future, and that can improve substantially the resource usage:

1. To increase the number of instructions in a trace, we can operate inter–procedural analysis and subroutine in–lining. In addition, if we can operate memory aliasing analysis, we also can apply unrolling techniques to simple recursive structures.

2. To disambiguate memory aliases, we need to apply pointer analysis techniques (Hendren and Nicolau 1990) to recursive data structures, or to force the front–end compiler to generate annotations when possible.

ABSOLUTE PERFORMANCE ANALYSIS

Processor Test Results

The VLSI processor has been tested in a "stand–alone" configuration, by means of a test board designed for the Tektronix LV500 Tester. The test machine has allowed to check the functionality and the timing performance of the ASIC. The test patterns have been generated by means of the compiler and a dedicated assembler, by forcing the tester to emulate the rest of the system in terms of timing relationships. The measured operating frequency is 30 MHz.

We have dedicated particular attention to the analysis of the delays of the fetch path since the pipeline structure of the board must allow to execute a 64–bit instruction at each cycle. However, all measured values have fallen within the range which was forecast by the simulation of the system performed during the design phase.

Benchmark Execution Time

Since the software host interface is still under development, we have obtained benchmarks results by means of an instruction emulator, and by applying the processor timing obtained from the test phase. However, the synchronous nature of the architecture makes the evaluation rather precise, since no dynamic synchronization overhead is added at runtime.

To compare our approach with other Prolog machines, we have computed the total execution time for the considered benchmarks.

The considered benchmarks do not include input/output operations and complex arithmetic (multiplication, division, etc.) as the prototype processor does not implement this class of instructions. The reason for this is that we are interested in comparing the basic processor features to accelerate the kernel of the Prolog abstract machine (stack management, unification, backtracking, tagged data, etc.). Arithmetic operations can be accelerated at any speed according to the required hardware cost.

Table 1. Absolute time comparison [msec]

	Quintus	PLM	KCM	BAM	Symbol-3
divide10	0.41	0.38	0.091	0.0387	0.0423
log10	0.15	0.109	0.039	0.0201	0.0146
mu	7.04	12.407	4.644	0.8557	1.2913
nreverse	1.62	2.1	0.65	0.2057	0.2401
ops8	0.24	0.214	0.059	0.0251	0.0274
prover	8.67	6.83	-	0.9722	1.2995
qsort	4.82	4.24	1.32	0.2253	0.2192
queens-8	21.2	28.8	1.205	1.2017	1.5492
sendmore	490	-	-	42.3364	44.0938
serialise	3.1	2.47	1.22	0.5133	0.6556
tak	1120	940	-	31.047	32.067
times10	0.345	0.247	0.082	0.0346	0.0363
zebra	423	-	-	86.8901	119.1841
average speedup	10.51	10.47	2.42	0.83	1

Table 1 shows the results on the considered benchmarks of a three–processor architecture and other Prolog machines. An analysis of the execution times shows that our architecture performs substantially better than Warren–based approaches (13 times faster than the PLM and 2.5 than the KCM), and we are very close to the BAM processor (85% of its performance), but with more implementation simplicity and a slower technology.

CONCLUSIONS

The SYMBOL project has demonstrated that we can achieve some gain in the execution of Prolog programs with the adoption of a VLIW architecture and through aggressive compiler optimizations. The project has produced a first prototype ASIC processor and a three-processor board is currently under test.

In this paper, we have presented a performance analysis of a real architecture devoted to the exploitation of instruction-level parallelism in symbolic programs. Some considerations emerge from the presented data.

First of all, we can see that, through the adoption of a RISC instruction set and the exploitation of fine grain concurrency in a VLIW style, absolute time performance of Prolog programs reaches a ten-fold improvement with respect to previous Warren-based implementations.

However, the degree of exploitable concurrency is still low when we assume a shared, single access memory. This is due to the dynamic data structure organization of Prolog data that prevents a simple memory aliasing analysis. There is still a lot of work to be done in this area, trying to apply more sophisticated techniques to the problem of memory aliasing in Prolog.

Finally, we remind that the optimization we propose are not in contrast with coarse grain parallelism (*or*-parallelism, *and*-parallelism), but can be considered complementary, and can give an additional speedup of a factor of 2-3.

References

Amdahl, G. Validity of the single processor approach to achieving large scale computing capabilities. In *Proc. AFIPS 1967 Spring Joint Computer Conference*, 1967.

De Gloria, A., and Faraboschi, P. Instruction level parallelism in prolog: Analysis and architectural support. In *Proc. 19th Annual International Symposium on Computer Architecture (ISCA19)*, 1992.

De Gloria, A., Faraboschi, P., and Guidetti, E. A parallel incremental architecture for prolog program execution. In Delgado-Frias, J., and Moore, W., editors, *VLSI for Artificial Intelligence and Neural Networks*. Plenum Press, New York, 1991.

Ellis, J. *Bulldog: A compiler for VLIW architectures.* The MIT Press, 1985.

Fisher, J. Trace scheduling: a technique for global microcode compaction. *IEEE Transactions on Computers*, C-30(7):478–490, July 1981.

Haygood, R. A prolog benchmark suite for aquarius. Technical Report UCB/UCSD 89/509, University of California at Berkeley, April 1989.

Hendren, L., and Nicolau, A. Parallelizing programs with recursive data structures. *IEEE Transactions on Parallel and Distributed Systems*, 1(1):35–47, January 1990.

Holmer, B., Sano, B., Carlton, M., Van Roy, P., Haygood, R., Bush, W., Despain, A., Pendleton, J., and Dobry, T. Fast prolog with an extended general purpose architecture. In *Proc. 17th Annual International Symposium on Computer Architecture (ISCA17)*, pages 282–291, 1990.

Van Roy, P. *Can Logic Programming Execute as Fast as Imperative Programming?* PhD thesis, University of California at Berkeley, December 1990.

Warren, D. An abstract prolog instruction set. Technical report, Artificial Intelligence Center, SRI International, 1983.

A PROLOG VLSI SYSTEM FOR REAL TIME APPLICATIONS

Pier Luigi Civera, Guido Masera and Massimo Ruo Roch

INTRODUCTION

VLSI technologies offer the capabilities to implement high performance processors for special applications; in the particular field of Artificial Intelligence, several hardware solutions to speed up the execution of PROLOG programs are today available. Fast execution of PROLOG programs suggests to exercise the programming language also in new application fields such as real-time controls. The programming benefits using Prolog or similar declarative languages instead of the procedural ones are evident: rules description is simpler and more effective, even if it imposes different methods and perspective to programmers.

The work described in the paper refers to an experience in rule based programming for a real time application, and point out weaknesses and missing parts of a PROLOG system in order to face real time problems. Computational load requirements, hardware and software aspects have been investigated for a future automotive environment, as depicted in the European PROMETHEUS Project, more precisely in PRO-CHIP activities (Prometheus 1987 and 1988). A Prolog system based on a VLSI Prolog processor, called PROXIMA, near to be operational, has been exploited for the new application. The first goal of the activity was pointing out the extensions to be introduced in the Prolog system in order to sustain real time requirements. These extensions were included in a very straightforward manner, avoiding to incorporate complex theoretical aspects. Extensions to sustain real time requirements will be reproduced on hardware and software in terms of new circuits and instructions to be included in future versions of the VLSI processor and compilers. The experiment has been arranged with simulation tools, developed for the sequential processor, with extensions for the new parts.

The paper starts with some considerations about the real time requirements and the extensions they imply on software and hardware; then it continues with a short description of the Prolog engine, in the actual version. The paper describes the solution adopted and the performance evaluations measured for the specific application, concluding with experience results and a short discussion on the future work.

A REAL TIME APPLICATION

Real time processing is a very general term and refers to many application fields. The common factor is that the software must sustain the dynamics of physical system by continuously sampling the input variables and supplying outputs or controls on time.

Almost each programming environment is suitable to develop some real time applications; the limit is represented by the whole computation speed or, alternatively, states if the environment is able to track the external system dynamics.

In order to make a given programming environment suitable for real time applications, some key features must be included and adequately supported:

- ability to perform fast input/output operations;
- ability to represent and manage time and events;
- ability to manage many processes, i.e. multitasking;
- ability to monitor processes at run time.

All the above mentioned capabilities usually require hardware supports such as timers, counters and interrupt controllers, and software support such as special operative system features for processes and resources management.

Real time systems comparison is carried out using three main performance parameters:

- the number of instructions per second (CPU speed);
- the number of interrupts serviced per second (interrupt handling capability);
- the number of I/O operations performed per second (I/O throughput).

Intrinsic real time systems must exhibit balanced and as highest as possible values of all the figures, as depicted in figure 1; on the contrary high performance RISC workstations tend to optimize just the CPU speed.

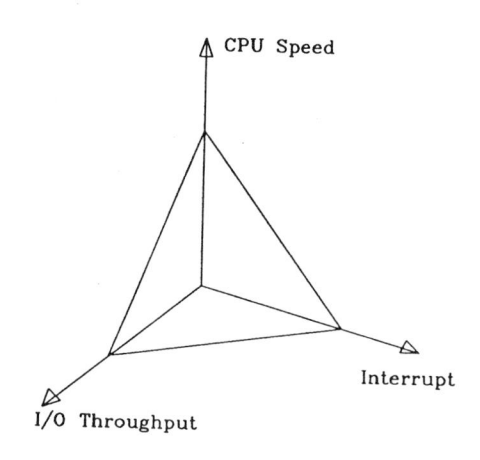

Figure 1. Real time performance domain

Real time computing and declarative programming styles, like Prolog, seem to be incompatible by definition. Prolog has no interrupt capability and provides a very poor input/output support. In addition, real time computing requires to the system the ability to track the external phenomena on time, following its dynamics anyhow.

Therefore the computation cycle must be solved within a finite period of time, strictly related to the sampling time (or the event arrival rate) of the external system.

On the opposite, declarative styles lowers programming effort, but they are no more able to cope with a fixed or deterministic response time.

The rule based program devolves almost all the effort on a general interpreter. The rule interpreter always performs less efficiently with respect to specific procedures, although noticeable improvements has been achieved by introducing the compilation techniques.

Assuming to use Prolog as a real time programming environment, these basic aspects must be accomodated.

The following design decisions have been adopted. The complete system is composed of several cooperating processes. Prolog clauses are limited to describe the behavior of monitoring and control processes. Processes are mapped on specific processors: it is assumed that input/output processes, and interrupt services are better managed on conventional processors (well suited to support interrupts and fast I/O), while the control and decision processes are directly managed by the Prolog environment.

The configuration allows the control part to be described through rules and be properly executed on dedicated hardware, without the burden of standard I/O and event processing.

The communication scheme among processes is based on events and messages; they are reported as database facts when they are exchanged with the declarative side of the system. The Prolog part will see the external world dynamics through the updated values of the data base facts.

Control part is computed in a time-independent form (like a snapshot), freezing the inputs and the system status at each step. The main activity of the new step computation is represented by the updating of the status and the output facts.

The external input variables are sampled and pre-processed, the processed version is transferred to the Prolog part. Output values are collected back from the I/O processes, post-processed (for instance translated into graphical formats or into PWM sequences to the actuators), and returned to the external world.

The control processes restart at each step to consult again the rules and the updated facts.

The control part is not a unique Prolog program, but it consists of a hierarchy of interacting processes, where each process is still an independent Prolog program, here referred as a *module*.

Each module runs considering only its input and status variables and computing the results. Inputs to, and results from a module may be directed to other modules (internal) or exchanged with the procedural part (external).

The modular method revealed very useful, as it allowed an easy design and refinement of the system, and also greatly simplified debugging and tuning phases.

Processor Extensions

Referring to the previous points two different issues (system variables and module management) had to be incorporated in the Prolog processor.

The first issue is related to an interprocess communication capability: the set of facts stored into the common database represents the system blackboard, where all the processes refer to communicate.

Prolog language is able to manage only one-time assignment variables, therefore global parameters, that are multiple assignment type, must be stored as facts. *Assert* and *retract* are the only available built-in's, in standard Prolog, able to modify the code clauses, but they are rather inefficient because they imply the interpretation or the incremental compilation (at run-time) of the modified code. To avoid this bottleneck, two new language primitives, the *statussave, statusload* have been introduced. The new built-in's do not create or delete any clauses (facts or rules), but they acts only on existent facts, reading or altering the value of their arguments.

The second point comes out considering that a large number of modules is always required for practical applications, therefore the Prolog processor must be able to switch from one task to another, simulating a set of virtual processors. Referring to the specific experiment, a simple multitasking capability was already present in the sequential version of PROXIMA for tracing and debugging. The two microcoded instructions to save and restore the internal processor status (*storereg, loadreg*) have been used for context switching.

THE PROXIMA PROLOG ENGINE

A sequential Prolog machine has been designed by the authors et al. (Civera et al. 1989), as the result of a precedent activity, developed inside the AWAI (Advanced Workstation for Artificial Intelligence) CNR project.

The PRolog eXecution MAchine (PROXIMA) is a VLSI implementation of a slightly modified Warren Abstract Machine (WAM). This computational model executes a compiled version of the Prolog programs, augmenting efficiency.

Two VLSI ASICs have been developed to build the core of the Prolog machine. The first one, named IPU (Instruction Processing Unit), reads the variable length byte code from the memory, and translates it to a fixed 32 bit format, better suited for high speed execution. It is able to automatically perform conditionless jumps, and to apply a very fast binary search algorithm to constant and structure tables, as required by some WAM instructions. Last, inside the chip a 24 words deep FIFO memory is realized, used as a prefetching buffer between the decoder and the execution unit, in order to reduce the apparent code memory access time. The IPU, containing about 40000 transistors, is realized in 0.8 μm double metal N-well CMOS technology, and its die size is 6.9×6.9 mm^2.

The second ASIC, named DPU (Data Processing Unit), operates on the data memory, fetching decoded instructions from the IPU, and executing them. It is a fully horizontal microcoded engine, with a 32 bit datapath, containing the structures required by the WAM computational model. The microcode has a high degree of internal parallelism, so that the execution efficiency is increased and the code memory bandwidth is reduced. The data and code memory interfaces are independent of each other (Harvard architecture), due to the high transfer speed required. Microinstructions are executed in a two stages pipeline, to optimize hardware resource utilization and to reduce the minimum clock cycle time. The circuit includes 128 kbit of ROM, 1 kbit of triple port RAM and about 80000 transistors in the datapath. The fabline is the same used for the IPU, and the resulting die size is about 8.1×8.9 mm^2.

A limited real time support is provided by a low priority interrupt input, which is sampled at each jump or fail operation. This constraint has been introduced to reach a consistent state of the Prolog engine before the activation of the interrupt handling procedure.

The two circuits introduced above are interfaced to a host computer via two banks of dual port memory, for code and data, respectively. The Prolog chip set acts as

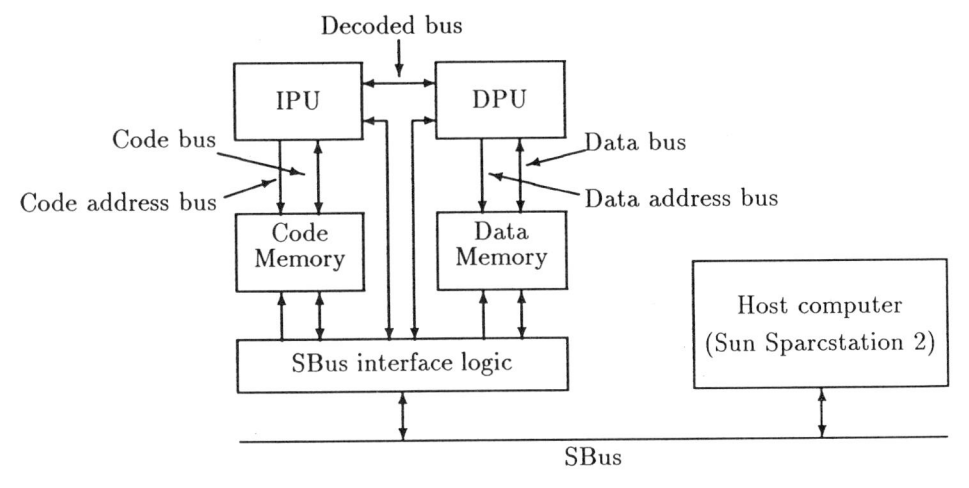

Figure 2. Sequential Prolog system architecture

an attached processor toward the host, which, in turn, provides I/O and file system handling support. The block diagram of the sequential system is depicted in figure 2.

SYSTEM SOLUTIONS AND ARCHITECTURE

The hardware implementation of a Prolog automotive system requires, as a preliminary step, an accurated analysis of the computational load of the involved modules. This evaluation task basically needs two points:

- the availability of realistic automotive modules, able to represent a meaningful part of the software environment to be introduced in the future vehicles;

- a way of measuring the performance of the choosed Prolog machine (PROXIMA), in terms of number of logic inferences per machine cycle.

As far as the first point is concerned, unfortunately no useful automotive applications, structured as rule systems and easy to be translated in Prolog, were available. So a set of software modules has been generated *ex novo*, in order to emulate some of the main computing functions in the vehicle. The following functions have been selected: road holding, overtaking monitoring and support, lane tracking, refuel control.

The whole software architecture, consisting of 5 Prolog control modules and 5 filtering programs, is given in figure 3.

The filtering programs select the correct inputs received from the sensors and prevent data contaminated by measurement errors from reaching the control modules; moreover they are able to make up for short periods of signal loss, through a moving average of the earlier values. Each filter receives from the sensors input pairs: the current measure and a second value representing its degree of confidence (this value is assumed to be produced by an intelligent sensor and ranges from 0 to 1). The filter outputs are the processed measures and a signal indicating the output reliability. Moreover some of the filters, those managing the most important and critical information, receive data from more than one sensor, for safety reasons, and dinamically give measurement values with better reliabilities.

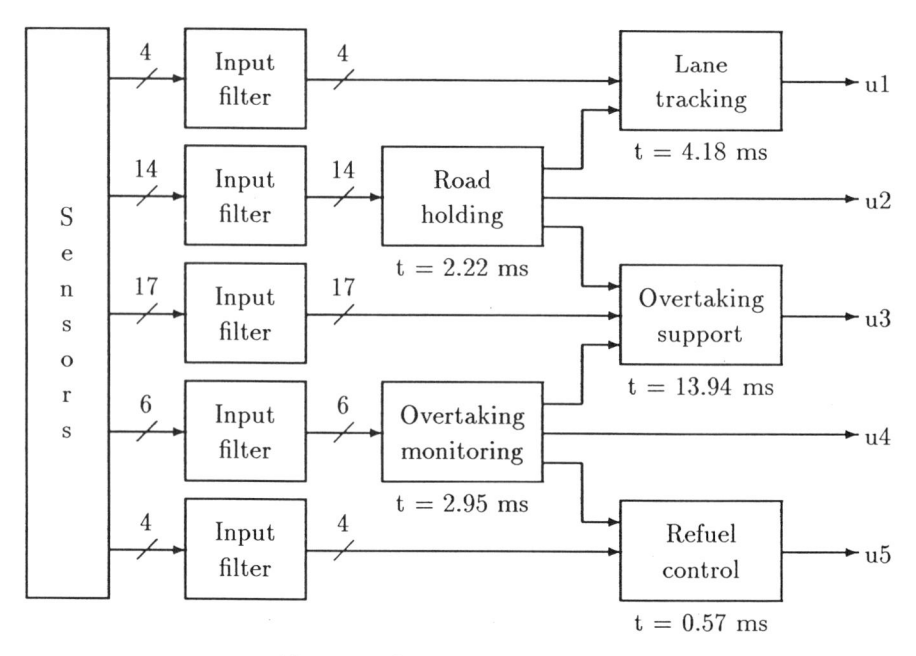

Figure 3. Software architecture

In the developed system, 45 sensors have been introduced, measuring several physical attributes related to the driving, for example: acceleration, velocity, brake fluid pressure, road surface condition, steering angle, relative position of other vehicles, recognized traffic signs, and so on.

While the filters are implemented in Pascal, as their function is simply procedural, the control modules are the kernel of the rule based automotive system and are written in Prolog. In figure 3, an example of a two levels hierarchy of control modules is reported: the *road holding* and *overtaking monitoring* modules receive just filter outputs and produce video warnings and inputs for the second level modules, while the *lane tracking*, *refuel control* and *overtaking support* modules receive data both from filters and first level modules, but return just video warnings.

In figure 4 the *road holding* module is depicted with its inputs. The module produces an output representing the attitude of the vehicle in maintaining a correct and safe running status, while external or internal events tend to modify it. This output ranges from the value "poor" to the values "satisfaying" and "good", and it is used as input information by the following modules. A second output is a velocity correction value, suggested to the driver.

In order to evaluate the performance of the PROXIMA engine when executing the described application, the implemented system has been exercised against a simulated sensor environment. All the data for the filters must be generated in a consistent way, that is preserving the correct correlations among them. The programs, written in Pascal, have been structured grouping the management of sensors depending on each other: for example, a road holding program generates all the data regarding the vehicle movement, like velocity, gear engaged, rounds per minute, and so on, while other programs give the vehicle position in relation with the road borders, velocity and position of the other vehicles, atmospheric and road surface conditions.

As the two PROXIMA ASICs were not yet available from the foundry, all the per-

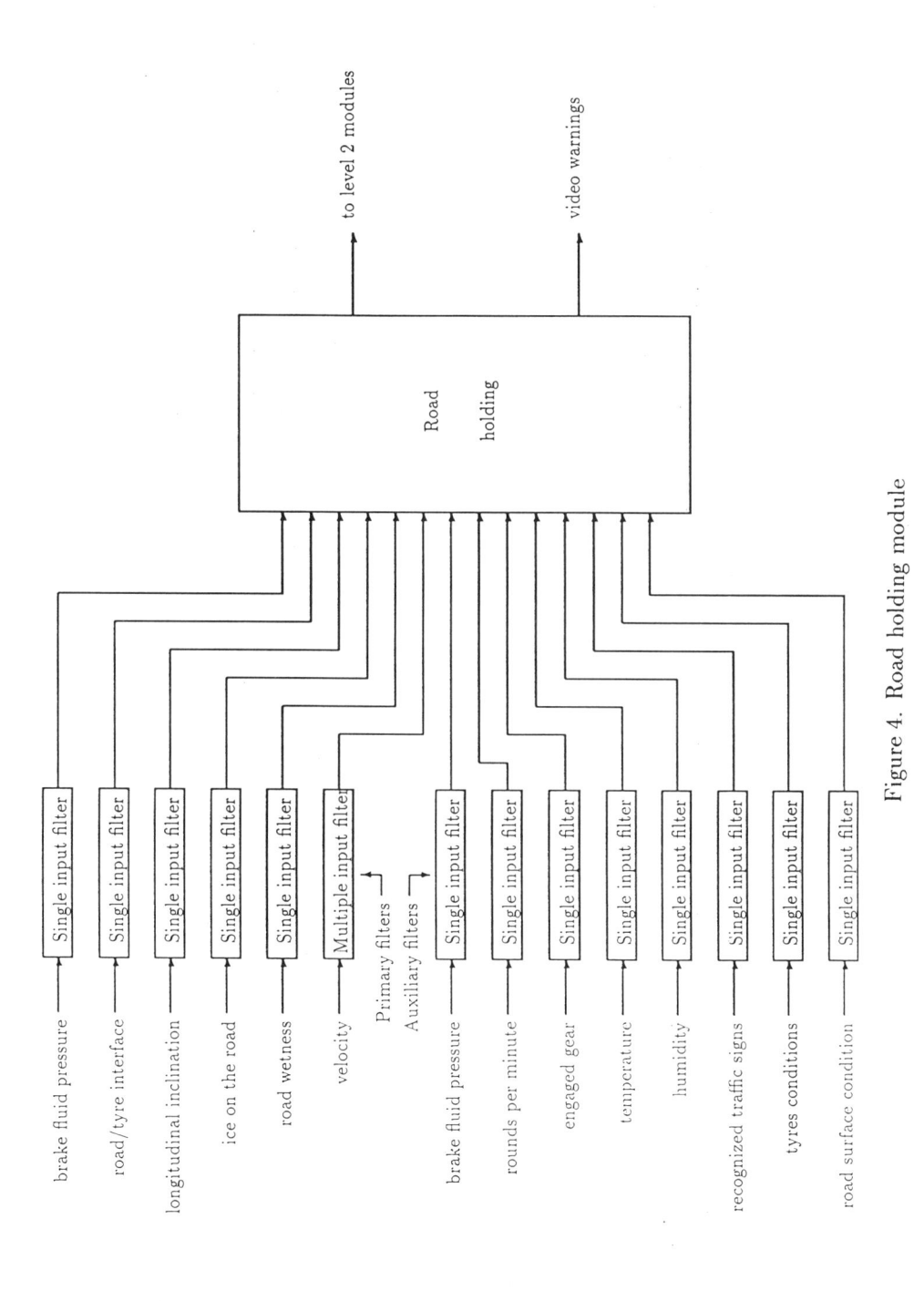

Figure 4. Road holding module

formance evaluations have been obtained using a software emulator of the PROXIMA engine (Civera et al 1989): this program is able to supply, for each simulation, the number of executed micro-operations and logic inferences.

COMPUTATIONAL EVALUATIONS

Given the described real time application, two opposite approaches can be followed in performance evaluation: either the time constraint is supposed to be fixed and the processor number required to meet it is calculated, or the maximum real time performance achievable for a planned system configuration is estimated.

As a starting point for the analysis, the system has been supposed to operate with a period of 10 ms. The maximum time required by the Prolog engine to execute the control modules, in the worst case, must be less then 10 ms, otherwise measures will be missed. In addition, the clock cycle time for the Prolog engine has been assumed equal to 100 ns (10 MHz clock frequency).

The most important information for the performance evaluation, is the time required to execute each of the control modules. As this time depends on the input data, a statistical evaluation has been performed, applying several sequences of simulation data to the developed system. Knowing the statistics of the execution times required by each module, it is possible to evaluate the processor number required to keep up with the frequency of the sensor data reception.

The execution time for each module can be obtained through two different vias: either as the ratio between the global logic inference number and the Lips (Logic inference per second) number, or as the microinstruction number times the machine cycle time.

As each module contains an input and an output part, managing reading and writing of file data, code modifications have been introduced in order to measure just the execution time required by the actually useful parts.

In the table, the maximum cycle execution time of the processor for the 5 descibed modules and the two types of filters (mono and multi-sensors) are given, assuming filtering tasks also executed on the Prolog part.

Table 1. Maximum cycle execution times

module name	micro-instruction number	maximum time per cycle [ms]
mono-sensor filter	975	0.09
multi-sensor filter	8729	0.8
road holding	9646	0.96
overtaking monitoring	2881	0.28
overtaking support	44259	4.4
refuel control	3148	0.31
lane tracking	16715	1.6

If just one processor is available, the time required to compute each of the system outputs can be evaluated knowing the number of filters that the modules need, as indicated in figure 3. For example, the output $u1$ requires a time equal to ≈ 4.2 ms, evaluated adding the delay introduced by the road holding module, with its 4 filters, and the lane tracking module, with its 14 filters.

As output $u3$ needs a time of $\approx$14 ms, one processor does not meet the real time constraint, and a multi-processor solution is required.

With no limitation on the processor number, the minimum time required to compute all the system outputs is equal to $\approx$5.5 ms. But in this case, a peak of 45 processors are needed and the system has a very poor efficiency, as it is shown by the execution profile in figure 5.

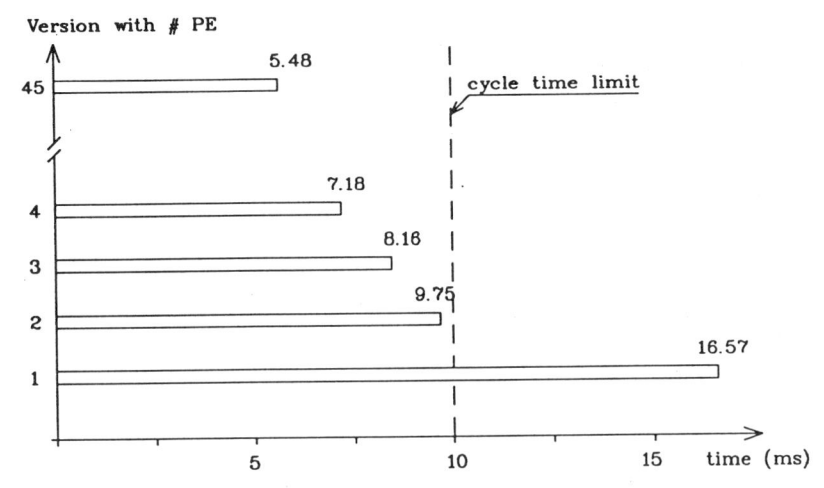

Figure 5. Execution profile with infinite resources

In figure 6, the execution profile is reported for different number of available processors: comparing the required time with the cycle period, the conclusion is that 2 processors should be sufficient, with a very restricted margin, and 3 processors seems offering the best trade-off between cost and performance.

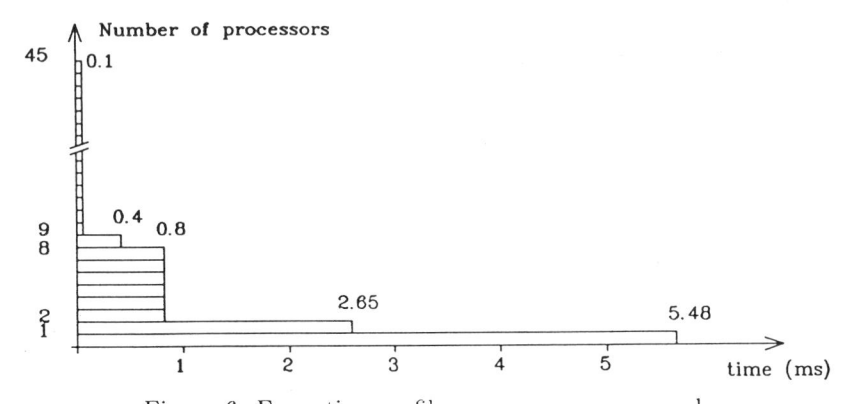

Figure 6. Execution profile versus processor number

CONCLUSIONS AND FUTURE WORK

The paper shows that real time applications can be feasible using fast Prolog processors combined with standard processor hardware. 144 kLIPS are demanded to sustain a 10ms cycle time, for a rather complex automotive application; unfortunately the computational load can be sustained only using more PROXIMA CPUs; however a 2 or 3-processors solution represents a reasonable tradeoff.

Dedicated VLSI processors play a basic role in real time applications using Prolog. However some enhancements must be considered for a Prolog processor. The proposed extensions seem to be effective at least within the frame of the presented application. More discussion is still required on the way to improve such a technique; a closer analysis of the programming technique adopted in the example gives the idea of a strong similarity between the technique and what is implemented using fuzzy set theory in control automation. Finite reaction time and robustness force the Prolog programming to use smooth and redundant rules. Actions must be enforced by a degree of confidence, input extimates must be evaluated starting from more independent sources, as values coming from a unique sensor could be critical.

Long searches and search failures must be absolutely avoided, better if more than one solution (per task) can be collected within the time step. The first solution should be the default one, e.g. reducing the car speed if no other solutions have been found... Therefore the experience shows that at least three different kind of parallelism must be considered for similar applications:

- Process parallelism: hardware must be able to switch among many modules to realize a set of virtual processors; however multiprocessor systems (more CPUs) must be considered in critical applications.

- Execution parallelism: more solutions per cycle with the selection of the best one by extra decision module is actractive. The opinion is that OR-parallel Prolog (Demarchi et al., 1992) execution seems very useful in this case; OR-parallel model is able to supply the first solution as a sequential case, but it can collect more solutions in a short time.

- Module parallelism: replicas of critical modules on different processors must also be considered, and outputs are evaluated in the same way as data coming from different sensors.

Up to now, only the first kind of parallelism has been evaluated, while the second and the third ones are left for future efforts. The behaviour exhibited on the simulated input data, concerning the control itself, was also encouraging and future work carried out by control experts is planned using field data.

The experience demonstrated that very sophysticated real time systems can be implemented leaving programmers free to focus on the playing rules, including those concerning anomalous situations, failures and error recovery. All these aspects, quite hard to be included in a classical and formal approach, will improve the system robustness. The weak point of the technique, beside the novelty and the hardware complexity, is that the exact system behaviour connot be predicted (non determinism), thus leaving the debugging phase very difficult, if not properly supported by adequate software tools.

REFERENCES

Bernard, J. A., "Use of a Rule-Based System for Process Control", in *IEEE Control System Magazine*, October 1988.

Civera, P.L., Masera, G., Ortelli, S., Piccinelli, P., Piccinini,G.L., Poluzzi, R., Ruo Roch, M. and Zamboni, M., "PROXIMA: an Integrated Prolog Machine" - Sixteenth European Solid-State Circuits Conference, Grenoble, Sept. 19-21 1990.

Civera, P.L., Masera, G., Piccinini, G.L. and Zamboni, M., "A Parallel Inference Engine for Automotive Applications" - *International Prometheus Workshop*, pp.221-230, Torino, April 1990.

Civera, P. L., Piccinini, G. and Zamboni, M., "A 32 Bit Processor for Compiled Prolog",in *VLSI for Artificial Intelligence*, J. Delgado-Frias and W. Moore (ed), Norwell, MA: Kluwer Academic, pp.13-26, 1989.

Civera, P. L., Piccinini, G. and Zamboni, M., "Implementation Studies for a VLSI Prolog Co-Processor",in *IEEE MICRO*, vol. 9, 1989.

Demarchi, D., Piccinini, G., Zamboni, M., "An Extended WAM Based Architecture for OR-Parallel Prolog Execution", *3rd International Workshop on VLSI for Neural Networks and Artificial Intelligence*, 1st-4th September 1992, Oxford.

"Eureka program: the Prometheus Project, a Contribution to Safety and Efficiency of Road Transport", ATA Events 1988, International Symposium, Torino 22 April 1988.

Nitao, J. J. and Parodi, A. M., "A Real-Time Reflexive Pilot for an Autonomous Land Vehicle", in *IEEE Control System Magazine*, February 1986.

"Topics of Research, Sensing Systems/Signal Processing", 2nd Prometheus Symposium, Bruxelles, 1987.

Warren, D. H. D., "An Abstract Prolog Instruction Set", Technical Note 309, SRI 1983.

AN EXTENDED WAM BASED ARCHITECTURE FOR OR-PARALLEL PROLOG EXECUTION

Danilo Demarchi, Gianluca Piccinini and Maurizio Zamboni

INTRODUCTION

The paper describes the performance evaluation and the VLSI design of a parallel architecture for high speed execution of Prolog programs. The achievement of very high performances makes sequential execution unsuitable, so parallel models have to be studied and adopted (Nakajima 1988). Our study starts from an existing VLSI sequential Prolog processor (PROXIMA), based on the Warren Abstract Machine; the idea is to extend it in order to support OR-parallel execution models, so that multiprocessor systems could be designed and tested.

The extension of the sequential execution model requires the new design of the parallel search strategy and the techniques for the multiple binding. In such a direction a detailed study has been carried out to compare different execution models (SRI, ARGONNE, VVWAM),(Warren 1987, Shen 1987, Hausman 1987). A parametric simulator has been developed to compare the costs of the different scheduling algorithms and binding techniques, considering three main system architecture topologies (shared memory, distributed memory, cross-bar switch).

Moreover the microarchitecture features of the VLSI processor (internal microconcurrency, memory accesses, etc...) have been analyzed to evaluate their influence on the system performance. A model has been selected with a distributed scheduling algorithm optimized to achieve high granularity processes, so reducing the overall startup costs for the processes initialization.

The model uses an extended instruction set built up on the original Warren Instruction Set. The added instructions implement a distributed scheduler where each processor is able to collect the available work from the other processors. This implies that the semantic of the Warren instructions for the choice point management has to be completely redefined. In order to optimize the process allocation, the standard compiler has been modified allowing the selection of the choice points to be managed as "parallel" ones. The "binding array technique" has been introduced to manage the multiple binding, saving the semantic of the sequential unification algorithm and instructions. A dedicated tag is used to reference data contained in the binding array, so that a dereferencing mechanism can be applied to retrieve the correct data component.

The VLSI implementation of the processor has been derived from the abstract ma-

---

[0]This work is partially supported by CNR *Progetto finalizzato calcolo parallelo* and by MURST *Progetto 40%*.

chine and from the simulations, following design criteria that minimize the prototype costs and the design time. With respect to the sequential VLSI Prolog processor PROXIMA, developed in our laboratory, the processing element has been designed by replacing the control microroms of the sequential processor with a large external Writable Control Storage, to make possible further optimizations on the computational model. The processor has been designed to maintain the compatibility with the prefetching unit (IPU) of the sequential Prolog processor.

EXECUTION MODELS: A QUANTITATIVE APPROACH

The choice of an execution model requires the definition of an appropriate metric to compare the different alternatives (Tick 1987 and Tick 1988). This is even more important in this case since we consider the physical implications of the models in terms of architectural solutions.

For this reason the costs associated to the physical operations are included in the simulator parameters; it is so possible to characterize the models according to the speed and the type of the hardware elements. For instance, the access to the physical memory has a cost that depends only on the system architecture. In such a way each model can be simulated by evaluating the operations needed to execute its primitives and then considering the costs. The main features of the OR models simulated are related to:

- the scheduling algorithm for the process activation;
- the binding techniques for shared variables.

The main architectural aspects that are considered in our simulation studies are:

- the number of processors;
- the system architecture (costs related to the access to the shared variables);
- the process installation costs;
- the processor efficiency (in terms of number of cycles required to execute a basic primitive of the algorithm).

The goal of the methodology used performing these simulations, is to find out the best execution model according to a system architecture with a fixed number of processors. The first problem that arises is to decide whether it is better to find the algorithm that best fits on a specific architecture or to optimize the system architecture for a selected algorithm. The solution can be mainly found by studying the influence of the architecture on the performance of an algorithm.

Moreover an architecture can be simply classified in terms of implementation costs. It could be better to select the algorithm that offers high performance with a low cost system architecture only if the performance are not too far from the maximum achievable. To reduce the number of the simulations, we used the SRI model as *reference model* to compare the different strategies and architectures: this choice seems reasonable since many subsequent execution models are based on the SRI one. It has been designed for a general shared memory multiprocessor and has been conceived as an extension of the sequential execution approach.

The SRI model keyfeatures are:

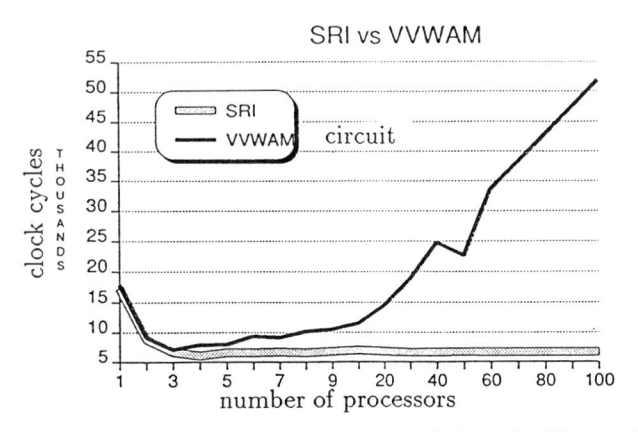

Figure 1. Execution times (in number of clock cycles) of the SRI mode vs VVWAM for a shared memory architecture (circuit benchmark)

1. The shared variables problem is solved by using the binding array technique. Each processor can bind a shared variable (conditional binding) in a local memory area. The shared variable is identified by an offset that each processor adds to the starting location of its own binding array area. For the non-shared variables (unconditional binding) the SRI model behaves exactly as the sequential WAM, binding the variable locally.

2. The scheduling algorithm is based on the *top-most* management of the choice points; in other terms the search strategy assigns high granularity choices to the idle processors, so reducing the overall installation costs.

Two models (ARGONNE and VVWAM) have been simulated comparing the execution times of some typical benchmarks with the SRI model execution. As an example the simulation results for the *"circuit"* (a very simple cmos gate level simulator) is reported in figure 1 and figure 2. It shows the execution times according to the number of processors for the SRI, VVWAM and ARGONNE models. The simulations have been performed for a shared memory architecture. The SRI reaches the optimal performances for 4-6 processors; then increasing the number of the processors no improvement is visible. On the contrary, the VVWAM has an optimum value of processors (3-6) with worse performances, then an increase of the processor number brings to a significant worsening of the results.

The ARGONNE model behaves as the SRI, with slightly improvements and saturates the performances for large amount of processing units. From the results obtained, the SRI model achieves better performances than VVWAM and just a little lower than ARGONNE. Since ARGONNE has a variable binding policy more complex than SRI, we have selected the SRI model as reference to compare the different architectural solutions.

The further simulations have been carried on focusing on the SRI model and varying the architecture; three basic architectures, shown in figure 3, have been selected: *shared memory*, *distributed memory* and *cross-bar switch*. The simulation times for the three architectures are compared in figure 4. It is important to analyze the results for systems with:

- a low number of processors (less than 4);

- a medium-high number of processors.

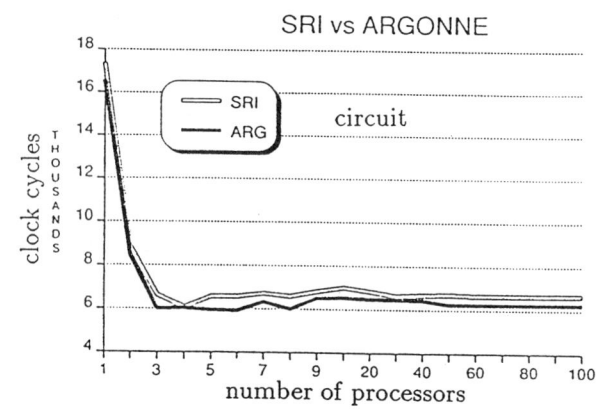

Figure 2 . Execution times (number of clock cycles) of the SRI model *vs* ARGONNE for a shared memory architecture (*circuit* benchmark)

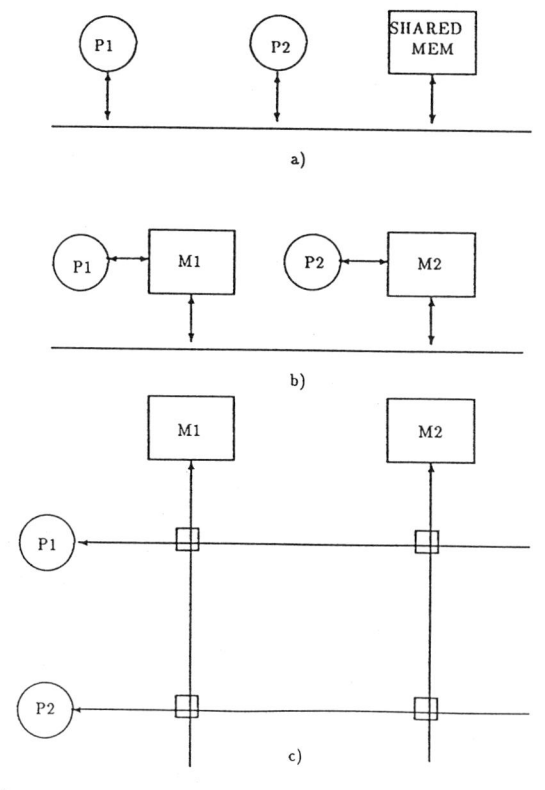

Figure 3 . Reference architectures block diagrams: a- shared memory, b- distributed memory, c- cross-bar switch

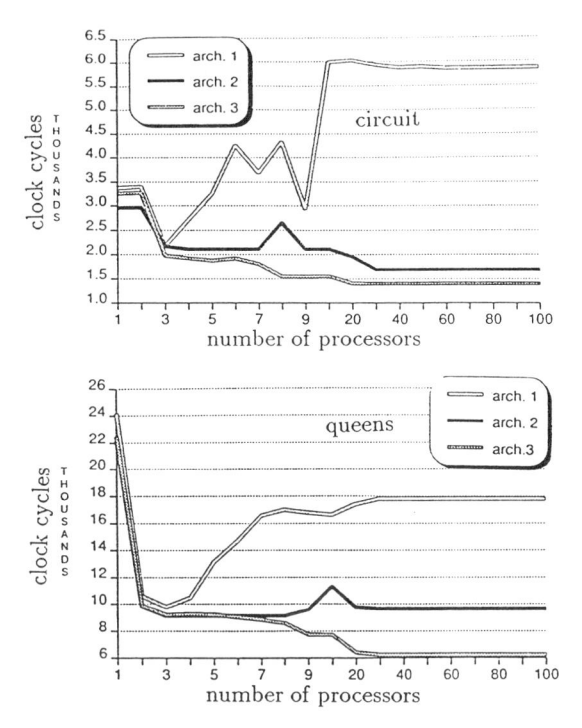

Figure 4. Execution times (number of clock cycles) of the SRI model for the three reference architectures (arch1-*shared memory*, arch2-*distributed memory*, arch3-*cross-bar switch*) for the *circuit* and *queens* benchmarks

The division in two classes is due to the analysis of the achievable performances; reasonably for a low number of processors the performance are almost the same for the three architectures. The shared memory architecture does not represent a limit to the performance since the conflicts on the bus are limited. For the same reasons the cross-bar switch architecture does not improve the performances.

If the number of processors increases, the differences become more evident. The global bus with distributed dual-port memory architecture shows good performance for a medium-high number of processors, by keeping limited the complexity of the hardware with respect to the cross-bar switch solution that has high implementation costs. An important feature of the distributed memory architecture is that saturation applies when the number of processors overcome the possible choices of the simulated benchmarks. Only the basic shared memory architecture shows a high degradation of the performance for more than 8-10 processors pointing out that memory access cost become the bottleneck of the system.

In the first step of the model definition, the processor loads have been evaluated by studying the time execution profiles.

As an example, the working and idle activities for a distributed memory system with 50 processors are reported in figure 5 for the *queens* problem benchmark. The upper profile shows the number of active processors. For very simple benchmarks the profiles show that a large number of alternatives are tried simultaneously at the beginning. So the load profile has an high busy/idle ratio, but quite early it decreases sharply when almost the possible open alternatives in the proof tree are closed. On the contrary, for the *queens* problem benchmark (that is a significant example of a large class of A.I. problems), a more realistic load profile has been obtained.

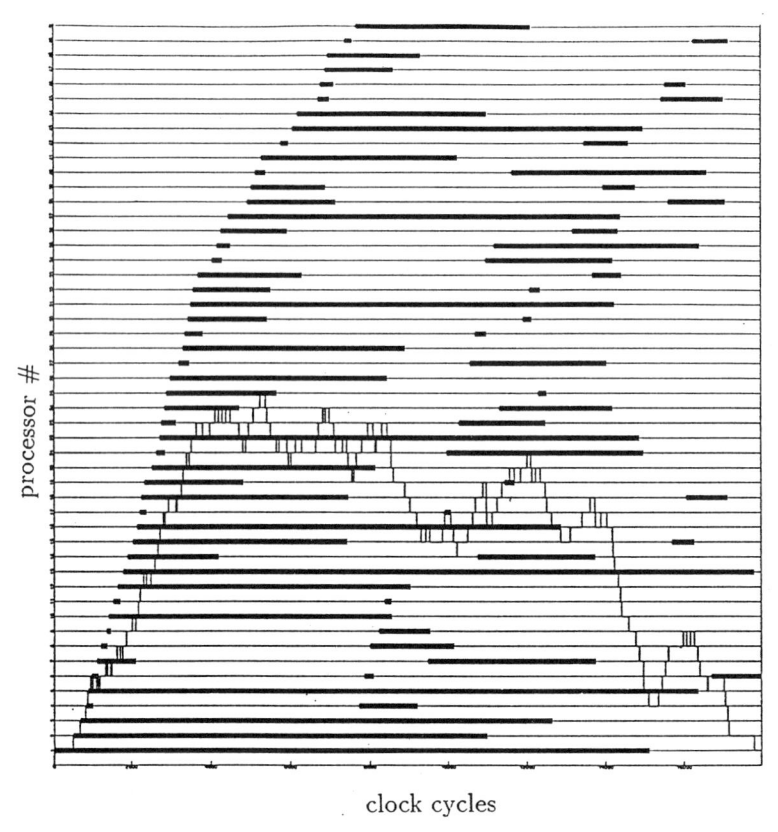

Figure 5. Execution profile of the SRI model with a distributed memory architecture (50 processors) for the *queens* benchmark

In this case the profile has a quite uniform behavior, showing that the SRI model with a distributed memory architecture is a good starting point for an high performance and low cost architecture.

THE EXTENDED WAM MODEL:P-PROX

The parallel model implemented has been designed assuming the WAM model and the Warren instruction set as starting point (Hermenegildo 1989).
The parallel primitives have been inserted in the Warren instructions by modifying their semantics.
They have to support the process scheduling and the variable binding. The most relevant constraint is to leave unchanged the data structures for the unification, and the procedural instructions for the search strategy.
Moreover to avoid the need of an additional unit for the management of the process scheduling, the scheduling operations have been distributed in the procedural instructions. This solution uses a *distributed scheduler* (Warren 1988 and Calderwood 1988) where each processor searches the available work when it has finished the search in its own sub-tree. The scheduling operations are performed in two main steps:

1. available work search;
2. available work transfer.

In the first phase each processor that has completely explored its own sub-tree starts the search procedure looking for available work at the other processors. The oldest (*top most*) open choice point is considered the candidate job.

The second phase starts with the request to the processing element (with the *top most* choice) for the transfer of the needed choice point. After that, the requesting processor can load the internal registers and also the sub-tree execution entry point address (the code is duplicated in each processor memory).

The binding array technique is the most effective in terms of computational overload if compared to the sequential case: it requires only an additional dereferencing. The binding array is stored in the processor local memory and the physical address of the variable is generated using the offset stored in the tagged variable. In such a way the unification instructions remain practically unaltered maintaining the high degree of efficiency of the sequential WAM.

The implemented model has the same instruction set of the sequential processor, with the only exception of the "Sequential-chpt" instruction. This instruction solves a problem that arises immediately when large programs are executed in parallel: the poor control of the program intrinsic parallelism. A first solution is to write programs with a limited degree of intrinsic parallelism, but this method reduces obviously the Prolog capabilities. A more effective approach is to introduce a compiler directive to mark the predicates that have not to be executed in parallel. As a consequence the compiler introduces the "sequential-chpt" instruction that causes the processor to mark the choice point created immediately after as sequential; i.e. it cannot be executed in parallel and therefore it is not stolen by any other processor.

The implemented model has been simulated by writing a parallel simulator in order to verify the model correctness and efficiency.

THE ABSTRACT MACHINE FOR THE EXTENDED MODEL

The parallel model implemented in the simulator has been obtained as an extension of the Warren Abstract Machine for the sequential execution of Prolog. The most relevant differences are related to the management of the new structures for the multiple binding and the parallel search strategy information. The data memory organization is similar to the sequential abstract machine. Only two areas are related to the parallel execution:

1. The Binding Array is the memory area where the "conditional bindings" are written: each variable is identified by an offset starting from the base address contained in the Base Address Pointer of each worker. This area is functionally "private" since only the owner can access it.

2. The Parallel Information Stack contains physically the additional fields of the choice point used to manage the parallelism of the model.

THE SYSTEM ARCHITECTURE

The target system architecture is a distributed memory system. Each processor has a "private" code memory and a local dual port memory that can be accessed by

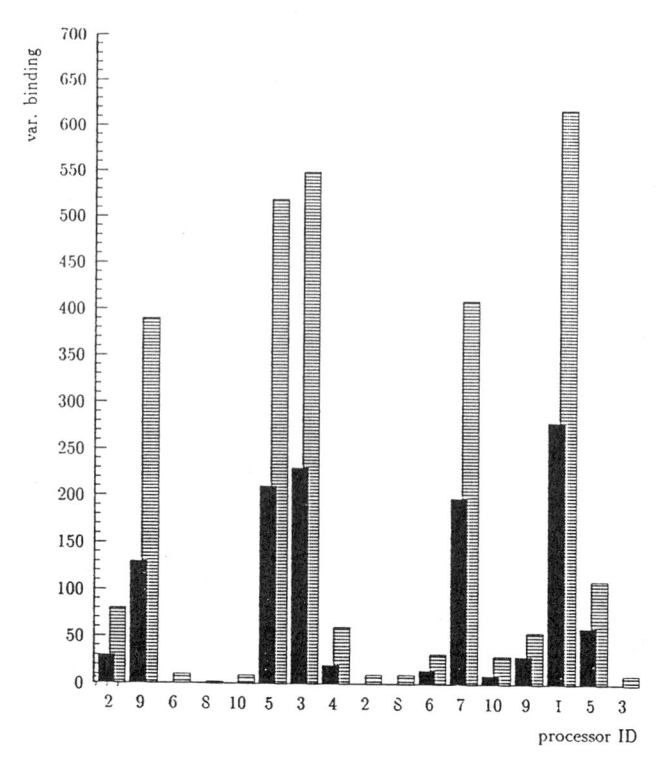

Figure 6 . Memory access distribution for the *queens* benchmark - dotted bars show the total number of variable bindings; black bars represent the local bindings

all the other processors via a global bus. The code is stored in a local memory to avoid additional traffic on the global bus; this implies the copy of program code to every processor board. The main advantage of using a distributed data memory with respect to the shared one is related to the locality of the memory accesses. Moreover the binding array technique keeps the accesses to the shared variables as local as possible. The benefits of storing the binding array in a distributed dual port memory without using the global bus are straightforward, as shown in figure 6. In this diagram the distribution of the accesses to the local variables and the shared ones (binding array) are indicated for each processor: more than 50% of them are in the binding array. The overall system architecture is shown in figure 7; the global bus is connected with the S-bus board interface to a workstation host (Sun Sparc Station 2).

Each processor board includes the P-PROX DPU (Data Processing Unit), the IPU (Instruction Processing Unit - the same of the PROXIMA sequential Prolog Machine), the dual port data memory and the code memory where the program code is downloaded in broadcasting for all the processors at the beginning of the computation.

THE PHYSICAL IMPLEMENTATION

The design of the VLSI processor has been carried out by considering the new constraints imposed by the parallel execution.

The main difference with respect to the sequential implementation is that the overall performances of the system are less dependent from the efficiency of the single processor. In fact the system capabilities have a greater influence than in the sequential case. To evaluate the relationships between the processor efficiency and the system speed-up, some simulations have been performed varying the microconcurrency factor of the

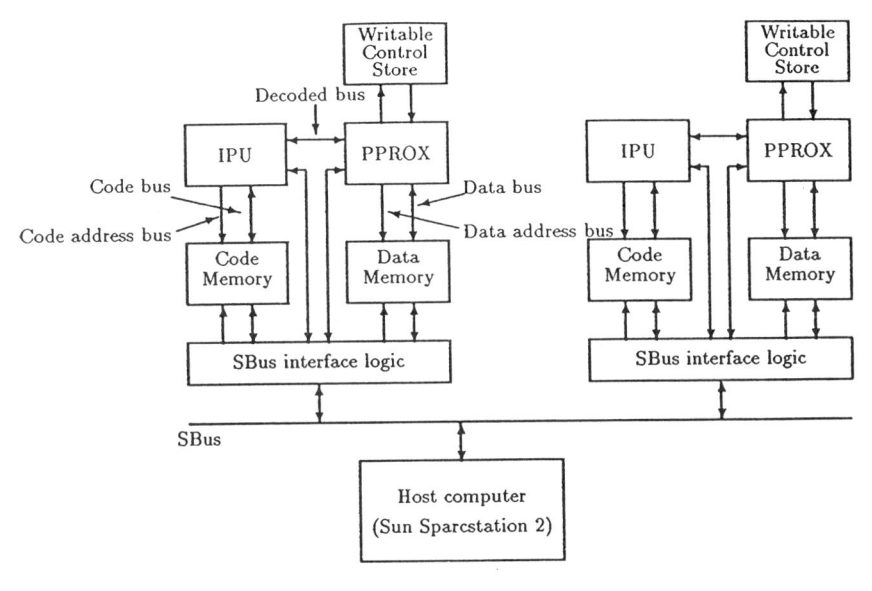

Figure 7. System architecture block diagram

processor (the number of single microinstructions that can be executed in parallel in a microcoded processor). The results, shown in figure 8, indicate that the microconcurrency for a distributed memory solution can be reduced with acceptable reduction in the performances, if k (number of microcycles per operation required by the architecture) increases from $k = 0.5$ (approximately the factor for the PROXIMA sequential processor) to $k = 2$. It is interesting to note that the microconcurrency factor affects marginally low performance system architectures (shared memory) but becomes much more important for the cross-bar switch solution. In our case this results justify some architectural choices in order to reduce the implementation costs of the processor. The control store has been implemented externally and the microcode is "vertical" to decrease the number of pads required in the processor to load the microinstructions. This solution is in accordance with the internal solution adopted for the large number of registers in the abstract machine. In fact, a large register file has been used (64 x 32) allowing an efficient coding of the operations. The sequencer of the control unit has been implemented inside the processor by using a dedicated data-path. The resulting integrated circuit, designed with the silicon compiler Genesil, in $1.2\mu m$ CMOS technology is 6.8×8.8 mm^2 and has been simulated at 20 Mhz.

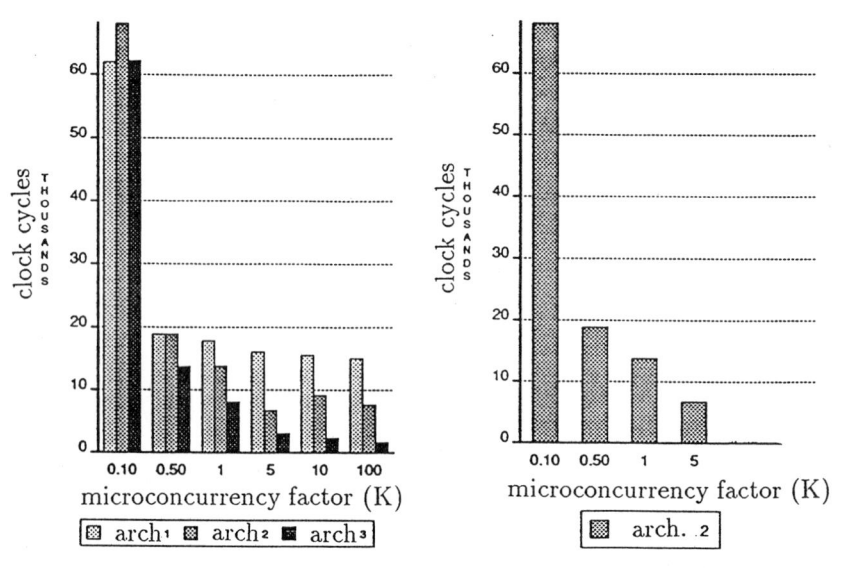

Figure 8 Execution times for the *queens* benchmark for the three reference architectures varying the microconcurrency factor

REFERENCES

Calderwood A. and Szeredi P. :"Scheduling OR-parallelism in Aurora: the Manchester Scheduler", *International Conference on Fifth Generation Computer Systems* Tokyo, 1988

Hausman B., Ciepielewski A. and Calderwood A. :"Cut and Side-Effects in OR-parallel Prolog", *SICS Technical Report* Swedish Institute of Computer Science (SICS), 1987

Hermenegildo M.: "High-Performance Prolog Implementation: the WAM and Beyond", *International Conference on Logic Programming*, Lisboa June 1989

Nakajima K., Inamura Y., Rokusawa K., Ichiyoshi, and T. Chikayama: "Distributed Implementation of KL1 on the Multi-PSI/V2", *Institute for New Generation Computer Technology*, pp. 436-451, 1988

Shen K. and Warren D.H.D.: " A simulation study of the ARGONNE model for OR-parallel execution of Prolog", *Int. Symposium on Logic Programming*, San Francisco, 1987

Tick E.:"A Performance Comparison of AND and OR-parallel Logic Programming Architectures", *Institute for New Generation Computer Technology*, pp. 452-467, 1988

Tick E.:"Studies in Prolog Architectures", *PhD thesis*, Stanford University June 1987

Warren D.H.D. et al. :"The Aurora OR-parallel Prolog System", *International Conference on Fifth Generation Computer Systems* pp. 819-830, 1988

Warren D.H.D.:"The SRI Model for OR-parallel Execution of Prolog: Abstract Design and Implementation Issues" *Proceeding of the 1987 Symposium on Logic Programming*, pp.92-102, 1987

ARCHITECTURE AND VLSI IMPLEMENTATION OF A PEGASUS-II PROLOG PROCESSOR

Takashi Yokota and Kazuo Seo

INTRODUCTION

Many new Prolog machines use the Warren abstract machine (WAM, Warren 1983a) as an execution model. Each WAM instruction can be broken down into a series of simple actions; yet its instruction level is high. Thus, even a RISC, which has a low-level instruction set, can be made to perform at high levels, if it has WAM oriented features.

The aim of Pegasus architecture is to offer a high-performance component for Prolog execution which uses a minimum of peripheral circuits (e.g., memories). The processor can be incorporated in a workstation as a backend processor, or can be used as a stand-alone processor in a manufacturing control system. RISC's essential features: a compact chip and fast execution, are suitable for such applications.

We first studied the WAM instruction set, and investigated what mechanisms were suitable and how to make them work efficiently. This effort resulted in the simple, powerful architecture called "Pegasus-I" (Seo and Yokota 1989); the forerunner to Pegasus-II. It has a simple instruction set as well as powerful hardware-supported features. Of particular note, is a special register file which directly supports backtracking. This architecture proved the architectural effectiveness of RISC applied to Prolog language.

Pegasus-I, however, had an inherent problem: low semantic density. This resulted because each WAM instruction corresponds to Prolog's flexible semantics; a mechanical translation into low-level actions decreases semantic density in objective codes. This also leads to inefficient execution. One solution to this problem came as result of breakthroughs in compilers and optimizers. For example, Aquarius (Van Roy 1990) does type inferences on all arguments, for all clauses; and then, tries to translate Prolog source programs into deterministic codes. This method can considerably reduce execution time in some cases.

We approached Pegasus-II architecture in another manner; we raised the instruction level. We made careful, detailed investigations of instruction sequencing and instruction-level parallelism in compiled WAM codes. Then, we introduced many ideas into new Pegasus-II architecture. We will describe these new concepts in the following sections.

ARCHITECTURE

Basic Architecture Inherited from Pegasus-I

The basic architecture of Pegasus is a tagged RISC. Each word consists of an 8-bit tag and a 32-bit value part. A tag represents data types utilizing the least significant six bits, and uses the remaining two bits for garbage collection. An instruction word has the same format, however, no distinction is made between tags and value parts. In principle, one instruction is executed at every cycle.

Addressing modes are limited to base register (or program counter) plus immediate offset. Each LOAD and STORE instruction has the capability to post-increment/-decrement a base register. This feature supports frequent stack access.

A multidirectional branch on a tag segment is supported. Two of the least significant six bits of the tag segment represent the four basic data types: variable, constant, list and structure. A tag-dispatching jump instruction concatenates basic type bits of two source registers. The concatenated four bits are used as an offset to loop up a table, which contains 16 destinations, and the instruction causes a branch. This 16-way branch is used in the unification procedure of two registers. If one of these is zero, a four-way branch is executed. This appears at head-unification.

Pegasus–I supports backtracking with a special register file. The register file contains 23 ordinary registers, a zero register, and 24 sets of paired registers. Paired registers (called *Main* and *Shadow* registers) have the ability to copy themselves into each other in one clock cycle. This special mechanism makes it faster and easier to create choice points and to backtrack. No other clocks are required for shallow backtracking.

Design Concept of Pegasus–II Architecture

One of the best, low-cost ways of achieving high performance is full utilization of on-chip resources. However, it seems impossible to attain 100% utilization. Performance degradation in Pegasus–I is caused by the following problems:

- limitations resulting from the primitiveness of the instruction set,
- redundant instruction words,
- interlocking,
- frequent and short branches,
- and lower memory bandwidth.

In designing Pegasus–II architecture, we adopted the following as solutions to these problems.

Compound Instructions

In Prolog processing, the dominant action is manipulation of data (load/store, register transfer, and tag/value manipulation). ALU operations are not frequent. Since a typical sequence of actions contains common registers, we can utilize parallelism which incorporates multiple pathways; rather than using multiple functional blocks (i.e., ALUs). For example, when allocating a variable in a heap area, a STORE-and-MOVE instruction can execute STORE, MOVE, and INCREMENT actions in parallel. Since these parallel actions do not interfere with each other, they can be combined. Possible combinations are:

- load/store and move,
- load/store and jump,

- load/store and add,
- and two sets of moves.

Compound instructions also reduce code size.

Dedicated Hardware

Several dedicated circuits, which contribute to high efficiency and code size reduction, are introduced for frequent operations. A *tag-comparator* minimizes dereferencing time-cost. A dereferencing action is specified by one bit of an instruction word. If the bit is activated and a source register has a reference tag, dereferencing cycles are inserted and performed before the original execution of the instruction.

Trail checking, which appears frequently, requires magnitude comparisons among three registers (B, HB and the variable register) and a stack-push operation. A *trailer* contains a complete copy of B and HB registers, and a set of comparators. It performs comparisons concurrently with a STORE operation to bind variables. A stack-push operation is inserted if the variable must be trailed.

Frequently used addresses are held in dedicated registers, called *vector* registers. Some instructions use inherent addressing using vector registers, a conditional fail instruction, for example.

TagRAM is a dual-port register file containing seven frequently-used tags.

A special instruction loads the instruction word into a special register, named X. Since the instruction needs no memory accesses, the X register can be accessed without interlocking in the succeeding instructions.

Dynamic Execution Switching Mechanism and Parallel Datapath Operation

Pegasus requires dynamic execution control; for example, control of branch penalty slots and dereferencing. We unified these features into the dynamic execution switch mechanism (DESM), and improved it to make it a powerful and flexible instruction set. The instruction decoder inputs various information including the instruction register, i.e., the processor status, the tag segment of the register file, the comparators added to the datapath, and others. It then determines actions corresponding to inputs.

DESM is employed to increase utilization of resources. Prolog execution contains frequent short branches. In an "IF <cond> THEN A ELSE B" procedure, if <cond>, A, and B are simple operations, and these three actions do not interfere with each other, they can be unified into one instruction using DESM. An action for <cond>, and a datapath arrangement for A and B begin at the same cycle. Throughout the cycle <cond> is being determined, and at the end of the cycle, the result is propagated to the instruction decoder. The decoder determines which action to execute, and the selected action continues in succeeding cycles.

The most significant example is WAM's "get_XXX" instruction. Details of this instruction are listed in figure 1. This procedure has three possible actions:

- memory read for dereferencing,
- memory write for variable-binding (write-mode),
- compare and branch to the fail/0 built-in predicate (read-mode).

These three actions start in parallel concurrent with instruction decoding (fig.2(a)–

(c)). With DESM capability, the decoder can determine which action to continue. For example, if the register has a reference tag, then dereferencing is encountered and execution of the same instruction is tried in the next cycle. Hence, "get_XXX" is executed in one instruction and datapath elements are used to their fullest (fig.2(d)).

Harvard Architecture

The necessity to access memory frequently is an undesirable aspect of Prolog. This is because execution efficiency is limited by a narrow memory bandwidth. To widen the memory bandwidth, Pegasus–II has incorporated Harvard architecture, which has two memory busses: an instruction bus and a data bus.

Harvard architecture can double memory bandwidth, and also influences pipelining (figure 3). Instructions are fetched in the *F(etch)* stage, and the *EX(ecute)* stage carries out instruction decoding, operand fetching, and the ALU operation. Write-back to register and memory referencing are executed in the *WM* stage. In cases of load instructions, an optional *XW* stage (extended register write) is employed.

The instruction set is summarized in table 1. The datapath is illustrated in Figure 4.

```
/* Get_Constant( Reg, Constant ) */
while( Reg.tag==Reference )        /* dereferencing */
  Reg := memory(Reg.val);
if( Reg.tag==Variable ){           /* write-mode */
  memory(Reg.val) := Constant;     /*   variable binding */
  if( is_trailed(Reg)==YES )       /*   trail checking */
    push_to_trail_stack( Reg );
} else {
  if( Reg != Constant )            /* read-mode */
    PC := fail/0;                  /*   if not identical, fail */
}
```

Figure 1 A Description of "get_constant" WAM Instruction

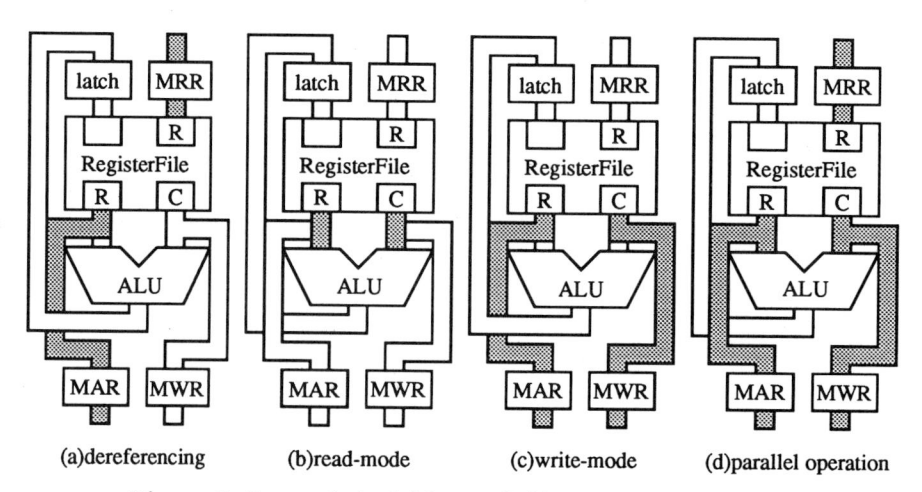

Figure 2 Datapath Activities Avaliable in "GET" Instruction

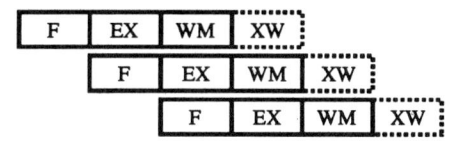

Figure 3 Pipelining

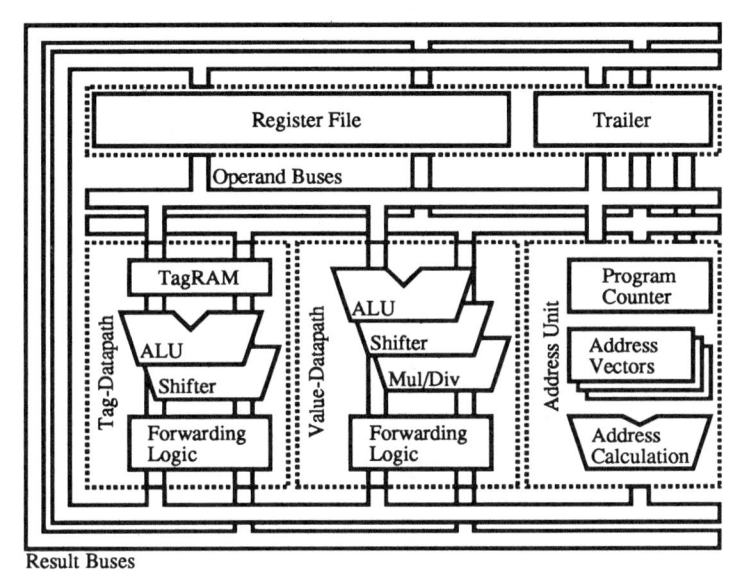

Figure 4 Datapath Diagram

IMPLEMENTATION OF A PROTOTYPE CHIP

Cell-Based Design for Rapid Implementation

For new architecture to prove its efficiency, quick prototyping is essential. Cell-based ASIC design methodologies are suitable for this purpose, because full-scale prototypes require large man-power outlays and take a long time. Therefore, only the special register file is custom-made in the Pegasus–II prototype. The register file, which has paired banks capable of copying one another, is a key component in Pegasus architecture. Originally, the layout of the Pegasus–I chip used a 2.0 μm CMOS rule. Since this layout can also accomodate a scalable lambda rule, enlargement and shrinking of the physical layout require no special effort. This layout has been shrunk in Pegasus–II architecture according to a 1.2 μm rule. The Pegasus–II design treats the register file as a large cell.

The remaining architecture was designed using high-level design tools:

Table 1 Pegasus–II Instruction Set Summary

Opcode	Function (*EA*=Effective Address)
Prolog Instructions	
UFY	Unify two registers
SW	Choice point creation, Main→Shadow register copy
SR	Choice point deletion, Main←Shadow register copy
GT	Get {nil,constant,list,structure}
JTD	Tag-dispatching jump, PC:=PC+dispatch(r_1,r_2)
STD	Tag-dispatching jump, PC:={PC+1,PC+Ofs1,PC+Ofs2,*FailVector*}
TSW	Switch on tag-type, PC:={PC+1,*EA*,*FailVector*}
Jump/Branch Instructions	
BV/BT	Conditional branch on Value/Tag segment, if TRUE then PC:=*EA*
FV/FT	Conditional fail on Value/Tag segment, if FALSE then PC:=*EA*
JP	Unconditional jump to any address
HJP	Hashed jump, PC:=PC+(Reg&Mask)
TRP	Software trap
Data-transfer Instructions	
LD/ST	Load/store a word
LDX/STX	Move a word from/to Coprocessor
M	Move registers and manipulate tag/value part
LA	Load effective address, r_d:=*EA*, (r_d is limitted)
MAD	Load effective address, r_d:=*EA*
Logical/Arithmetic Operation Instructions	
OV/OT	Operate on Value/Tag, $r_d:=r_1$<op>r_2
MUL/DIV	Integer multiplication/division
Compound Instructions	
LRJ/SRJ	Load/Store and jump
LDM/STM	Load/Store and move
Miscellaneous	
NOP	No Operation
HLT	Halt processor
STI/CLI	Set/Clear interrupt mask
STM/CLM	Set/Clear GC-tag mask
X	Set special purpose registers/vectors

- state machine compiler,
- datapath module generator,
- and cell placement and router.

The state machine compiler inputs boolean equations and state transition diagrams, and generates a standard cell netlist. It contains a powerful logic synthesizer, and its results are nearly optimal for given constraints (input/output delay, output driving capability, input capacity).

The datapath module generator inputs a bit-slice expression of schematics. It stacks primitive functional cells to a specified number of bits. This tool can also generate a standard cell netlist. The place/router organizes standard cell blocks and composes the entire chip (including I/O pads).

Obviously, Pegasus–II architecture requires more complex control logic than its earlier version. Hardware design language allows designers to translate strictly from architectural specifications to implementation circuits. Since the state machine compiler inputs low-level, redundant syntax, we used a simple translator which converts the state machine compiler syntax to a higher level description. This design method is used mainly to eliminate human mistakes.

Most of the component blocks are designed using the state machine compiler and the datapath module generator. Table 2 lists the percentage of gates compiled by these design tools. Exceptions are the carry look-ahead circuit, the clock generator, the I/O pads, and miscellaneous small combinatorial circuits. Over 2,300 lines from 17 state machine descriptions were compiled into standard cell netlists containing about 2,400 gates. Datapath schematics were compiled into standard cell netlists containing 3,500 gates and layout blocks containing 17,000 gates.

Fabrication

The prototype chip was fabricated using a double-metal 1.2μm CMOS process. It contains about 144,000 transistors in a 9.3mm × 9.3mm die. Figure 5 is a photomicrograph of the chip. The master control circuit is physically divided into three blocks. A wide horizontal channel near the center contains operand busses and control lines. The other horizontal channel, at the bottom, holds two memory busses (instruction and data). The chip works using a 100ns clock cycle.

EVALUATION

We assumed two imaginary architectural models for comparison. The *P1* model has a Pegasus–I instruction set executed on the same hardware and pipelining as Pegasus–II. The *P0* model is similar to *P1*, however, it uses a primitive instruction set of RISCs. The Pegasus–II model is abbreviated as *P2*.

We evaluated these architectural models using Warren's benchmarks (Warren 1983b). The execution cycles, showing both GET and compound instructions in each program, are summerized in figure 6. The figure reveals that high-level instructions introduced in Pegasus–II contribute greatly to improved performance.

Static code sizes are shown in figure 7. As is evident, Pegasus–II architecture drastically reduces code sizes; which leads to a reduction in the cache mishit ratio.

CONCLUSIONS

We described Pegasus–II architecture and its VLSI implementation. Primitive RISC architecture increases the size of codes and makes Prolog execution inefficient. To make Pegasus–II most effective for Prolog execution, it is necessary to carefully select instructions and to raise the instruction level of Pegasus–II. Benchmark evaluations revealed that the instruction set was high-level, and the architecture significantly improved performance and semantic density of machine codes.

The prototype chip was fabricated using cell-based ASIC methodologies for rapid implementation. The chip works using a 100ns clock cycle, and, by our estimation, proper implementation reduces the clock cycle to 50 ∼ 60 ns.

Table 2 Percentages of Gates Compiled by Design Tools

Block	#Gates	%Compile	Tool
Datapath (tag)	2,462	97.7	d.p. generator
Datapath (value)	6,395	100.0	d.p. generator
Address Unit	8,172	100.0	d.p. generator
Master Control	7,675	82.3	state mach. comp. & d.p. generator
Others	374	100.0	state mach. comp.
(total/weighted mean)	25,134	92.8	

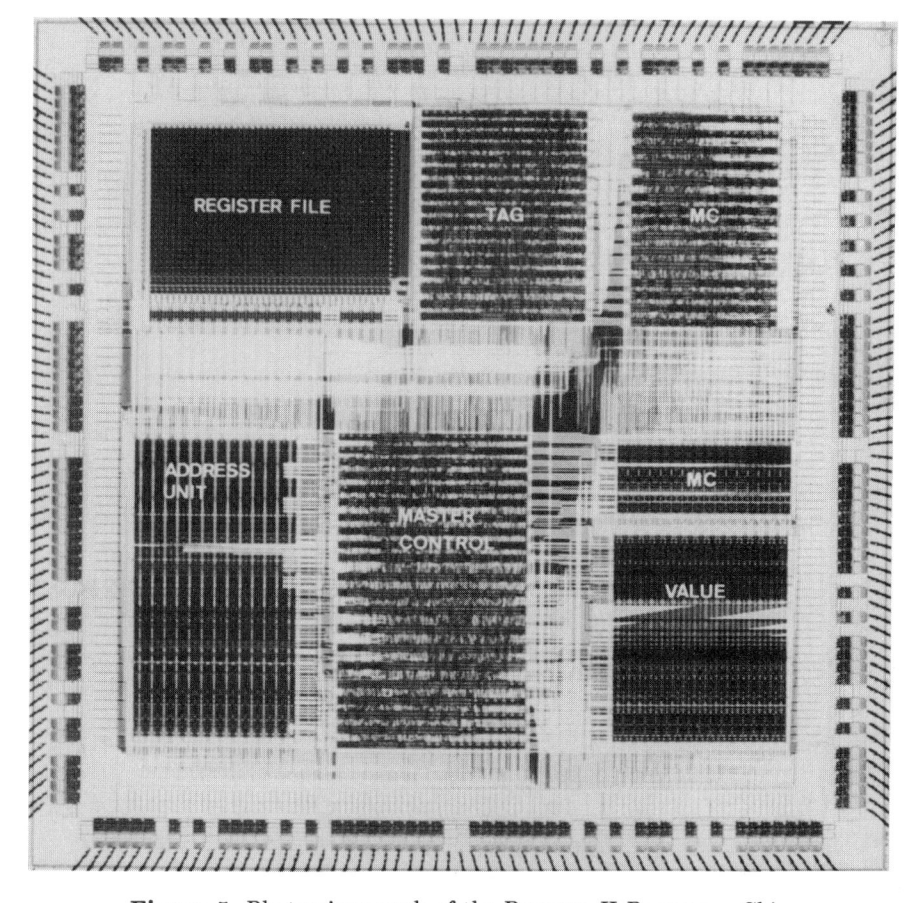

Figure 5 Photomicrograph of the Pegasus–II Prototype Chip

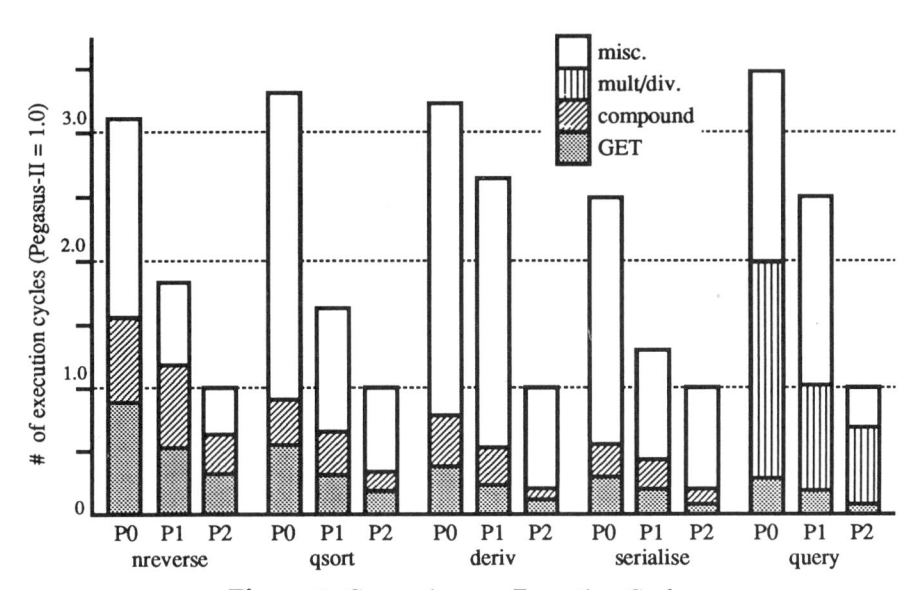

Figure 6 Comparison on Execution Cycles

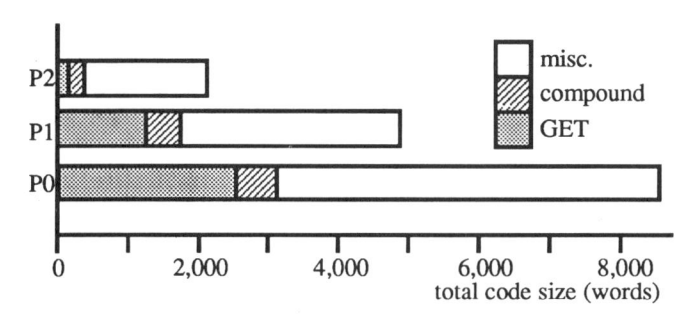

Figure 7 Comparison on Static Code Size

References

Seo, K. and Yokota, T., "Design and Fabrication of Pegasus Prolog Processor," in *VLSI '89*, pp.265–274, North Holland, 1989.

Van Roy, P. L., "Can Logic Programming Execute as Fast as Imperative Programming," Report No. UCB/CSD 90/600, University California Berkeley, 1990.

Warren, D. H. D., "An Abstract Prolog Instruction Set," Technical Note 309, Artificial Intelligence Center, SRI International, 1983.

Warren, D. H. D., "Applied Logic – Its Use and Implementation as Programming Tool," Technical note 290, Artificial Intelligence Center, SRI International, 1983.

CONTRIBUTORS

Tadashi Ae, *Hiroshima University, Japan*
Reiji Aibara, *Hiroshima University, Japan*
Nigel M. Allinson, *University of York, UK*
Krste Asanovic´, *Int. Computer Science Institute / UC-Berkeley, USA*
Steven Barber, *State University of New York at Binghamton, USA*
James Beck, *Int. Computer Science Institute / UC-Berkeley, USA*
E. Belhaire, *University of Paris, France*
Giacomo M. Bisio, *University of Genoa, Italy*
François Blayo, *Swiss Federal Institute of Technology, Switzerland*
Hamid Bolouri, *University of Hertfordshire, UK*
Rüdiger W. Brause, *J. W. Goethe University, Germany*
Erik L. Brunvand, *University of Utah, USA*
Mario Cannataro, *CRAI, Italy*
Howard C. Card, *University of Manitoba, Canada*
Daniele D. Caviglia, *University of Genoa, Italy*
Pier Luigi Civera, *Politecnico di Torino, Italy*
Alessandro De Gloria, *University of Genoa, Italy*
Eric Delaunay, *Ecole Polytechnique, France*
José G. Delgado-Frias, *State University of New York at Binghamton, USA*
Danilo Demarchi, *Politecnico di Torino, Italy*
Hui Ding, *State University of New York at Binghamton, USA*
Paolo Faraboschi, *University of Genoa, Italy*
William Fornaciari, *Politecnico di Milano, Italy*
P. Garda, *University of Paris, France*
Jean-Dominique Gascuel, *Ecole Polytechnique, France*
Manfred Glesner, *Darmstadt University, Germany*
Raymond J. Glover, *Brunel University, U.K.*
Karl Goser, *University of Dortmund, Germany*
Douglas M. Green, *The Johns Hopkins University, USA*
Kevin Gurney, *Brunel University, UK*
Denis B. Howe, *Imperial College, UK*
Terence Hui, *University of Hertfordshire, UK*
John F. Hurdle, *University of Utah, USA*
Paolo Ienne, *Swiss Federal Institute of Technology, Switzerland*
Giacomo Indiveri, *University of Genoa, Italy*
Lüli Josephson, *University of Utah, USA*
N. Kasabov, *University of Essex, U.K.*
Brian E. D. Kingsbury, *Int. Computer Science Institute / UC-Berkeley, USA*
Kazumasa Kioi, *Hiroshima University, Japan*
Phil Kohn, *Int. Computer Science Institute / UC-Berkeley, USA*
Aleksander R. Kolcz, *University of York, UK*
Andreas König, *Darmstadt University, Germany*
V. Lafargue, *University of Paris, France*
Oliver Landolt, *CSEM, Switzerland.*
Simon H. Lavington, *University of Essex, U.K.*
Adrian Lawrence, *University of Oxford, U.K*
Christian Lehmann, *Swiss Federal Institute of Technology, Switzerland*
S. Lin, *University of Essex, U.K.*
Wei Lin, *State University of New York at Binghamton, USA*
Vincent Lok, *University of Oxford, U.K*
Wayne Luk, *University of Oxford, U.K*
Guido Masera, *Politecnico di Torino, Italy*
Jill R. Minick, *Texas A&M University, USA*
Lionel Montoliu, *Ecole Polytechnique, France*
Bahram Moobed, *Ecole Polytechnique, France*
Nelson Morgan, *Int. Computer Science Institute / UC-Berkeley, USA*

Paul Morgan, *University of Hertfordshire, UK*
E. Naroska, *University of Dortmund, Germany*
Mauro Olivieri, *University of Genoa, Italy*
Ian Page, *University of Oxford, U.K*
G. Palm, *University of Ulm, Germany*
Gerald G. Pechanek, *IBM, USA*
Gianluca Piccinini, *Politecnico di Torino, Italy*
Luigi Raffo, *University of Genoa, Italy*
Darren P. Rodohan, *Brunel University, U.K.*
Ulrich Rückert, *University of Dortmund, Germany*
Massimo Ruo Roch, *Politecnico di Torino, Italy*
S. Rüping, *University of Dortmund, Germany*
Thomas F. Ryan, *State University of New York at Binghamton, USA*
Silvio P. Sabatini, *University of Genoa, Italy*
Fabio Salice, *Politecnico di Milano, Italy.*
Kazuo Seo, *Mitsubishi Electric Corporation, Japan*
Giandomenico Spezzano, *CRAI, Italy*
Richard Stamper, *University of Oxford, U.K.*
M. A. Styblinski, *Texas A&M University, USA*
Domenico Talia, *CRAI, Italy*
Edward P. K. Tsang, *University of Essex, UK*
A. Ultsch, *University of Dortmund, Germany*
Maurizio Valle, *University of Genoa, Italy*
Stamatis Vassiliadis, *IBM, USA*
Marc A. Viredaz, *Swiss Federal Institute of Technology, Switzerland*
Chang J. Wang, *University of Essex, UK*
John Wawrzynek, *Int. Computer Science Institute / UC-Berkeley, USA*
Michel Weinfeld, *Ecole Polytechnique, France*
Takashi Yokota, *Mitsubishi Electric Corporation, Japan*
Maurizio Zamboni, *Politecnico di Torino, Italy*

INDEX

3D-VLSI, 109

adaptive architectures, 159
analog circuits, 1, 25, 35, 45
analog VLSI architecture, 61
AND/OR parallelism, 253
application specific integrated circuits (ASIC),
    311, 285
architecture, 285
artificial intelligence, 253, 285
associative memory, 231, 159, 167, 207
associative processors, 243
asynchronous computing, 151
asynchronous design, 129
automotive, 285

backpropagation learning, 35, 81, 71, 151
bit-serial processors, 45, 71
boltzmann machines, 45

circuit macro-modeling, 35
classification, 141
CLIPS, 231
cluster transformation, 53
CMAC, 129, 177
complex-node architectures, 177
constrained optimization, 217
constraint satisfaction, 187
content addressable memory, 243, 231
correctness-preserving transformations, 197
cortical cells, 61
cortical map, 61

datadriven computations, 151
dataflow computers, 151

digital electronics, 93, 159

feed-forward networks, 81
field programmable arrays, 177, 197, 141
fully-connected neural network, 71

hardware compilation, 197
hash coding, 177
heuristic search, 307
HLL processors, 297
hybrid knowledge processing, 207

IFS/2, 231
image processing, 61
instruction level parallelism, 275, 307

Kohonen network, 25, 109
Kuhn-tucker conditions, 217

latency, 81
learning algorithms, 1, 53, 187
linear equations, 217
linear programming, 217
locality of computation, 35
logic programming systems, 253

minimum entropy, 53
mixed implementation, 45
multilayer networks, 71, 81
multiport memory, 109

networks cooperation, 159
neural associative memories, 141
neural coprocessor, 141
neuron architecture, 81
noise reduction, 177

on-chip learning, 1, 25, 71, 93
overflow algorithm, 167

parallel architecture, 253, 297, 187
parallel associative processing, 265
parametrised design, 197
perceptron networks, 197
performance measurements, 275
pipeline design, 71
principal component analysis, 53
probabilistic, 119
production systems, 231
prolog machine, 253 , 265 , 275, 285, 299, 307

real-time processing, 141,287
receptive field, 476
reconfigurable architecture, 159
reduced instruction set computers (RISC), 169, 307
reward-penalty, 119
ruby, 197
rule extraction, 207

scalable SIMD architecture, 141
segmentation, 61
self-organization, 25, 109
self-timed circuits, 129

semantic density, 307
sequential emulation of neurocomputing, 71
sets, 265
SIMD architectures, 93, 103, 231
similarity join, 109
sparse coding, 207
stochastic search, 187
surface approximation, 53
synapse, 25
synaptic multiplier, 35
synchronous update, 45
system implementation, 141
systolic arrays, 93

Tank/Hopfield network, 217
texture analysis, 61
trace-scheduling compilation, 275
transputer, 253
unconstrained optimization, 217

visual system, 61
VLIW Architectures, 275, 103
VLSI design methodology, 35

Warren abstract machine (WAM), 297, 307
weight normalization circuit, 53
weightless systems, 119
winner-take-all, 25